Pollution Prevention

Strategies and Technologies

Mark S. Dennison

 **Government Institutes, Inc.**
Rockville, Maryland

Government Institutes, Inc., 4 Research Place, Suite 200, Rockville, Maryland 20850

Copyright ©1996 by Government Institutes. All rights reserved.

99 98 97 96 95   5 4 3 2 1

No part of this work may be reproduced or transmitted in any form or by any means, electronic or mechanical, including photocopying, recording, or any information storage and retrieval system, without permission in writing from the publisher. All requests for permission to reproduce material from this work should be directed to Government Institutes, Inc., 4 Research Place, Suite 200, Rockville, Maryland 20850.

The reader should not rely on this publication to address specific questions that apply to a particular set of facts. The author and publisher make no representation or warranty, express or implied, as to the completeness, correctness or utility of the information in this publication. In addition, the author and publisher assume no liability of any kind whatsoever resulting from the use of or reliance upon the contents of this book.

Printed on acid-free, recycled paper (50% recycled fiber and 10% post-consumer waste).

Manufactured in the United States of America.

ISBN: 0-86587-480-8

For Tracey, my love and inspiration

Table of Contents

Detailed Contents vii
Preface xv
About the Author xvii
Acknowledgements xix

Chapter 1: Introduction to Pollution Prevention 1

Chapter 2: Pollution Prevention Regulatory Requirements 6

Chapter 3: Solid Waste Reduction Programs 55

Chapter 4: Hazardous Waste Minimization Programs 94

Chapter 5: Water Pollution Prevention Programs 111

Chapter 6: Industry-Specific Pollution Prevention Strategies .. 162

Chapter 7: Pollution Prevention Case Studies:
        Industrial Facilities 269

Chapter 8: Pollution Prevention Case Studies:
        Federal Facilities 296

Appendix A: Pollution Prevention Act of 1990 307

Appendix B: U.S. EPA Pollution Prevention
　　　　　　Program Contacts 315

Appendix C: State Pollution Prevention Program Contacts ... 333

Appendix D: University Pollution Prevention
　　　　　　Research Centers 358

Appendix E: State Pollution Prevention Laws 384

Glossary of Terms 435
Index 455

Detailed Contents

PREFACE xv
ABOUT THE AUTHOR xvii
ACKNOWLEDGEMENTS xix

PART ONE—POLLUTION PREVENTION FRAMEWORK

Chapter 1: Introduction to Pollution Prevention 1
IMPORTANCE OF POLLUTION PREVENTION   1
TYPES OF POLLUTION PREVENTION   1
  Source Reduction   1
  Recycling   3
  Treatment   4
BENEFITS OF POLLUTION PREVENTION   4

Chapter 2: Pollution Prevention Regulatory Requirements 6
INTRODUCTION   6
REGULATORY POLICY SHIFT TO POLLUTION PREVENTION   6
  Pollution Prevention Act of 1990   8
  EPA's Pollution Prevention Strategy   10
  EPA's Pollution Prevention Policy Statement   12
  EPA's Five-Year Strategic Plan   14
FEDERAL POLLUTION PREVENTION REGULATION   23
  RCRA Waste Minimization Regulatory Guidance   24
  RCRA Regulation of Recycled Materials   25
  EPCRA Toxic Chemical Release Inventory   32
  CWA Best Management Practices   33
  Storm Water Pollution Prevention Plans   37
STATE POLLUTION PREVENTION REGULATION   41

Mandatory State Requirements   42
  Voluntary State Requirements   53

PART TWO—POLLUTION PREVENTION PROGRAMS

Chapter 3: Solid Waste Reduction Programs 55
  INTRODUCTION   55
  BENEFITS OF WASTE REDUCTION   55
  WASTE REDUCTION APPROACHES   56
    Waste Prevention   56
    Recycling   59
    Composting   59
    Purchasing   60
  PROGRAM PLANNING AND ORGANIZATION   60
    Management Support   61
    Waste Reduction Team   61
    Program Objectives   62
    Employee Involvement   63
  CONDUCTING WASTE ASSESSMENTS   64
    Purpose   64
    Waste Assessment Approaches   65
    Records Examination   68
    Facility Walk-Through   68
    Waste Sort   70
    Documenting the Waste Assessment   71
    Sample Waste Assessment   72
  EVALUATING WASTE REDUCTION OPTIONS   73
    Compiling and Screening Options   74
    Analyzing and Selecting Options   75
    Waste Prevention Options   76
    Recycling Options   76
    Questions to Ask Recycling Companies   76
    Composting Options   79
    Purchasing Options   80
  IMPLEMENTING THE WASTE REDUCTION PROGRAM   81
    Employee Training and Education   81
    Employee Participation   82
    Program Evaluation   83
  COMMON SOLID WASTE REDUCTION PRACTICES   84
  COMMON RECYCLABLE MATERIALS   88

Chapter 4: Hazardous Waste Minimization Programs 94
  INTRODUCTION   94
  BENEFITS OF WASTE MINIMIZATION   95
  ELEMENTS OF A HAZARDOUS WASTE MINIMIZATION PROGRAM   96
    Management Support   97
    Waste Generation and Management Costs   99
    Waste Minimization Assessments   99
    Cost Allocation   100
    Technology Transfer   100
    Program Implementation and Evaluation   101
  HAZARDOUS WASTE MINIMIZATION CASE STUDY:
   AGENT REGENERATION   101
    Problem Identification   102
    Strategic Alternatives   105
    Electrochemical Option   106
    Ozone Gas Option   107
    Air Oxidation Option   108
    Option Selection; Results   109

Chapter 5: Water Pollution Prevention Programs 111
  INTRODUCTION   111
  POLLUTION PREVENTION AND BEST MANAGEMENT PRACTICES   111
    Types of BMPs   112
    Factors Affecting BMP Selection   114
  COMPONENTS OF BMP PLANS   115
  PLANNING PHASE   116
    BMP Committee   116
    BMP Policy Statement   121
    Release Identification and Assessment   124
  DEVELOPMENT PHASE   132
    Good Housekeeping   133
    Preventive Maintenance   137
    Inspections   142
    Security   146
    Employee Training   149
    Recordkeeping and Reporting   154
  PLAN EVALUATION AND REEVALUATION   159

PART THREE—POLLUTION PREVENTION STRATEGIES

Chapter 6: Industry-Specific Pollution Prevention Strategies ... 162
    INTRODUCTION   162
    PAINT MANUFACTURING INDUSTRY   162
        Equipment Cleaning Wastes   164
        Off-Site Specification Paint   169
        Bags and Packages   169
        Air Emissions   170
        Spills   171
        Filter Cartridges   171
        Obsolete Products/Customer Returns   172
    AUTOMOTIVE REPAIR INDUSTRY   172
        Shop Cleanup   175
        Parts Cleaning   177
        Automotive Maintenance   183
    FIBERGLASS AND PLASTICS INDUSTRY   185
        Equipment Cleaning Wastes   187
        Scrap Solvated and Partially Cured Resins   192
        Gelcoat Resin and Solvent Overspray   193
        Rejected and/or Excess Raw Materials   196
        Empty Bags and Drums   197
        Air Emissions   198
        Miscellaneous Waste Streams   199
    MARINE MAINTENANCE AND REPAIR YARDS   200
        Chemical Stripping Wastes   202
        Abrasive Blast Wastes   204
        Paint and Solvent Wastes   206
        Equipment Cleaning Wastes   210
        Machine Shop Wastes   210
        Engine Repair Shop Wastes   211
        Vessel Cleaning Wastes   212
        Spill Control   212
    COMMERCIAL PRINTING INDUSTRY   214
        Material Handling and Storage   216
        Image Processing   217
        Plate Processing   220
        Makeready   221
        Printing and Finishing   223

PRINTED CIRCUIT BOARD INDUSTRY   227
    Product Substitution   231
    Cleaning and Surface Preparation   232
    Pattern Printing and Masking   235
    Electroplating and Electroless Plating   236
    Etching   247
    Wastewater Treatment   248
AUTOMOTIVE REFINISHING INDUSTRY   250
    Body Repair   253
    Paint Application   253
    Shop Cleanup Wastes   259
PESTICIDE FORMULATION INDUSTRY   260
    Equipment Cleaning Wastes   262
    Spills and Area Washdowns   263
    Off-Specification Products   264
    Containers   264
    Air Emissions   266
    Miscellaneous Waste Streams   268

**Chapter 7: Pollution Prevention Case Studies:
Industrial Facilities** 269
INTRODUCTION   269
NUCLEAR-POWERED ELECTRICAL GENERATING STATION   270
    Existing Waste Management Activities   270
    Waste Minimization Opportunities   271
    Additional Options Identified   272
MANUFACTURER OF FINISHED LEATHER   272
    Existing Waste Management Activities   272
    Waste Minimization Opportunities   273
LOCAL SCHOOL DISTRICT   274
    Existing Waste Management Activities   274
    Waste Minimization Opportunities   275
    Additional Options Identified   275
STATE DEPARTMENT OF TRANSPORTATION MAINTENANCE
  FACILITY   276
    Oil Waste Reduction Opportunities   276
    Antifreeze Waste Reduction Opportunities   277
    Freon/CFC Waste Reduction Opportunities   277
    Paint Waste Reduction Opportunities   277
    Recycling of Used Tires   278

LEGAL SUPPLY PRINTING COMPANY   278
    Waste Reduction Opportunities   278
    Engraving Process   279
    Printing Process   280
AUTOMOTIVE AIR CONDITIONING CONDENSER MANUFACTURER   280
    Waste Generation and Management Activities   280
    Waste Minimization Opportunities   281
RAILCAR REFURBISHER   282
    Waste Generation and Management Activities   282
    Waste Minimization Opportunities   283
PERMANENT-MAGNET DC ELECTRIC MOTOR MANUFACTURER   284
    Waste Generation and Management Activities   284
    Waste Minimization Opportunities   285
METAL BANDS, CLAMPS, AND TOOLS MANUFACTURER   286
    Waste Generation Activities   286
    Existing Waste Minimization Practices   287
    Waste Minimization Opportunities   288
ALUMINUM EXTRUSIONS MANUFACTURER   288
    Waste Generation Activities   289
    Existing Waste Minimization Practices   289
    Waste Minimization Opportunities   290
COMMERCIAL ICE MACHINE AND ICE STORAGE BIN
  MANUFACTURER   290
    Waste Generation Activities   291
    Existing Waste Minimization Practices   292
    Waste Minimization Opportunities   293
ADDITIONAL CASE STUDIES   293
    Printed Circuit Board Manufacturer   293
    Specialty Chemical Manufacturer   294
    Aluminum Can Manufacturer   294
    Treated Wood Products Manufacturer   295

Chapter 8: Pollution Prevention Case Studies: Federal Facilities   296
    INTRODUCTION   296
    SCOTT AIR FORCE BASE   297
        Existing Waste Management Activities   297
        Waste Minimization Opportunities   298
        Recommendations   299
    FITZSIMMONS ARMY MEDICAL CENTER OPTICAL FABRICATION
      LABORATORY   299

      Existing Waste Management Activities   299
      Waste Minimization Opportunities   300
      Recommendations   301
  FORT RILEY, KANSAS   301
      Existing Waste Management Activities   301
      Waste Minimization Opportunities   302
      Recommendations   303
  CINCINNATI DEPARTMENT OF VETERANS AFFAIRS HOSPITAL   303
      Existing Waste Management Activities   303
      Research and Development Opportunities   304
      Recommendations   305
  AIR FORCE PLANT NO. 6   305
      Plant Background   306
      Cleaning Solvent Substitutions   306

Appendix A: Pollution Prevention Act of 1990 307

Appendix B: U.S. EPA Pollution Prevention Program Contacts . 315

Appendix C: State Pollution Prevention Program Contacts 333

Appendix D: University Pollution Prevention Research Centers . 358

Appendix E: State Pollution Prevention Laws 384
      Mandatory State Programs   384
          Arizona   384
          California   396
          Minnesota   406
          Texas   412
      Voluntary State Programs   419
          Colorado   419
          Delaware   425
          Florida   430
          Iowa   431

GLOSSARY OF TERMS 435
INDEX 455

Preface

Pollution Prevention Strategies and Technologies has been written as an all-purpose guide to help business and industry toward an understanding of the pollution prevention policies, regulatory initiatives, and technologies designed to reduce wastes at their source. To achieve this purpose, the book has been divided into three parts, plus appendices.

Part One provides a helpful introduction to the subject of pollution prevention, explaining the meaning and benefits of pollution prevention to waste generators, and offering a detailed discussion of the policy shift from traditional "end-of-pipe" pollution controls to reduction of wastes at the source. The most important federal and state pollution prevention programs and laws are examined in-depth. Part Two outlines the components of different types of pollution prevention programs, with practical treatment of solid waste reduction programs, hazardous waste minimization programs, and water pollution prevention programs. Part Three gives concrete examples of pollution prevention strategies through presentation of specific waste reduction technologies for eight different industries. Case studies for fifteen different commercial/industrial facilities and five different federal facilities are provided to illustrate proven ways of implementing pollution prevention. Finally, the appendices contain helpful listings of federal, state, and university-sponsored pollution prevention program contacts. Copies of the federal Pollution Prevention Act of 1990 and various representative state pollution prevention laws are also provided for handy reference purposes.

Pollution prevention, commonly referred to as P^2 by the environmental community, is the new environmental compliance frontier, at both the federal and state levels, as voluntary pollution prevention practices are steadily becoming regulatory mandates. It is hoped that this book will help companies to meet the challenges of source reduction and recycling by providing businesses, plant managers, attorneys, environmental consulting firms, environmental managers, and other environmental professionals with an easy-to-read, comprehensive guide to all legal and technical issues concerning pollution prevention.

Mark S. Dennison
Ridgewood, New Jersey

About the Author

Mark S. Dennison is an attorney and author or co-author of numerous books and articles dealing with environmental law and regulatory compliance issues. His books include *Environmental Reporting, Recordkeeping, and Inspections: Compliance Guide for Business and Industry* (1995); *Environmental Due Diligence for Lenders: Practical Guide to Implementing an Environmental Risk Program*, with R. Kenneth Keim and Philip Lee (1995); *Storm Water Discharges: Regulatory Compliance and Best Management Practices* (1995); *OSHA and EPA Process Safety Management Requirements: A Practical Guide for Compliance* (1994); *Hazardous Waste Regulation Handbook: A Practical Guide to RCRA and Superfund* (1994); *Understanding Solid and Hazardous Waste Identification and Classification* (1993); *Wetlands and Coastal Zone Regulation and Compliance*, with Steven Silverberg (1993); *Wetlands: Guide to Science, Law, and Technology*, with James Berry (1993).

Mr. Dennison is also Editor-In-Chief of the monthly newsletter, *Environmental Strategies for Real Estate* and is a regular columnist for the American Planning Association's bi-monthy newsletter, *Environment & Development*. He is in private practice in Ridgewood, New Jersey, where he specializes in environmental, land use, real estate, and zoning law. He is admitted to practice in New Jersey and New York. Mr. Dennison holds a B.A., *magna cum laude*, from the State University of New York (Oswego), an M.A. from Syracuse University, and a J.D. from New York Law School.

Acknowledgements

I wish to express my appreciation to everyone who played a role in publication of this book. I thank the editorial and production staff at Government Institutes, especially my editor, Alex Padro, who helped to develop the concept for the book and used his editorial leadership to guide me through the publishing process.

I also wish to thank the many helpful individuals working at different federal and state regulatory offices who answered my questions and provided copies of various forms and government documents.

Additional thanks must go to family and friends for providing moral support and inspiration, especially Erin Carrather, Jonathan Huff, Jessie Ann Huff, Bobbie Waits, Harry Huff, Jr., Mrs. J. Howard Gould, Keith Dennison, and Chelsea.

But, as always, my greatest appreciation is reserved for my dear Tracey, whose constant love and support empowered me to get to the finish line.

M.S.D.

Chapter 1

Introduction to Pollution Prevention

IMPORTANCE OF POLLUTION PREVENTION

Previous regulatory efforts focused on controlling and cleaning up the most immediate environmental problems. Although those efforts yielded major reductions in pollution, regulators have learned that traditional "end-of-pipe" approaches not only can be expensive and less than fully effective, but sometimes transfer pollution from one medium to another. Thus, additional improvements to environmental quality require measures to prevent pollution from occurring in the first place.

Preventing pollution offers important economic benefits, as pollution never created avoids the need for expensive investments in waste management or cleanup. Pollution prevention has the exciting potential for both protecting the environment and strengthening economic growth through more efficient manufacturing and raw material use.

TYPES OF POLLUTION PREVENTION

Pollution prevention, in its broadest sense, refers to the reduction in volume and/or toxicity of waste prior to discharge or disposal. Pollution prevention techniques generally consist of source reduction and recycling activities. Although treatment may be used to reduce the toxicity of some waste streams, it is not generally thought of as pollution prevention in its truest sense.

Source Reduction

Source reduction means the reduction or elimination of waste at its source. Facilities seeking to implement source reduction techniques need to evaluate their manufacturing, production, and general waste generating operations for opportunities to reduce wastes before they are generated. Each process or manufacturing operation must be closely examined to determine material inputs, transformations that occur as part of production processes, and material outputs. The impact of quality control parameters, product specifications, and production goals must also be considered.

Good operating practices are key in achieving source reduction goals. These practices typically have been used to improve efficiency and reduce production costs. Improving yields by reducing production losses has long been a common practice by industries where raw materials account for a significant portion of operating costs. Good operating practices generally require little or no capital investment, are easily implemented, and result in significant savings. Good operating practices might include the following measures:

- Waste reduction programs
- Management and personnel practices
- Material handling and inventory practices
- Loss prevention
- Waste segregation
- Production scheduling

A waste reduction program should be formalized to indicate management support for the program. Management gives strong recognition to company progress in meeting program goals by identifying employees or departments that make significant strides toward achieving the goals of the company's waste reduction program. Certain personnel practices may be used to encourage employee participation on pollution prevention. Incentives may be used as an effective means of rewarding employees who make contributions to company's pollution prevention efforts.

Materials handling and inventory practices include programs to reduce loss of input materials caused by mishandling, expired shelf life, and poor storage conditions. The proper control over materials as they are handled or transferred from one location to another reduces the chances of spills. Simple procedures may be implemented to ensure proper materials handling, such as properly training employees in the operation of each type of transfer equipment, allowing adequate spacing of containers, stacking containers so as to avoid the chances of punctures and breaks, and proper labeling of containers to indicate the name and type of substance inside. Computerized systems are the most efficient method of inventory control and materials tracking. Poor inventory control can result in overstocking or disposal of expired materials.

Loss prevention minimizes wastes by avoiding spills and leaks from production equipment and storage areas. The most effective way to minimize wastes that are needlessly generated by spills and leaks is to take precautionary measures to ensure that spills and leaks don't occur in the first place.

Waste segregation reduces the volume of wastes by preventing the mixing of hazardous and nonhazardous components of the company's waste streams. Mixing of hazardous and nonhazardous waste may cause the entire mixture to become classified as a hazardous waste. Thus, by separating nonhazardous from hazardous waste, the overall volume of hazardous waste requiring disposal will be reduced, resulting in significantly reduced hazardous waste management and disposal costs.

Production scheduling changes, especially where batch processes are used, can also be an effective waste reduction technique. Schedule changes in batch production runs can reduce the frequency of equipment and tank cleaning that result in large amounts of solvent waste. To reduce cleaning frequency, batch sizes should be maximized or batches of one material should be followed with a similar product, which may not necessitate cleaning between batches.

Recycling

If a waste stream or component of the waste stream cannot be reduced or eliminated through source reduction, recycling presents the next best option for pollution prevention. Recycling can take two basic forms: preconsumer and postconsumer recycling. As the term implies, preconsumer recycling involves raw materials, products, and by-products that have not yet reached consumers for intended end-use, but are typically reused within the original production process. On the one hand, postconsumer recycled materials are those that have served their intended end-use, and have then been separated from the municipal solid waste stream. Although postconsumer recycling is beneficial toward recovery of some discarded materials for reuse, it is clearly not as effective as preconsumer recycling in achieving pollution prevention because of the high waste management costs and regulatory burdens associated with recovery of postconsumer wastes. Preconsumer recycling, by contrast, enables facilities to reclaim or reuse certain components of process waste streams for a beneficial purpose in a different process.

Facilities can implement recycling to help eliminate waste disposal costs, reduce raw material costs, and provide income from saleable waste. A material is recycled if used, reused, or reclaimed. Recycling through use and/or reuse involves returning the waste material either to the original process as a substitute for an input material, or to another process as an input material. Recycling through reclamation is the processing of waste for recovery of valuable material or for regeneration.

Treatment

Unfortunately, not all wastes can be entirely eliminated at the source, nor recovered for recycling. When these pollution prevention opportunities are unavailable, effective treatment methods should be used to reduce the toxicity of remaining wastes. According to the US EPA's definition, treatment is "any practice, other than recycling, designed to alter the physical, chemical, or biological character or composition of a hazardous substance, pollutant, or contaminant, so as to neutralize said substance, pollutant, or contaminant or to render it nonhazardous through a process or activity separate from the production of a product or the providing of a service." Treatment can generally be divided into three categories: (1) chemical treatment; (2) biological treatment; or (3) physical treatment. These methods are basically available for treating wastes after they have been generated and, as such, are not true methods of pollution prevention. Accordingly, the chapters that follow will not focus on the use of treatment technologies as a pollution prevention alternative.

BENEFITS OF POLLUTION PREVENTION

Pollution prevention is very important as an environmental management tool. It provides a means of reducing wastes that could give rise to environmental liabilities. By implementing source reduction and waste minimization measures, a company reduces the volume and toxicity of wastes normally generated by a given chemical manufacturing process or business operation. Pollution prevention results in economic benefits, competitive advantages, enhanced public image, and reduced environmental risks. Obviously, by having to manage fewer wastes, a company has fewer wastes to dispose of, fewer wastes to report, and reduced risk of spills and releases that contaminate the environment (and less money spent on cleanup and environmental penalties). The goal of this book is to provide practical pollution prevention strategies that companies can apply to their business operations.

Chapter 2

Pollution Prevention Regulatory Requirements

INTRODUCTION

By now, most companies that generate hazardous and solid wastes are aware of the environmental regulatory trend toward pollution prevention. This chapter outlines the current regulatory policy toward pollution prevention and explains the various federal and state regulatory initiatives designed to promote source reduction, waste minimization, and recycling activities.

REGULATORY POLICY SHIFT TO POLLUTION PREVENTION

About a quarter century ago, our nation decided to make a concerted effort toward achieving a cleaner environment. Since passage of the National Environmental Policy Act (NEPA) in 1969, and establishment of the U.S. Environmental Protection Agency (USEPA) in 1970, public environmental advocacy, good corporate citizenry, and committed governmental action have helped to improve the quality of the environment. Emissions of air pollutants from cars and industrial facilities has been reduced, over 5,000 wastewater treatment facilities have been constructed, ocean-dumping of wastes has been prohibited, production and use of hazardous substances, such as asbestos, PCBs, and CFCs, have been banned or are being phased out.[1]

Nevertheless, progress has been slow toward realization of many environmental goals. Forty percent of the country's rivers and lakes are remain unsuitable for fishing or swimming. One in five Americans lives in areas where the air quality does not meet federal Clean Air Act

---

[1] *The New Generation of Environmental Protection: EPA's Five-Year Strategic Plan,* EPA 200-B-94-002 (July 1994), p.1.

standards. One in four Americans lives within four miles of a hazardous waste site. Tellingly, only about 200 Superfund sites have been fully remediated since the Comprehensive Environmental Response, Compensation and Liability Act (CERCLA or Superfund) became law back in 1980. Clearly, it is going to take a long time to cleanup an environment spoiled by previously unabated pollution-generating activities.

Acknowledging this slow progress and the limited success of various "command-and-control" measures to regulate pollution, Congress, the U.S. EPA, and the states have been incorporating new pollution prevention requirements into the environmental regulatory framework. Until recently, the predominant waste management practice had been "end of pipe" treatment or land disposal of hazardous and nonhazardous wastes. While this approach provided substantial progress in improving the quality of the environment, obvious limits exist on how much environmental improvement can be achieved using methods that manage pollutants after they have been generated. The following passage from *EPA's Five-Year Strategic Plan*, released in July 1994, conveys this sentiment:[2]

> The media-specific (i.e., air, water, land) nature of environmental laws and EPA's resulting administrative structure have fragmented EPA's response to environmental protection. Too often, our piecemeal approach to pollution has ended up simply moving contaminants around—from air, to water, to land—rather than reducing and preventing pollution. "Command-and-control" approaches to mitigating pollution have proven sometimes to be blunt instruments—overregulating in some areas, undercontrolling in others....
>
> EPA is at a crossroads. To continue leading national efforts to preserve the earth's environment, the Agency must pursue a new generation of environmental protection—one founded on renewed commitment to broad environmental goals combined with common sense, innovation, and flexibility.... EPA must move beyond strategies that react to today's environmental problems to strategies that anticipate and prevent pollution.

---

[2] *Id.* at pp. 1-2.

Pollution Prevention Act of 1990

In 1990, Congress passed the Pollution Prevention Act of 1990 (PPA)[3] to focus more attention on reducing the volume and toxicity of wastes at the source. Thus, source reduction, recycling, and other waste minimization strategies are fast becoming a significant environmental regulatory compliance issue. In Section 6602(b) of this law, Congress declared it a national policy to prevent or reduce pollution at the source whenever feasible.[4]

This statute is basically an enabling act which states congressional commitment to waste reduction and recycling activities and which mandates that U.S. EPA implement pollution prevention strategies and regulations. The PPA requires that U.S. EPA provide grants to the states to implement their own pollution prevention programs, and requires that U.S. EPA set up an information clearinghouse and conduct pollution prevention research. The PPA also requires that companies report their pollution prevention practices under SARA Title III, also known as the Emergency Planning and Community Right-to-Know Act (EPCRA).[5]

The PPA sets forth a hierarchy of waste management options in descending order of preference: prevention/ source reduction, environmentally sound recycling, environmentally sound treatment, and environmentally sound disposal. Pollution should be prevented or reduced at the source wherever feasible, while pollution that cannot be prevented should be recycled in an environmentally safe manner. In the absence of feasible prevention or recycling opportunities, pollution should be treated. Disposal or other release into the environment should be used as a last resort.

Pollution prevention is explained in the PPA to mean source reduction and other practices that reduce or eliminate the generation of pollution. Section 6602 notes that:

> "There are significant opportunities for industry to reduce or prevent pollution at the source through cost-effective changes in production, operation, and raw materials use.... The opportunities

---

[3] Pub. L. 101-508, 42 USC 13101 et seq.
[4] PPA 6602(b), 42 USC 13101(b).
[5] 42 USC 11001-11050. EPCRA was originally enacted as part of the Superfund Amendments and Reauthorization Act of 1986.

for source reduction are often not realized because existing regulations, and the industrial resources they require for compliance, focus upon treatment and disposal, rather than source reduction.... Source reduction is fundamentally different and more desirable than waste management and pollution control."[6]

Source reduction is defined in the law as any practice which reduces the amount of any hazardous substance, pollutant or contaminant entering any waste stream or otherwise released into the environment (including fugitive emissions) prior to recycling, treatment or disposal; and which reduces the hazards to public health and the environment associated with the release of such substances, pollutants, or contaminants.[7]

The Pollution Prevention Act mandates establishment of an EPA office to carry out various functions required by the Act.[8] The EPA office must function independently of the Agency's other "single-medium" programs; it must promote a multi-media approach to source reduction. The law contains a listing of more than a dozen specific measures that EPA must develop and implement, including:[9]

- Facilitate the adoption of source reduction techniques by businesses and by other federal agencies;

- Establish standard methods of measurement for source reduction;

- Review regulations to determine their effect on source reduction;

- Investigate opportunities to use federal procurement to encourage source reduction;

- Develop improved methods for providing public access to data collected under federal environmental statutes;

---

[6] PPA 6602(a), 42 USC 13101(a).

[7] PPA 6603(5)(A), 42 USC 13102(5)(A).

[8] PPA 6604(a), 42 USC 13103(a). Although not specifically referred to as such in the PPA, this independent office is the Office of Pollution Prevention and Toxics, headquartered in Washington, D.C.

[9] PPA 6604(b), 42 USC 13103(b).

- Develop a training program on source reduction opportunities, model source reduction auditing procedures, a source reduction clearinghouse, and an annual award program.

The PPA authorizes a grant program for state technical assistance programs to promote source reduction techniques by businesses. Approved state programs may receive up to 50% in federal matching funds for each year that a state participates in the program.[10]

Under the PPA, facilities required to report releases to EPA for the annual Toxic Release Inventory (TRI) must now additionally provide information on pollution prevention and recycling for each facility and for each toxic chemical. The information includes: the quantities of each toxic chemical entering the waste stream and the percentage change from the previous year, the quantities recycled and percentage change from the previous year, source reduction practices, and changes in production from the previous year.[11]

The PPA also requires that EPA report to Congress within 18 months (and biennially afterwards) on actions needed to implement a strategy to promote source reduction, and an assessment of the clearinghouse and the grant program.[12]

EPA's Pollution Prevention Strategy

Under the PPA, EPA must develop a pollution prevention strategy that reduces pollution at the source. In February 1991, EPA published its pollution prevention strategy, commonly referred to as the Industrial Toxics Project or the "33/50" Initiative.[13] Under the plan, EPA is seeking to reduce releases and off-site transfers of 17 high-volume EPCRA Section 313 toxic chemicals. These industrial chemicals include known and potential carcinogens, developmental toxins, chemicals that bioaccumulate, ozone-depleting chemicals, and chemicals contributing to

---

[10] PPA 6605, 42 USC 13104.
[11] See Section 2.3.3 below.
[12] PPA 6608, 42 USC 13107.
[13] 56 Fed. Reg. 7849 (Feb. 26, 1991). Copies of the 33/50 Initiative are available free of charge by calling the EPCRA hotline at (800) 535-0202. Ask for Pub. No. EPA/741/R-92/001.

ozone at ground level. EPA set a goal to reduce the releases of these chemicals by 33% by the end of 1992, and 50% by the end of 1995.

The 17 target chemicals are:

- Benzene
- Cadmium and Cadmium Compounds
- Carbon Tetrachloride
- Chloroform (Trichloromethane)
- Chromium and Chromium Compounds
- Cyanide and Cyanide Compounds
- Lead and Lead Compounds
- Mercury and Mercury Compounds
- Methyl Ethyl Ketone
- Methyl Isobutyl Ketone
- Methylene Chloride (Dichloromethane)
- Nickel and Nickel Compounds
- Tetrachloroethylene (Perchloroethylene)
- Toulene
- 1,1,1-Trichloroethane (Methyl Chloroform)
- Trichloroethylene
- Xylene (all xylenes)

Under the program, EPA encourages companies to reach the 33 and 50% goals by using the pollution prevention hierarchy outlined in the PPA. Since the PPA accords prevention/source reduction the highest value, EPA correspondingly makes source reduction the preferred method of pollution prevention. Recycling is considered the next best method, and treatment is the least preferred method of pollution prevention. Although participation in the Industrial Toxics Project is completely voluntary and the program goals are not enforceable, companies should take steps to implement pollution prevention techniques because such measures are destined to become mandatory components of pollution control laws.

EPA is seeking reductions primarily through pollution prevention practices that go beyond regulatory requirements. EPA also will be encouraging industry to develop a preventive approach, seeking continuous environmental improvements beyond these reductions and the 17 priority chemicals.

Success in the program will be measured by nationwide reductions, rather than results at each company or facility. This approach provides flexibility and allows participating companies to develop reduction strategies that are the most cost-effective for their facilities.

In numeric terms, the goal is to reduce the amount of releases and off-site transfers from the 1.4 billion pounds reported in 1988 to 700 million pounds by 1995. EPA's Toxic Release Inventory (TRI) will be used to track these reductions using 1988 data as a baseline. As of mid-November 1992, more than 1000 companies were participating in the Program.

Each of the 17 chemicals was selected from the TRI, based on a number of factors, including high production volume, high releases and off-site transfers of the chemical relative to total production, opportunities for pollution prevention, and potential for causing detrimental health and environmental effects.

The 33/50 Program is part of EPA's overall Pollution Prevention Strategy and the first of the Agency's pollution prevention initiatives. It is also a major component of the Office of Pollution Prevention and Toxics' Existing Chemicals Revitalization Program.

All of the 33/50 Program chemicals are regulated under one or more existing environmental statutes, and the 33/50 Program is intended to complement, not replace, ongoing EPA programs. All 17 targeted chemicals will be subject to the Maximum Achievable Control Technology (MACT) standards of the Clean Air Act Amendments of 1990. EPA believes that the incentive for early reductions offered by the MACT provisions will further the progress of the 33/50 Program.

EPA has been contacting companies to provide them with information on the 33/50 Program and to solicit their participation. Companies are being asked to identify and implement cost-effective pollution prevention practices related to the 17 chemicals and to develop written commitments stating their reduction goals and how they plan to achieve them. Access to these written commitment statements is available at a public docket at EPA Headquarters.

EPA's Pollution Prevention Policy Statement

On June 13, 1993, EPA Administrator, Carol Browner issued the agency's Pollution Prevention Policy Statement. In the policy statement, she offered guidance on how the agency could achieve the main goal of the PPA—"that pollution should be prevented or reduced at the source whenever feasible"—by making pollution prevention the guiding principle for all EPA programs. The main points of this policy statement are summarized here.

The statement identifies various commitments already made to further the Clinton-Gore Administration's commitment to environmental solutions that reduce pollution at its source. For example:

- The Administration's budget request for the 1994 fiscal year included a $33 million increase in spending for pollution prevention programs at EPA;

- On Earth Day, 1993, the President announced a commitment to an Executive Order establishing voluntary source reduction goals for procurement, and requiring federal agencies to comply with Emergency Planning and Community Right-to-Know reporting requirements for toxic chemical wastes;

- On May 25, 1993, Carol Browner released new Pollution Prevention Act data on the type and amount of toxic chemicals generated as waste, and announced her intention to expand Emergency Planning and Community Right-to-Know to include different chemicals and sources of pollution.

While acknowledging these accomplishments, the policy statement directs EPA to go further by building pollution prevention into the framework of EPA's mission to protect human health and the environment. With the new focus on pollution prevention in mind, the policy outlines the following guiding principles for the Agency to follow:

- *Regulation and Compliance*: The mainstream activities at EPA, such as regulatory development, permitting, inspections, and enforcement, must reflect our commitment to reduce pollution at the source, and minimize the cross-media transfer of waste.

- *State and Local Partnerships*: Increasingly, state and local agencies are the "face of government" for the general public. We will strengthen the national network of state and local prevention programs, and seek to integrate prevention into state and local regulatory, permitting, and inspection programs supported with federal funds.

- *Private Partnerships*: We will identify and pioneer new cooperative efforts that emphasize multi-media prevention strategies, reinforce the mutual goals of economic and environmental well-being, and represent new models for government/private sector interaction.

- *Federal Partnerships*: We must work closely with our counterparts in other agencies to ensure that pollution prevention guides our management and procurement decisions, and to pursue opportunities for reducing waste at the source in the non-industrial sector.

- *Public Information/The Right-to-Know*: We will collect and share useful information that helps identify pollution prevention opportunities, measure progress, and recognize success.

- *Technological Innovation*: We will try to meet high priority needs for new pollution prevention technologies that increase competitiveness and enhance environmental stewardship, through partnerships with other federal agencies, universities, states, and the private sector.

- *New Legislation*: Where justified, we must not hesitate to seek changes in federal environmental law that will encourage investment in source reduction.

EPA's Five-Year Strategic Plan

In July 1994, EPA released its "Five-Year Strategic Plan," which has pollution prevention as one of its primary goals.[14] In the plan, EPA set forth a strategy to reorient efforts toward reduction and elimination of pollution at the source. The plan recognizes that pollution prevention—anticipating problems and stopping them before they occur—is far more cost-effective and protective of the environment than focusing efforts on solving environmental problems long after they have

---

[14] *The New Generation of Environmental Protection: EPA's Five-Year ~~gic Plan*, EPA 200-B-94-002 (July 1994).

been created—when solutions are more likely to be costly and less likely to be effective. Consequently, the plan states that "pollution prevention should be the strategy of choice in all that the Agency does."

Strategies. The plan identifies the following strategies designed to further the agency's commitment to pollution prevention:

1. Incorporating multi-media prevention principles into the Agency's mainstream environmental programs;

2. Strengthening partnerships with state, tribal, and local governments;

3. Developing new cooperative efforts with the private sector;

4. Promoting prevention with other federal agencies;

5. Providing information to the public;

6. Encouraging technological innovation and diffusion; and

7. Working to change existing environmental legislation, where necessary.

These strategies are consistent with the principles outlined in the pollution prevention policy statement issued by the agency the year before.[15]

Measures of Progress. The plan identifies four complementary ways to measure progress in pollution prevention: measures of Agency activity, measures of state activity, measures of activity undertaken by the regulated community and the public, and measures of environmental improvements that result from pollution prevention.

■ *Agency Activity*: As the EPA reorients its efforts toward prevention, it will move beyond measuring resource commitments and identify specific activity measures, for example:

---

[15] See discussion beginning on p. 12.

—The number and environmental value of Supplemental Environmental Projects (SEPs), permits, and regulations that include prevention approaches; and

—More qualitative measures of progress, such as the identification of pollution prevention options in the development of all major program and regional strategies, and changes in requirements governing state programs.

- *State Activity*: The states have been at the forefront of pollution prevention outreach and experimentation. Possible measures of progress by states will include:

—The amount of grant funds redirected through flexibility agreements to pollution prevention efforts;

—Changes in base operations such as permitting, inspections, and enforcement settlements that encourage pollution prevention; and

—Direct outreach to the regulated community and the public through workshops and facility visits that promote pollution prevention.

- *Activity of the Regulated Community and Public*: Progress will be measured by assessing the adoption of pollution prevention approaches in the private sector, including:

—The number of facilities that enroll in Agency-sponsored voluntary programs;

—Voluntary changes in industry and consumer behavior, such as the development of prevention plans, the development of prevention technologies, and the marketing of environmentally preferable products; and

—Pollution prevention planning and implementation among federal and state agencies.

- *Toward Environmental Results*: Reductions in the volume and toxicity of pollutants generated will be measured across the entire range of programs that the Agency administers. While the Toxics Release Inventory (TRI) is EPA's most visible measure, TRI can be complemented with information from other environmental programs to create a more complete picture of releases from facilities.

The Agency also will develop better methods for evaluating the effectiveness of particular prevention approaches and better methods for showing how prevention contributes to environmental results.

Key Activities. The Agency is pursuing innovative pollution prevention approaches throughout the regions and programs in support of its objectives.

- *Mainstream Programs*: EPA will expand all its compliance tools inspections, and enforcement regulations, permitting, to explicitly promote prevention.

 -Common Sense Initiative—Focusing on selected industry sectors, EPA will create cross-Agency teams to identify and implement environmental management solutions that provide greater environmental benefits at reduced compliance cost.

 -Regulations and Permitting—EPA will incorporate pollution prevention into ongoing activities, including effluent guidelines, Maximum Available Control Technology (MACT) standards, and permitting under the Clean Air Act and Clean Water Act.

 -Compliance and Enforcement—EPA will develop auditing policies that encourage consideration of pollution prevention by private parties conducting environmental audits and exploring facility-wide pollution prevention options during the settlement process, both as injunctive relief and as Supplemental Environmental Projects.

- *State, Tribal, and Local Partnerships*: EPA will support state, tribal, and local efforts to reorient ongoing activities to promote prevention, consistent with the goal of building prevention into mainstream environmental protection programs.

 -Resources—EPA will provide states and tribes with more flexibility in administering federal programs. For example, Region 10 is working with the Alaska Department of Environmental Conservation to direct state grant funds to multi-media pollution prevention projects. The State Pollution Prevention Office received three percent of the program grant resources in the first year, and four percent in the second year. The goal for the third year is five percent.

-Innovation—EPA will support innovation at all levels. The Agency will strengthen its support for compliance assistance and other forms of technical assistance at the state, tribal, and local level, and not simply channel resources into support of traditional regulatory programs.

-Flexibility—EPA will reassess the required state commitments to federal programs if these commitments interfere with pollution prevention.

- *Private Partnerships*: EPA is initiating voluntary programs in all environmental media to complement its traditional regulatory programs. In implementing these programs, EPA will work to better integrate delivery systems, avoid multi-media tradeoffs, and focus on customer needs.

-Common Sense Initiative—EPA will work with industry, environmental groups, states, tribes, and local governments to identify and encourage regulatory flexibility and innovative nonregulatory approaches that enhance environmental protection while reducing compliance costs for selected industries. EPA plans to expand the program beyond the pilot industries in the future.

-Coordinated Voluntary Programs—EPA will explore the feasibility of concurrently marketing its various voluntary programs to particular businesses and facilities.

-MERIT Partnership—EPA Region 9, for example, will continue to support the MERIT (Mutual Efforts to Reduce Industrial Toxics) partnership, involving industry, federal, state, and local governments, with the goal of reducing toxic emissions in southwest Los Angeles County.

- *Federal Partnerships*: Federal partnerships extend across a wide range of federal activities management of government facilities, in acquisition, and in developing new cooperative arrangements for working with the private sector.

    -Facilities—Executive Order 12856 establishes a framework for federal management of its own facilities; it requires TRI reporting, goals for reducing releases and off-site transfers, and pollution prevention plans. Source reduction is the preferred approach for attaining these reductions.

    -Acquisition—Executive Order 12856 requires plans and goals for eliminating or reducing the unnecessary acquisition of products containing extremely hazardous substances or toxic chemicals, and Executive Order 12873 mandates EPA to develop guidance for other federal agencies to use in acquiring environmentally preferable products.

    -Cooperative Ventures—The President's Climate Change Action Plan establishes energy efficiency programs that are implemented jointly by the Department of Energy and EPA. EPA has signed cooperative agreements with the National Institute of Standards and Technology (NIST) Manufacturing Extension Partnership, and with the Small Business Administration (SBA). The NIST agreement will increase its capacity for bringing innovative pollution prevention technologies and technical assistance to small and medium-sized manufacturers, and the SBA partnership will help SBA provide technical assistance and assess various financial mechanisms that support prevention technology. The Agency for International Development-funded Environmental Pollution Prevention Project helps EPA foster pollution prevention through specific projects all over the world.

- *Public Information and the Right to Know*: Information is a tool for both the public and industry. It fosters more knowledgeable advocacy by the public, better understanding of opportunities to reduce waste and improve efficiency, and advances in professional expertise that can inform consumer, citizen, and corporate decision making.

-Facility Database—EPA will support the development of a facility database that will consolidate all the information developed under EPA, state, and tribal programs concerning pollutants and releases from a particular facility.

-Environmental Justice—EPA will embark on a new effort to provide grants to strengthen the capacity of minority and low-income communities to use environmental information to advance pollution prevention.

-Education—Region 1, for example, will support the *Pollution Prevention Consortium of New England Universities,* which includes 11 member universities with a mission of facilitating joint pollution prevention research, education, and training projects.

- *Technological Innovation and Diffusion*: In addition to promoting technology as an adjunct to specific regulatory or response activity, EPA will promote good design and clean technology to reduce pollution overall.

-Environmental Technology Initiative (ETI)—ETI funds research and demonstration projects, technology diffusion activities, and the assessment of regulatory barriers to the adoption of prevention technologies. Through ETI, Office Research Development is focusing nearly $5 million in research and technical assistance on the metal fabrication industry, which is dominated by small businesses. Future funding will be focused on industries participating in EPA's Common Sense Initiative.

-Regulatory Barriers—EPA will work to eliminate barriers to the adoption of prevention technologies that are caused by the regulatory process, and it will investigate opportunities to allow industry to coordinate regulatory compliance with capital development cycles.

-Green Chemistry—The *Design for the Environment* (DfE) program will continue to foster a "green chemistry" ethic that encourages industry to avoid the generation of toxic waste when synthesizing chemicals.

- *Legislation*: EPA will work to reorient fundamental statutory mandates toward prevention, particularly in the Clean Water Act, the Safe Drinking Water Act, food safety and pesticide legislation, and environmental technology legislation.

    -Coordination—Potential changes include both substantive provisions that can advance prevention, and efforts to help align specific federal, state, and tribal activities under these statutes. Better coordination of activities under different environmental statutes can encourage prevention, rather than shifting waste from one environmental medium to the next, and provide industry with opportunities to identify, implement, and finance prevention alternatives to end-of-pipe treatment. For example, EPA will promote the coordination of permitting activities under different environmental statutes.

    -Superfund—The Agency will work to promote settlements and preserve liability through the Superfund reauthorization process. Liability has been a major incentive for pollution prevention, since industry realizes that the surest way to avoid future liability is to avoid generating wastes in the first place.

Office of Prevention, Pesticides and Toxic Substances. The Office of Prevention, Pesticides and Toxic Substances (OPPTS) is experiencing dynamic change as it considers its agenda for the next five years. While basic chemical and pesticide regulatory programs and information reporting requirements continue as mandated by existing legislation, in the future OPPTS will devote increased attention to a new and expanded mission. The new mission for OPPTS is (1) to promote safer designs, wiser use of materials, products, processes, practices and technologies, and disposal methods using pollution prevention as the guiding principle of first choice; and (2) to provide information, education, and technical assistance to empower the public to make informed decisions on the risks associated with pesticides and toxic substances.

To obtain these ends, OPPTS plans to make greater use of the scientific data and information used for long-standing regulatory programs for pesticides and industrial chemicals. These new risk reduction and pollution prevention opportunities fall into eight major categories:

1. Reducing the use of pesticides by removing higher-risk pesticides from the market, and accelerating regulatory decisions on safer pesticides and promoting increased use of integrated pest management techniques.

2 Improving the safety of the nation's food supply by implementing health-based tolerances for pesticide residues.

3. Identifying sectors of the general population that are at greater risk from pesticides (children and farm workers, for example) and taking action to reduce exposures to these groups.

4. Expanding the use of pollution prevention guiding principles in the Agency, the federal government, the states, and the private sector, including the development of safer chemicals, products, and technologies in the industrial sector.

5. Empowering the public with information about chemicals that will help identify significant risks and define pollution prevention opportunities for reducing those risks.

6. Reducing exposure to chemicals of national interest such as lead, other heavy metals, PCBs, and other complex organic compounds.

7. Providing expertise in chemical risk assessment, using tools such as structure activity relationship assessments to provide highly accurate predictions about the toxicity of chemicals and making information available both within and outside the Agency.

8. Meeting statutory responsibilities to register pesticides and ensure that chemicals in commerce do not pose unacceptable risks to the public or the environment.

OPPTS is the leading advocate for promoting pollution prevention as the guiding principle of first choice to achieve environmental stewardship. This advocacy applies within the Agency to other programs and outside the Agency to the states, the general public, and constituent groups.

OPPTS now has reoriented its chemical programs to support the emphasis on source reduction in the Pollution Prevention Act. The new chemicals program, the Agency's first opportunity to practice pollution prevention guiding principles, keeps dangerous chemicals from entering commerce. It also is being used to instill a "green chemistry" ethic in industry, aimed at selecting chemical synthesis pathways that avoid the generation of toxic waste. OPPTS will continue to reform the existing chemicals program to promote the use of safer chemicals, materials and technologies in the industrial sector. The Design for the Environment (DfE) program, which is an outgrowth of the existing chemical program, will become a key component in the Agency's Environmental Technology Initiative.

The pesticides program's pollution prevention priorities will focus on promoting sustainable agriculture. Increased emphasis will be given to promoting the wider application of alternative agricultural practices including Integrated Pest Management (IPM), education and outreach to growers and other pesticide users, and recycling of pesticide containers to encourage pollution prevention. Initiatives to reduce the use of high-risk pesticides are also an important ongoing part of OPPTS pollution prevention activities.

The TRI program will take a more prominent role as one of the cornerstones of the Agency's pollution prevention activities. The program will be expanded to include federal facilities, additional chemicals, and new categories of reporting facilities. Providing usable information from the TRI to many diverse groups of customers is essential to assist and enhance participation in environmental decision making.

FEDERAL POLLUTION PREVENTION REGULATION

The Pollution Prevention Act of 1990 essentially formalized a national policy and commitment to waste reduction, functioning primarily as an enabling act that mandates and authorizes pollution prevention measures at the federal government level. As indicated in the section on Regulatory Policy Shift to Pollution Prevention, EPA has issued strong policy initiatives to further nationwide pollution prevention activities. At this time, very few pollution prevention measures are considered mandatory; rather EPA has tried to encourage business and industry to make voluntary commitments to pollution prevention, largely through participation in various EPA-sponsored administrative programs. Some mandatory pollution prevention requirements do, however, exist. These federal pollution prevention regulatory requirements are reviewed in this section.

RCRA Waste Minimization Regulatory Guidance

Before the Pollution Act of 1990 became law, some early consideration was given to waste reduction activities. In fact, with the passage of the Hazardous and Solid Waste Amendments (HSWA) to RCRA in 1984, Congress established a significant new policy concerning hazardous waste management. Specifically, Congress declared that the reduction or elimination of hazardous waste generation at the source should take priority over the management of hazardous wastes after they are generated. In RCRA Section 1003(b), Congress declared it to be national policy that, whenever feasible, the generation of hazardous waste is to be reduced or eliminated as expeditiously as possible. Waste that is nevertheless generated should be treated, stored, or disposed of so as to minimize the present and future threat to human health and the environment.[16]

In furtherance of this national policy toward pollution prevention, the 1984 amendments to RCRA added a significant new waste minimization requirement. Under RCRA Section 3002(b), hazardous waste generators who transport their wastes off-site are required to certify on their hazardous waste manifests that they have programs in place to reduce the volume or quantity and toxicity of hazardous waste generated to the extent economically practicable. Certification of a waste minimization "program in place" is also required as a condition of any permit issued under section 3005(h) for the treatment, storage, or disposal of hazardous waste at facilities that generate and manage hazardous wastes on-site.

On May 28, 1993, EPA issued interim final guidance to assist hazardous waste generators and owners and operators of hazardous waste treatment, storage, or disposal (TSD) facilities to comply with the waste minimization certification requirements of RCRA Sections 3002(b) and 3005(h), as amended by the HSWA.[17] The guidance document fulfills a commitment made by EPA in its 1986 report to Congress entitled *The*

---

[16] 42 USC 6902(b).

[17] "Guidance to Hazardous Waste Generators on the Elements of a Waste Minimization Program," 58 Fed. Reg. 31114 (May 28, 1993).

Minimization of Hazardous Waste[18] to provide additional information to generators on the meaning of the certification requirements added by the HSWA.[19] In addition, the interim guidance set forth a detailed plan for waste generators to develop and implement hazardous and solid waste minimization programs. The waste minimization program guidance is discussed in detail in Chapter 4.

RCRA Regulation of Recycled Materials

After source reduction, recycling is considered the next preferable method of pollution prevention according to the pollution prevention hierarchy set forth in the Pollution Prevention Act, as well as in the 1984 amendments to RCRA. Unfortunately, it is not always clear exactly which activities qualify as recycling since EPA's RCRA regulations concerning recycled materials are very complex, and oftentimes confusing.[20] This section endeavors to clarify which the solid and hazardous waste management activities qualify as "recycling" within the meaning of these RCRA regulations.

In order to fully understand these recycling regulations, one must first possess a working knowledge of RCRA's solid and hazardous waste classification system. Although a discussion of solid and hazardous waste identification and classification cannot be provided here, it is important to note a basic rule of thumb: that a substance must be both a "solid" and a "hazardous" waste to be subject to regulation under RCRA Subtitle C (the provisions relating to hazardous waste management).[21] "Recycled materials" are regulated under RCRA Subtitle C.

In general terms, the regulation of recycled materials consists of the following components:

---

[18] EPA/530-SW-86-033 (Oct. 1986).

[19] See 51 Fed. Reg. 44683 (December 11, 1986).

[20] See 40 CFR 261.10 (definition of "recycled"); 260.30 through 260.33, 260.40, 260.41; 261.1 through 261.6; 40 CFR Part 266, Subparts C-G; 50 Fed. Reg. 614 (1985).

[21] For a full explanation of how to determine whether a waste is subject to regulation as a RCRA hazardous waste, see M. Dennison, *Understanding Solid and Hazardous Waste Identification and Classification: Practical Guide for the Waste Generator* (John Wiley & Sons, Inc., 1993).

1. Certain legitimate recycling activities will remove all but a few special substances from classification as solid wastes. Thus, it will not be regulated under RCRA Subtitle C if it is recycled in one of the approved manners.

2. Certain substances that are not solid wastes but are either characteristic or listed hazardous wastes will become solid wastes (and therefore regulated) if they are recycled in certain specified manners. These substances are considered "recyclable materials" (hazardous wastes that are recycled).

3. Certain materials known as "inherently waste-like materials" are considered solid wastes regardless of how they are recycled, even if the recycling process would otherwise render them exempt from regulation. These materials are "listed" hazardous wastes and are always subject to RCRA regulation.

Each of these situations are addressed below.

Recycling Activities That Remove Certain Materials from RCRA Regulation. Materials will not be classified as solid wastes (and therefore outside the scope of RCRA Subtitle C regulation) when they can be shown to be recycled by one of the following methods:[22]

1. Used or reused as ingredients in an industrial process to make a product, provided the materials are not being reclaimed;[23] or
2. Used or reused as effective substitutes for commercial products; or
3. Returned to the original process from which they are generated, without first being reclaimed. The material must be returned as a substitute for raw material feedstock, and the process must use raw materials as principal feedstocks.

---

[22] 40 CFR 261.2(e)(1).
[23] 40 CFR 261.1(c)(4).

However, certain materials will be considered solid wastes, regardless of whether the recycling involves use, reuse, or return to the original process. They are:[24]

1. Materials used in a manner constituting disposal, or used to produce products that are applied to the land; or

2. Materials burned for energy recovery, used to produce a fuel, or contained in fuels; or

3. Materials accumulated speculatively; or

4. Inherently waste-like materials

Proof Against Claims of "Sham" Recycling. Even if a company examines a process it is using and determines that the particular recycling activity exempts certain materials from classification as a solid waste, it must be prepared to document that the approved recycling process was actually used to recycle the materials. EPA included a provision to guard against "sham" recycling claims in the recycling regulations.[25]

Defendants in actions brought by EPA to enforce regulations implementing Subtitle C of RCRA who raise a claim that a certain material is not a solid waste by virtue of a recycling process, must demonstrate that there is a known market or disposition for the material, and that they meet the terms of the recycling exclusion or exemption. In doing so, they must provide appropriate documentation to demonstrate that the material is not a waste, such as contracts showing that a second person uses the material as an ingredient in a production process, or is otherwise exempt from regulation. In addition, owners or operators of facilities claiming that they actually are recycling materials must show that they have the necessary equipment to do so.

The best approach to recycling materials, which could otherwise be considered solid or hazardous waste, is to maintain careful records of the recycling activity to the greatest extent possible.

---

[24] 40 CFR 261.2(e)(2).
[25] 40 CFR 261.2(f).

"Inherently Waste-Like" Materials. Inherently waste-like materials are considered solid wastes and subject to RCRA regulation regardless of how they may be recycled.[26] These materials are:

1. Materials with hazardous waste Nos. F020, F021 (unless used as an ingredient to make a product at the site of generation), F022, F023, F026, and F028; or

2. Secondary materials fed to a halogen acid furnace that are characteristic hazardous waste or listed hazardous wastes, except brominated material that meets several criteria.

EPA may list additional materials as inherently waste-like by following specified procedures and listing criteria.

"Recyclable Material." The "recyclable material" provisions of the regulations are especially complex and difficult to understand.[27] A "recyclable material" is the term used by EPA to connote a hazardous waste that is recycled.[28] These materials are regulated separately because somewhat different regulations apply to recycled hazardous wastes than to RCRA hazardous wastes in general.

Determination of whether a material is a "recyclable material" starts with the same process used to identify all materials subject to RCRA Subtitle C regulation. Initially, a determination must be made regarding whether the material is a "solid waste." In the case of recycled materials, this analysis requires a two step process. First, the material must be included within the range of recycled materials governed by the regulations. The list of generally described materials is produced as Table 1 to 40 CFR Section 261.2.

These materials are:

- Spent Materials
- Sludges Listed in 40 CFR Section 261.31 or 261.32
- Sludges Exhibiting Hazardous Waste Characteristics

---

[26] 40 CFR 261.2(d).
[27] 40 CFR 261.6.
[28] 40 CFR 261.6(a).

- By-Products Listed in 40 CFR 261.32 or 261.32
- By-Products Exhibiting Hazardous Waste Characteristics
- Commercial Chemical Products Listed in 40 CFR 261.33
- Scrap Metal

If the material is included in the list, it will be a solid waste only if it is recycled in one of several prescribed ways. If the material is also hazardous, it will be a "recyclable material." Some of the substances on the list are hazardous by definition because they are "listed" or characteristic hazardous wastes under RCRA. Others may need further analysis to determine whether they are hazardous and, therefore, "recyclable materials."

The materials listed above are solid wastes if they are *recycled* or accumulated, stored, or treated before recycling and they are:

1. *Used in a manner constituting disposal.*[29] Materials are used in a manner constituting disposal if applied to or placed on the land in a manner that constitutes disposal, or used to produce products that are applied to or placed on land or otherwise contained in products that are applied to or placed on the land (in which case the product itself remains a solid waste.) Commercial chemical products listed in 40 CFR Section 261.33, however, are not solid wastes if they are applied to the land and that is their ordinary manner of use.

2. *Burned for energy recovery.*[30] The materials are solid wastes if recycled by being burned to recover energy, or used to produce a fuel or are otherwise contained in fuels (in which case the fuel itself remains a solid waste.) Again, commercial chemical products listed in 40 CFR Section 261.33, however, are not solid wastes if they are themselves meant to be fuels.

3. *Reclaimed.*[31] Spent materials, listed sludges and by-products, and scrap metal are solid wastes if they are recycled by being reclaimed. Characteristic sludges and by-products and commercial chemical products are not. A material is reclaimed if it is processed to recover

---

[29] 40 CFR 261.2(c)(1)(I)-(ii).
[30] 40 CFR 261.2(c)(2)(I)-(ii).
[31] 40 CFR 261.2(c)(3).

a usable product or if it is regenerated. Examples are recovery of lead values from spent batteries and regeneration of spent solvents.

4. *Accumulated speculatively.*[32] All of the materials listed above, except commercial chemical products are classified as solid wastes when they are "accumulated speculatively." A material is accumulated speculatively if it is accumulated before being recycled. A material is not accumulated speculatively, however, if the person accumulating it can show that the material is potentially recyclable and has a feasible means of being recycled; and that during the calendar year the amount of material that is recycled, or transferred to a different site for recycling equals at least 75% by weight or volume of the amount of that material accumulated at the beginning of the period. Refer to 40 CFR Section 261.1(c)(8) for specific guidance on calculating the 75% requirement.

Recycling activities should be checked to determine whether any recycled materials are solid wastes by virtue of being included on the list of recyclable materials and by being recycled in one of the four prescribed manners. Initially, however, it may be helpful to check the short list of recycled materials exempt from RCRA regulation listed in 40 CFR Section 261.6(a)(3).

If the material is already considered hazardous by virtue of being a characteristic or listed hazardous waste, it is a recyclable material. In the case of spent materials, and scrap metal, however (which are not already hazardous by definition), an independent analysis will need to be made to determine whether they are listed or characteristic hazardous wastes. If they are identified as hazardous wastes, they will, of course, be classified as recyclable materials. If they are not listed or characteristic hazardous wastes, but only solid wastes, they will not be subject to RCRA Subtitle C regulation.

Once a material has been identified as a recyclable material, it is necessary to determine how it is regulated. This is not a particularly easy task since recyclable materials are not regulated in precisely the same manner as hazardous waste in general. Refer to 40 CFR Section 261.6 for specific requirements if you determine that you have recyclable materials.

---

[32] 40 CFR 261.2(c)(4).

Certain recyclable materials are, however, altogether exempt from RCRA regulation. Check 40 CFR Section 261.6(a)(3) for this list of exemptions..

Variances from Classification of Recyclable Materials as Solid Wastes. On a case-by-case basis, subject to specific standards and criteria, EPA may grant variances from classification of recyclable materials as solid wastes (in which case the material would be excluded from RCRA Subtitle C regulation). In order to obtain a variance, an application must be submitted to the EPA Regional Administrator for the Regional Office with jurisdiction where the recycling operation is located. The application must meet address certain relevant regulatory criteria.[33] The application will be evaluated and the Regional Administrator will issue a draft notice tentatively granting or denying the variance. An informal public notice and comment period will follow and public notice of a tentative decision will be provided by newspaper advertisement or radio announcement. The public comment period lasts for thirty days, and a public hearing may be held at the discretion of the Regional Administrator. Following the notice and comment period, EPA issues a final decision on the application. the decision is final and not subject to administrative appeal to EPA.[34]

Variances may only be sought to exclude the following three types of materials from classification as a solid waste:

1. Materials that are accumulated speculatively without sufficient amounts being recycled;[35]

2. Materials that are reclaimed and then reused within the original primary production process in which they were generated;

3. Materials that have been reclaimed but must be reclaimed further before the materials are completely recovered.

The criteria that need to be addressed in the variance application vary depending on the type of material for which the variance is being sought.

---

[33] 40 CFR 260.33(a).
[34] 40 CFR 260.33(b).
[35] See 40 CFR 261.1(c)(8) (definition of accumulated speculatively).

EPCRA Toxic Chemical Release Inventory

In 1986, the Emergency Planning and Community Right-to-Know Act (EPCRA),[36] also known as SARA Title III, was added as a freestanding part of the Superfund Amendments and Reauthorization Act (SARA), which amended CERCLA. The primary goal of EPCRA is to facilitate public awareness and emergency response planning for chemical hazards. To fulfill this purpose, EPCRA requires that certain companies that manufacture, process, and use chemicals in specified quantities must file written reports, provide notification of spills/releases, and maintain toxic chemical inventories.

Certain companies must submit an annual report of releases of listed "toxic chemicals" pursuant to EPCRA Section 313, known as the Toxic Chemical Release Inventory (Form R).[37]

Regulated companies must be file this report on July 1 with the EPA and the relevant state authority. On Form R, the company reports any releases made during the preceding twelve months. Form R must be filed if the business has ten or more full-time employees, has a Standard Industrial Classification Code 20-39, and the business manufactures, stores, imports, or otherwise uses designated toxic chemicals at or above threshold levels. Generally, a company must file an annual Form R if it manufactures, imports, or processes at least 25,000 pounds of a listed "toxic chemical," or if it uses at least 10,000 pounds of a listed "toxic chemical" during the previous calendar year. "Toxic chemicals" subject to the Form R reporting requirements and their respective threshold quantities are listed at 40 C.F.R. Section 372.65. Certain exemptions from the Form R reporting of toxic chemical releases may apply.[38]

EPA is also considering expanding the list of chemicals subject to Form R reporting. Other substances being considered for addition to the list include Clean Water Act priority pollutants; extremely hazardous substances (EHSs) listed for purposes of EPCRA Section 312; RCRA "listed" hazardous wastes from 40 C.F.R. Sections 261.33(e), (f); and air toxins under Sections 112(b) and 602(a), (b) of the Clean Air Act Amendments of 1990.

---

[36] 42 USC 11001-11050.

[37] Violations of EPCRA Section 313 reporting are punishable by fines of up to $25,000 per day. 40 CFR 372.18.

[38] See 40 CFR 372.38.

With the passage of the Pollution Prevention Act of 1990 (PPA),[39] new requirements were added to the Form R. Section 6607 of the PPA expands and makes mandatory source reduction and recycling information on the EPCRA list of toxic chemicals. These requirements have been added by EPA through modifications to sections 6, 7, and 8 of the Form R. Section 6 has been modified so that off-site location and transfer amounts are reported together, including amounts sent off-site for recycling. Section 7 has been modified to include detailed information about on-site recycling activities, as well as changes to the information provided for treatment activities. Section 8 contains the majority of the new source reduction and recycling reporting requirements. For more information about the Form R, call the EPCRA hotline at (800) 535-0202.

CWA Best Management Practices

The Clean Water Act (CWA)'s primary objective is the restoration and maintenance of the chemical, physical, and biological integrity of the Nation's waters.[40] To achieve its objective, the CWA sets forth a series of goals, including attaining fishable and swimmable designations and eliminating the discharge of pollutants into navigable waters. As part of the CWA strategy to eliminate discharges of pollutants to receiving waters, National Pollutant Discharge Elimination System (NPDES) permit limitations have become more stringent.

The principal mechanism for reducing the discharge of pollutants from point sources is through implementation of the NPDES program, established by Section 402 of the CWA. All facilities with point source discharges must apply for and obtain a NPDES permit. EPA has delegated authority to most states to issue NPDES permits. Where State NPDES authorization has not yet occurred, EPA Regions issue NPDES permits.

A NPDES permit is essentially a license that allows a facility to discharge contaminated water. On the NPDES permit application, the facility provides information about the type of facility and type of discharge being requested. If the facility's application is approved, the permitting authority will issue a permit that contains various conditions related to the facility's pollutant discharges. The permit will generally

---

[39] 42 USC 13101-13109.
[40] CWA 101(a), 33 USC 1251(a).

contain specific limitations on contamination levels or specific actions that the facility must take, such as sampling or inspections.

Four minimum elements are typically included in each permit issued:

1. Effluent discharge limitations;

2. Monitoring and reporting requirements;

3. Standard conditions; and

4. Special conditions.

The numeric effluent discharge limits contained in a NPDES permit are based on the most stringent value among technology-based effluent guidelines limitations, water quality-based limitations, and limitations derived on a case-by-case basis. Permits also contain standard conditions that prescribe administrative and legal requirements to which all facilities are subject. Finally, permits may contain any supplemental controls, referred to as special conditions, that may be needed in order to ensure that the regulations driving the NPDES program and, ultimately, the goals of the CWA are met. Best management practices (BMPs) are one such type of supplemental control.

Section 304(e) of the CWA authorizes the EPA Administrator to publish regulations to control discharges of significant amounts of toxic pollutants listed under Section 307 or hazardous substances listed under Section 311 from industrial activities that the Administrator determines are associated with or ancillary to industrial manufacturing or treatment processes.

In 1978, EPA proposed regulations addressing the use of procedures and practices to control discharges from activities associated with or ancillary to industrial manufacturing or treatment processes. The proposed rule indicated how BMPs would be imposed in NPDES permits to prevent the release of toxic and hazardous pollutants to surface waters.[41] While this Subpart (40 CFR Part 125, Subpart K) never became effective, it

---

[41] 43 Fed. Reg. 37078 (Aug. 21, 1978) (40 CFR Part 125, Subpart K, Criteria and Standards for Best Management Practices Authorized Under Section 304(e) of the CWA).

remains in the Code of Federal Regulations and can be used as guidance by permit writers.

Although these regulations were never finalized, EPA and States continue to incorporate BMPs into permits based on the authority contained in Section 304(e) of the CWA and the regulations set forth in 40 CFR 122.44(k). While Section 304(e) of the CWA restricts the application of BMPs to ancillary sources and certain chemicals, the regulations contained in 40 CFR 122.44(k) authorize the use of BMPs to abate the discharge of pollutants under the following circumstances:

1. They are developed in accordance with Section 304(e) of the CWA;

2. Numeric limitations are infeasible, or

3. The practices are necessary to achieve limitations/standards or meet the intent of the CWA.

Thus, permit writers are afforded considerable latitude in employing BMPs as pollution control mechanisms.

As defined by CWA Section 304(e), the discharges to be controlled by BMPs are plant site run off, spillage or leaks, sludge or waste disposal, and drainage from raw material storage. These activities have historically been found to be amenable to control by BMPs. Some examples include the following:

- *Material storage areas* for toxic, hazardous, and other chemicals including raw materials, intermediates, final products, or by-products. Storage areas may be piles of materials or containerized substances. Typical storage containers could include liquid storage vessels ranging in size from large tanks to 55-gallon drums; dry storage in bags, bins, silos and boxes; and gas storage in tanks and vessels. The storage areas can be open to the environment, partially enclosed, or fully contained.

- *Loading and unloading operations* involving the transfer of materials to and from trucks or rail cars, including in-plant transfers. These operations include pumping of liquids or gases from truck or rail car to a storage facility or vice versa, pneumatic transfer of dry chemicals during vehicle loading or unloading, transfer by mechanical conveyor systems, and transfer of bags, boxes, drums, or other containers from vehicles by fork-lift, hand, or other materials handling methods.

- *Facility run off* generated principally from rainfall on a plant site. Run off can become contaminated with harmful substances when it comes in contact with material storage areas, loading and unloading areas, in-plant transfers areas, and sludge and other waste storage/disposal sites. Fallout, resulting from plant air emissions that settle on the plant site, may also contribute to contaminated run off. In addition to BMPs, facility run off from industrial sites may also be directly regulated under the NPDES storm water permitting program.

- *Sludge and waste storage and disposal areas* including landfills, pits, ponds, lagoons, and deep-well injection sites. Depending on the construction and operation of these sites, there may be a potential for leaching of toxic pollutants or hazardous substances to groundwater, which can eventually reach surface waters. In addition, liquids may overflow to surface waters from these disposal operations.

Many facilities currently implement successful measures to reduce and control environmental releases of all types of pollutants. These measures have been successfully implemented both formally as part of best management practice (BMP) plans and informally as part of unwritten standard operating procedures. In the context of the NPDES permit program, permittees are required to develop BMP plans to address specific areas of concern. The BMP plan developed by the permittee becomes an enforceable condition of the permit. The EPA believes that to ensure the continuing and greater successes of these programs, pollution prevention measures should be incorporated into a written company-wide plan.

BMPs may apply to an entire site or be appropriate for discrete areas of an industrial facility. Many of the same environmental controls promoted as part of a BMP plan may currently be used by industry in storm water pollution prevention plans, spill prevention control and countermeasure (SPCC) plans, Occupational Safety and Health Administration (OSHA) safety programs, fire protection programs, insurance policy requirements, or standard operating procedures. Additionally, where facilities have developed pollution prevention programs, controls such as source reduction and recycling/reuse may be similar to those promoted as part of a BMP plan.

With the increase in awareness of pollution prevention opportunities, as well as the increase in legislation and regulatory policies directing efforts towards pollution prevention, much of the traditional focus of BMP activities is being redirected from ancillary activities to industrial manufacturing processes. This redirection is resulting in the integrated application of traditional BMPs and pollution prevention practices into cohesive and encompassing plans that cover all aspects of industrial facilities. Specific guidance for implementing a BMP plan to control water pollution discharges is provided in Chapter 5.

Storm Water Pollution Prevention Plans

In addition to the general best management practices commonly incorporated in NPDES permits, EPA specifically mandates that all permits for storm water pollution discharges contain storm water pollution prevention plans. For each of the three types of storm water permits (i.e., individual, group, general), a storm water pollution prevention plan must be developed for each facility covered by the permit. Storm water pollution prevention plans must be prepared in accordance with good engineering practices and in accordance with factors outlined in 40 CFR 125.3(d) (2) or (3), as appropriate. The plan must identify potential sources of pollution which may reasonably be expected to affect the quality of storm water discharges associated with activities at the facility. In addition, the plan must describe and ensure the implementation of practices that are to be used to reduce the pollutants in storm water discharges associated with activities at the facility and to ensure compliance with the terms and conditions of the permit. Facilities must implement the provisions of the storm water pollution prevention plan as a condition of permit issuance.[42]

---

[42] For a comprehensive discussion of storm water pollution prevention plan requirements, see M. Dennison, *Stormwater Discharges: Regulatory Compliance and Best Management Practices* (Lewis Publishers, 1995).

Pollution Prevention Plan Requirements for Industrial Activities. The following checklist summarizes the pollution prevention plan requirements for industrial facilities covered by a storm water permit:

☐ **Pollution Prevention Team** - Each facility will select a Pollution Prevention Team from its staff, and the Team will be responsible for developing and implementing the Plan.

☐ **Components of the Plan** - The permit requires that the Plan contain a description of potential pollutant sources, and a description of the measures and controls to prevent or minimize pollution of storm water. The description of potential pollutant sources must include:

☐ A map of the facility indicating the areas which drain to each storm water discharge point

☐ An indication of the industrial activities which occur in each drainage area

☐ A prediction of the pollutants which are likely to be present in the storm water

☐ A description the likely source of pollutants from the site

☐ An inventory of the materials which may be exposed to storm water

☐ The history of spills or leaks of toxic or hazardous materials for the past 3 years

The measures and controls to prevent or minimize pollution of storm water must include:

☐ Good housekeeping or upkeep of industrial areas exposed to storm water

☐ Preventive maintenance of storm water controls and other facility equipment

- ☐ Spill prevention and response procedures to minimize the potential for and the impact of spills

- ☐ Test all outfalls to ensure there are no cross connections (only storm water is discharged)

- ☐ Training of employees on pollution prevention measures and controls, and recordkeeping

The permit also requires that facilities:

- ☐ Identify areas with a high potential for erosion and the stabilization measures or structural controls to be used to limit erosion in these areas

- ☐ Implement traditional storm water management measures (oil/water separators, vegetative swales, detention ponds, etc) where they are appropriate for the site

☐ **Inspection/Site Compliance Evaluation** - Facility personnel must inspect the plant equipment and industrial areas on a regular basis. At least once every year a more thorough site compliance evaluation must be performed by facility personnel to:

- ☐ Look for evidence of pollutants entering the drainage system

- ☐ Evaluate the performance of pollution prevention measures

- ☐ Identify areas where the Plan should be revised to reduce the discharge of pollutants

- ☐ Document both the routine inspections and the annual site compliance evaluation in a report.

☐ **Consistency** - The Plan can incorporate other plans which a facility may have already prepared for other permits including Spill Prevention Control and Countermeasure (SPCC) Plans, or Best Management Practices (BMP) Programs.

40 / Pollution Prevention Strategies and Technologies

☐ **Signature** - The plan must be signed by a responsible corporate official such as the president, vice president or general partner.

☐ **Plan Review** - The plan is to be kept at the permitted facility at all times. The plan should be submitted for review only when requested by EPA.

☐ **Semi-Annual Monitoring/Annual Reporting** is required of the following facilities:

- EPCRA Section 313 facilities

- Primary metal industries Standard Industrial Classification (SIC) 33

- Land disposal units / incinerators / Boilers and industrial furnaces (BIFs)

- Wood treatment facilities

- Facilities with coal pile run off

- Battery reclaimers

☐ **Annual Monitoring** is required of the following types of facilities (although there are no reporting requirements for these facilities):

- Airports with at least 50,000 flight operations per year

- Coal-fired steam electric facilities

- Animal handling/meat packing facilities

- Additional facilities, including:

    - SIC 30 and 28 with storage piles for solid chemicals used as raw materials that are exposed to precipitation

    - Certain automobile junkyards

- Lime manufacturing facilities where storm water comes into contact with lime storage piles

- Oil handling sites at oil fired steam electric power generating facilities

- Cement manufacturing and cement kilns

- Ready-mix concrete facilities

- Shipbuilding and repairing facilities

☐ **Additional Monitoring Requirements**

☐ Testing parameters for facilities are listed in the general permits.

☐ At a minimum, all dischargers must conduct an annual site inspection of the facility.

☐ **Alternative Certification**

☐ A discharger is not subject to the monitoring requirements for a given outfall if there is no exposure of industrial areas or activities to storm water within the drainage area of that outfall within a given year.

☐ The discharger must certify, on an annual basis, that there is no exposure to storm water, and such certification must be retained in the storm water pollution prevention plan. Facilities subject to semi-annual monitoring requirements must submit this certification to EPA in lieu of monitoring data.

STATE POLLUTION PREVENTION REGULATION

In response to the new regulatory trend toward pollution prevention, in contrast to the traditional "end-of-pipe" pollution controls, numerous states have been incorporating source reduction and recycling provisions into their regulatory regimes. Many states have made these requirements mandatory through enactment of various types of pollution prevention laws. Others have initiated pollution prevention activities through voluntary programs.

Mandatory State Requirements

A number of states have imposed mandatory pollution prevention requirements designed to reduce wastes generated by commercial and industrial facilities.[43] Table 2.1 outlines the pollution prevention laws of these states.

Table 2.1. Mandatory State Pollution Prevention Programs

Arizona:	Ariz. Rev. Stat. Ann. 49-961 to -73.
California:	Cal. Health & Safety Code 25244.12 to .24.
Georgia:	Ga. Code Ann. 12-8-60 to -83.
Louisiana:	La. Rev. Stat. Ann. 30.2291 to .2295.
Maine:	Me. Rev. Stat. Ann., tit. 38, 2301 to 2312.
Massachusetts:	Mass. Ann. Laws ch. 211, 1 to 23.
Minnesota:	Minn. Stat. Ann. 115D.01 to .12.
Mississippi:	Miss. Code Ann. 49-31-1 to -27.
New Jersey:	N.J. Stat. Ann. 13:1D-35 to -50.
New York:	N.Y. Envtl Conserv. Law 27-0900 to -0925.
Oregon:	Or. Rev. Stat. 465.003 to .037.
Tennessee:	Tenn. Code Ann. 68-212-301 to -312.
Texas:	Tex. Health & Safety Code Ann. 361.501 to .510.
Washington:	Wash. Rev. Code 70.95C.010 to .240

A brief description of some of these laws is provided here.

Arizona

Arizona's pollution prevention law applies to facilities that (1) must file the annual Toxic Chemical Release Inventory Form R required by

---

[43] See, generally, Rabe, Barry G., "From Pollution Control to Pollution Prevention: The Gradual Transformation of American Environmental Regulatory Policy," 8 Envtl & Plan. L.J. 226 (1991) (examining the mandatory pollution prevention programs developed in Massachusetts, Minnesota, and New Jersey).

Pollution Prevention Regulatory Requirements / 43

EPCRA Section 313 or (2) during the preceding 12 months, generated an average of one kilogram per month of an acutely hazardous waste (as defined in 40 C.F.R. Part 261) or an average of 1,000 kilograms per month of hazardous waste.[44] These facilities must file an annual toxic data report with the state environmental protection agency and implement a pollution prevention plan designed to reduce the use of toxic substances and the generation of hazardous wastes.[45] The toxic data report must contain a copy of the Form R filed pursuant to EPCRA Section 313[46] and an annual progress report concerning the facility's pollution prevention plan.[47]

The pollution prevention plan must include the following components:[48]

1. The name and location of and principal business activities at the facility.

2. The name, address and telephone number of the owner or operator of the facility and of the senior official with management responsibility at the facility.

3. A certification by the senior official with management responsibility at the facility that he has read the plan and that it is to the best of his knowledge true, accurate and complete.

4. Specific performance goals for the prevention of pollution, including an explanation of the rationale for each performance goal. The plan must include a goal for the facility and may include goals for individual production processes.

---

[44] Ariz. Rev. Stat. Ann. 49-962(A).

[45] Facilities subject to the reporting requirements of the state's pollution prevention act must file the toxic data report annually until (1) the facility ceases operation or (2) it did not have to file a Form R for the preceding calendar year or (3) for two consecutive years, it did not generate enough hazardous wastes to meet the prescribed threshold quantities. Ariz. Rev. Stat. Ann. 49-962(B).

[46] 42 U.S.C. 13106.

[47] The pollution prevention plan must, at a minimum, cover a two-year time period. Ariz. Rev. Stat. Ann. 49-963(K).

[48] Ariz. Rev. Stat. Ann. 49-963(J).

5. A written policy setting forth management and corporate support for the pollution prevention plan and a commitment to implement the plan to achieve the plan goals.

6. A statement of the plan's scope and objectives.

7. An analysis identifying pollution prevention opportunities to reduce or eliminate toxic substance releases and hazardous waste generation.

8. An analysis of pollution prevention activities that are already in place and that are consistent with the requirements of this article.

9. Employee awareness and training programs to involve employees in pollution prevention planning and implementation to the maximum extent feasible.

10. Provisions to incorporate the plan into management practices and procedures in order to ensure its institutionalization.

11. A description of the options considered and an explanation of why the options considered were not implemented.

In addition to preparing and implementing this pollution prevention plan, each facility must also file an annual progress report. The annual progress report must "analyze the progress made, if any, in pollution prevention including toxics use reduction, source reduction and hazardous waste minimization relative to each performance goal established and relative to the prescribed plan contents. Pollution prevention achieved under previously implemented activities may also be included. In the progress report, the facility also must set forth amendments to the pollution prevention plan and explain the need for any amendments.[49]

If a facility that is required to submit a pollution prevention or an annual progress fails to do so, the state environmental agency must order the facility to submit an adequate plan or report within a reasonable time period of at least 90 days. If the facility fails to comply with this order, the agency may take action against the facility, including inspection of the

---

[49] Ariz. Rev. Stat. Ann. 49-963(L).

facility, gathering necessary information and preparing a plan or progress report at the facility's expense, or entering an administrative compliance order that is enforceable in a proceeding.[50]

In addition to these mandatory requirements, Arizona's law also directs the state Department of Environmental Quality to establish a pollution prevention technical assistance program, which includes a hazardous waste reduction clearinghouse, hazardous waste minimization workshops and training, on-site technical assistance to hazardous waste generators, and incentives for innovative hazardous waste management.[51]

Arizona has also developed a parallel pollution prevention program for state agencies.[52] State agencies that produce hazardous wastes or use toxic substances in excess of the threshold quantities and time limits applicable to other facilities must also file pollution prevention plans. The pollution prevention plan must have a goal of 20 % reduction in hazardous waste within two years, 50 % reduction in hazardous waste within five years and a 70 % reduction in hazardous waste in ten years. The pollution prevention plan must address a reduction in the use of toxic substances and the generation of hazardous wastes. The plan must initially be filed on or before January 1, 1993 and every five years thereafter.[53] Just like other regulated facilities, these state agencies must also submit annual progress reports concerning their pollution prevention plans[54] and file annual toxic data reports.[55]

California

In 1989, California passed the Hazardous Waste Source Reduction and Management Review Act.[56] The primary purpose of the Act is to:[57]

1. Reduce the generation of hazardous waste.

---

[50] Ariz. Rev. Stat. Ann. 49-964(F).
[51] Ariz. Rev. Stat. Ann. 49-965.
[52] Ariz. Rev. Stat. Ann. 49-972, 49-973.
[53] Ariz. Rev. Stat. Ann. 49-972.
[54] Ariz. Rev. Stat. Ann. 49-972(I).
[55] Ariz. Rev. Stat. Ann. 49-973.
[56] Cal. Health & Safety Code 25244.12 to .24.
[57] Cal. Health & Safety Code 25244.13(b).

2. Reduce the release into the environment of chemical contaminants which have adverse and serious health or environmental effects.

3. Document hazardous waste management information and make that information available to state and local government.

The California Department of Toxic Substances Control is required to establish a program, in coordination with other state agencies, that promotes hazardous waste source reduction. The Act promotes the reduction of hazardous waste at its source, and wherever source reduction is not feasible or practicable, encourages recycling. The goal of the Act is to reduce the generation of hazardous wastes in the state by five percent per year from the year 1993 through the year 2000.[58]

The Act only applies to hazardous waste generators who produce more than 12,000 kilograms of hazardous waste in a calendar year, or more than 12 kilograms of extremely hazardous waste[59] in a calendar year.[60] Until December 31, 1997, however, generators of more than 5,000 kilograms in a calendar year of certain hazardous wastes[61] are also subject to the requirements of the Hazardous Waste Source Reduction and Management Review Act.[62]

Each generator regulated under the Act must conduct a source reduction evaluation review and plan every four years, commencing on or before September 1, 1991.[63] The source reduction evaluation review and plan must be conducted and completed for each site according to a specified format and include information concerning each of the following components:

1 The name and location of the site.

2. The SIC Code of the site.

---

[58] Cal. Health & Safety Code 25244.15(e).

[59] Hazardous waste refers to the those wastes considered hazardous under the state's Hazardous Waste Control Law. See Cal. Health & Safety Code 25110.

[60] Cal. Health & Safety Code 25244.15(d)(1).

[61] These hazardous wastes are listed at Cal. Health & Safety Code 25179.7(a)(1)-(3).

[62] Cal. Health & Safety Code 25244.15(d)(3)(A).

[63] Cal. Health & Safety Code 25244.19.

Pollution Prevention Regulatory Requirements / 47

3. Identification of all routinely generated hazardous waste streams which result from ongoing processes or operations that have a yearly volume exceeding five percent of the total yearly volume of hazardous waste generated at the site, or, for extremely hazardous waste, five percent of the total yearly volume generated at the site.

4. For each hazardous waste stream, the review and plan shall include all of the following information:

   -An estimate of the quantity of hazardous waste generated.

   -An evaluation of source reduction approaches available to the generator which are potentially viable, including consideration of input change, operational improvement, production process change, and product reformulation.

5. A specification of, and a rationale for, the technically feasible and economically practicable source reduction measures which will be taken by the generator with respect to each hazardous waste stream. The review and plan shall fully document any statement explaining the generator's rationale for rejecting any available source reduction approach.

6. An evaluation, and, to the extent practicable, a quantification, of the effects of the chosen source reduction method on emissions and discharges to air, water, or land.

7. A timetable for making reasonable and measurable progress towards implementation of the selected source reduction measures.

8. Certification by a registered professional engineer.

9. Four-year numerical goals for reducing the generation of hazardous waste streams, based upon its best estimate of what is achievable in that four-year period.,

10. A progress report as part of its submittal of a biennial report on March 1 of each even-numbered year pursuant to Section 66262.41 of Title 22 of the California Code of Regulations. If the generator is

not required to submit a biennial report pursuant to that regulation, it shall prepare a separate progress report on the same time schedule required for the biennial report. Generators not required to submit a biennial report shall not be required to submit their prepared progress report. Progress reports must address plan implementation activities undertaken by the generator during the two years preceding the year in which the biennial report is required to be submitted.

11. The progress report must briefly summarize and, to the extent practicable, quantify, in a manner which is understandable to the general public, the results of implementing the source reduction methods identified in the generator's source reduction evaluation, review, and plan for each waste stream addressed by that plan. The progress report due on March 1, 1994, and every other progress report thereafter, shall also include an estimate of the amount of reduction the generator anticipates will be achieved by the implementation of source reduction methods during the period between the preparation of the progress report and the preparation of its next progress report.

For the generators who generate less than 12,000 kilograms per year, the Act requires that the California Department of Toxic Substances Control modify the review and plan requirements of Section 25244.19 by substituting a compliance checklist approach for source reduction evaluation reviews and plans.[64] The purpose of the compliance checklist is to provide a simple, understandable method for small generators to comply with the waste reduction requirements of the Act in an inexpensive, convenient manner.

The Act also directs the California Department of Toxic Substances Control to establish a technical and research assistance program to assist generators in identifying and applying source reduction approaches.[65]

---

[64] Cal. Health & Safety Code 25244.15(d)(3)(B).
[65] Cal. Health & Safety Code 25244.17.

Minnesota

Minnesota has also enacted mandatory pollution prevention legislation.[66] The Minnesota Toxic Pollution Prevention Act identifies the "preferred means of preventing toxic pollution as techniques and processes that are implemented at the source and that minimize the transfer of toxic pollutants from one environmental medium to another."[67]

Minnesota - much like the Arizona and California programs discussed in the preceding subsections - seeks to achieve its goals through mandatory toxic pollution prevention plans.[68] Facilities that are required to file toxic chemical release reporting forms pursuant to EPCRA or Minnesota's emergency planning and community right-to-know law[69] must prepare a toxic pollution prevention plan for the facility.[70] Each toxic pollution prevention plan must establish a program identifying the specific technically and economically practicable steps that could be taken during at least the three years following the date the plan is due, to eliminate or reduce the generation or release of toxic pollutants reported by the facility. Toxic pollutants resulting solely from research and development activities need not be included in the plan.

The plan must be updated every two years and contain the following information:[71]

1. A policy statement articulating upper management support for eliminating or reducing the generation or release of toxic pollutants at the facility;

---

[66] Minn. Rev. Stat. Ann. 115D.01 to .12.

[67] Minn. Rev. Stat. Ann. 115D.02(a).

[68] Minn. Rev. Stat. Ann. 115D.07.

[69] Minn. Rev. Stat. Ann. 299K.08.

[70] The Minnesota Toxic Pollution Prevention Act contains different deadlines for plan completion, depending on the type of facility. Minn. Rev. Stat. Ann. 115D.07, subd.1(b)-(d). However, for all facilities that become subject to the requirements of the Act after July 1, 1993, the plan must be completed by six months after the facility's first submittal of a toxic chemical reporting form pursuant to EPCRA or the state's emergency planning and community-right-to-know law. Minn. Rev. Stat. Ann. 115D.07, subd. 1(e).

[71] Minn. Rev. Stat. Ann. 115D.07, subd. 2.

2. A description of the current processes generating or releasing toxic pollutants that specifically describes the types, sources, and quantities of toxic pollutants currently being generated or released by the facility;

3. A description of the current and past practices used to eliminate or reduce the generation or release of toxic pollutants at the facility and an evaluation of the effectiveness of these practices;

4. An assessment of technically and economically practicable options available to eliminate or reduce the generation or release of toxic pollutants at the facility, including options such as changing the raw materials, operating techniques, equipment and technology, personnel training, and other practices used at the facility. The assessment may include a cost benefit analysis of the available options;

5. A statement of objectives based on the assessment and a schedule for achieving those objectives. Wherever technically and economically practicable, the objectives for eliminating or reducing the generation or release of each toxic pollutant at the facility must be expressed in numeric terms. Otherwise, the objectives must include a clearly stated list of actions designed to lead to the establishment of numeric objectives as soon as practicable;

6. An explanation of the rationale for each objective established for the facility;

7. A listing of options that were considered not to be economically and technically practicable; and

8. A certification, signed and dated by the facility manager and an officer of the company attesting to the accuracy of the information in the plan.

The Act requires that regulated facilities prepare and submit annual progress reports on October 1 of each year.[72]

---

[72] Minn. Rev. Stat. Ann. 115D.08, subd. 1.

Minnesota's Toxic Pollution Prevention Act also requires that the state's pollution control agency establish a pollution prevention technical assistance program[73] and provide grants to study or demonstrate the feasibility of applying specific pollution prevention technologies.[74]

Texas

Texas has also passed a mandatory pollution prevention law.[75] Like the Arizona, California, and Minnesota laws discussed in the preceding subsections, the primary goal of the Texas law is to reduce pollution at its source.[76] Pollution that cannot be reduced at the source should be minimized wherever possible. Thus, the law sets forth a two-tier system of pollution prevention, with source reduction as the primary goal and waste minimization as secondary in preference. To carry out these goals, the law requires that regulated waste generators prepare a source reduction and waste minimization plan, which contains components to address source reduction and separate components to address waste minimization activities.[77]

Except for "conditionally exempt small-quantity generators,"[78] all other hazardous waste generators must prepare a source reduction and waste minimization plan, as must facilities subject to toxic chemical release reporting under EPCRA Section 313 that have releases of certain pollutants in excess of quantities established by the state commission.[79]

Regulated waste generators must conduct an initial survey that identifies activities that generate hazardous waste. Facilities subject to toxic chemical release reporting must conduct an initial survey that

---

[73] Minn. Rev. Stat. Ann. 115D.04.
[74] Minn. Rev. Stat. Ann. 115D.05.
[75] Tex. Health & Safety Code Ann. 361.501-509.
[76] Tex. Health & Safety Code Ann. 361.502.
[77] Tex. Health & Safety Code Ann. 361.505(a).
[78] "Conditionally exempt small-quantity generator" means a generator that does not accumulate more than 1,000 kilograms of hazardous waste at any one time on his facility and who generates less than 100 kilograms of hazardous waste in any given month. Tex. Health & Safety Code Ann. 361.501(3).
[79] Tex. Health & Safety Code Ann. 361.504(a). 361.503(c) requires that the state commission and advisory board jointly develop a common list of pollutants or contaminants and the level of releases of those pollutants or contaminants subject to source reduction and waste minimization planning.

identifies activities that result in the release of designated pollutants or contaminants.[80]

In addition to the initial survey, the source reduction and waste minimization plan must contain the following information:[81]

1. Based on the initial survey, a prioritized list of economically and technologically feasible source reduction and waste minimization projects;

2. An explanation of source reduction or waste minimization projects to be undertaken, with a discussion of technical and economic considerations, and environmental and human health risks considered in selecting each project to be undertaken;

3. An estimate of the type and amount of reduction anticipated;

4. A schedule for the implementation of each source reduction and waste minimization project;

5. Source reduction and waste minimization goals for the entire facility, including incremental goals to aid in evaluating progress;

6. An explanation of employee awareness and training programs to aid in accomplishing source reduction and waste minimization goals;

7. Certification by the owner of the facility, or, if the facility is owned by a corporation, by an officer of the corporation that owns the facility who has the authority to commit the corporation's resources to implement the plan, that the plan is complete and correct;

8. An executive summary of the plan; and

9. Identification of cases in which the implementation of a source reduction or waste minimization activity designed to reduce risk to human health or the environment may result in the release of a

---

[80] Tex. Health & Safety Code Ann. 361.505(a)(1).
[81] Tex. Health & Safety Code Ann. 361.505(a)(2)-(10).

different pollutant or contaminant or may shift the release to another medium.

Facilities required to develop source reduction and waste minimization plans must a submit an annual report and a current executive summary of the facility's progress in implementing the plan.[82]

Like the Arizona, California, and Minnesota pollution prevention programs, the Texas law requires that the state office of pollution prevention provide source reduction and waste minimization assistance to regulated facilities and waste generators.[83]

Voluntary State Requirements

In addition to the mandatory pollution prevention requirements discussed in the preceding subsection, other states have chosen, thus far, to promote pollution prevention through voluntary programs, as opposed to the mandatory requirements discussed in the preceding subsection. Although these voluntary state provisions set forth goals to encourage source reduction, waste minimization, and recycling activities, these state programs attempt to achieve these goals through voluntary practices. The voluntary programs have generally adopted educational outreach and technical assistance mechanisms to promote pollution prevention technologies and techniques. Table 2.2 outlines the pollution prevention programs of these states.

---

[82] Tex. Health & Safety Code Ann. 361.506(a)-(b).
[83] Tex. Health & Safety Code Ann. 361.509.

Table 2.2. Voluntary State Pollution Prevention Programs

Alaska:	Alaska Stat. 46.06.021 to .041.
Colorado:	Colo. Rev. Stat. Ann. 25-16.5-101 to -110.
Connecticut:	Conn. Gen. Stat. Ann. Appendix Pamphlet, P.A. 91-376.
Delaware:	7 Del. Code Ann. 7801 to 7805.
Florida:	Fla. Stat. Ann. 403.072 to .074.
Illinois:	Ill. Ann. Stat. ch. 111½, 7951 to 7957.
Indiana:	Ind. Code Ann. 13-9-1 to -7.
Iowa:	Iowa Code Ann. 455B.516 to .518.
Kentucky:	Ky. Rev. Stat. Ann. 224.46-310 to -325.
Rhode Island:	R.I. Gen. Laws 37-15.1-1 to .11
South Carolina:	S.C. Code Ann. 68-46-301 to -312.
Wisconsin:	Wis. Stat. Ann. 144.955.

Chapter 3

Solid Waste Reduction Programs

INTRODUCTION

Until recently, you might not have paid much attention to the solid waste your company produces. Many businesses have been content simply to establish and manage an efficient system for removing trash. Times have changed, however, and so has waste management. In many areas of the country, companies are seeing a dramatic increase in the complexity and costs of managing their waste. At the same time, public concern over the effects of all this waste has grown significantly. Today, more and more consumers are taking environmental considerations into account when purchasing products and services.

In response, innovative companies are incorporating waste reduction principles into their daily operations. What exactly is waste reduction? Waste reduction includes all actions taken to reduce the amount and/or toxicity of waste requiring disposal. It includes waste prevention, recycling, composting, and the purchase and manufacture of goods that have recycled content or produce less waste. Some companies are adopting simple waste reduction options such as reducing paper consumption through the use of electronic mail. Other businesses are reviewing their entire operation to identify and implement as many opportunities for reducing waste as possible. Whether simple alterations or large-scale initiatives, companies are finding that waste reduction offers impressive dividends.

BENEFITS OF WASTE REDUCTION

In addition to saving money through lower waste removal costs—sometimes thousands of dollars annually—waste reduction makes good business sense in other ways, too. Waste reduction can help reduce expenditures on raw materials, office supplies, equipment, and other purchases. Streamlining operations to reduce waste often can enhance overall efficiency and productivity as well. Furthermore, waste reduction measures can help demonstrate concern for the environment, increasing customer loyalty. For many companies, therefore, waste reduction is rapidly becoming an important component of their long-term business planning.

Waste reduction can help protect the environment, too. Waste reduction slows the depletion of natural resources, helps reduce pollution associated with the extraction of raw materials and the manufacture of products, and conserves valuable landfill space. Some waste reduction efforts also serve to reduce hazardous constituents in solid waste.

WASTE REDUCTION APPROACHES

Several different approaches can be taken to reducing the amount and toxicity of soil wastes. The four primary approaches to waste reduction are:

1. *Waste prevention, or source reduction,* is the design, manufacture, purchase, or use of materials and products to reduce the amount and/or toxicity of discarded waste.

2. *Recycling* is the collection and use of materials that would otherwise have been discarded as the raw materials in the manufacture of new products.

3. *Composting* is a natural process by which food scraps, yard trimmings, and other organic materials are collected and allowed to decompose under controlled conditions into a rich, soil-like substance called compost.

4. *Purchasing* is the procurement of products made from recycled materials and/or designed to result in less waste after their useful life.

Each of these approaches is discussed in detail in the next several subsections.

Waste Prevention

The most effective way to reduce wastes is for the company to generate less wastes in the first place. Companies can adopt a wide range of waste prevention strategies, including:

- *Using or manufacturing minimal or reusable packaging.* Encourage suppliers to minimize the amount of packaging used to protect their products or seek new suppliers who offer products with minimal packaging. Work with suppliers to make arrangements for returning shipping materials, such as crates, cartons, and pallets for reuse. In restaurants and company cafeterias, using bulk food and beverage dispensers instead of individual-serving containers also will help prevent waste. When opting for reusable containers, be sure to take steps to ensure proper hygiene is maintained. In addition, examine the packaging used for your own products to determine if it is possible to use fewer layers of materials or to ship merchandise in returnable or reusable containers.

- *Using and maintaining durable equipment and supplies.* Purchase quality, long-lasting supplies and equipment that can be repaired easily, and establish regular maintenance schedules for them. These items will stay out of the waste stream longer, and the higher initial costs are often justified by lower maintenance, disposal, and replacement costs. In addition, these items are replaced far less frequently, offering further cost savings.

- *Reusing products and supplies.* Using durable, reusable products rather than single-use materials is one of the most effective waste prevention strategies. Consider adopting simple, cost-effective measures, such as washing and reusing ceramic mugs in place of disposable cups. Another idea is to reuse common items such as file folders and interoffice envelopes.

- *Reducing the use of hazardous constituents.* Often, substitutes for the standard cleaning solvents, inks, paints, glues, and other materials used by graphics and maintenance departments are available which are free of the hazardous ingredients that otherwise could end up being disposed of with the rest of your company's solid waste. Ask suppliers to direct you toward reformulated products, such as toners with no heavy metals and water-based paints and cleaning solutions.

- *Using supplies and materials more efficiently.* There are many strategies that your company can adopt to reduce waste and conserve materials, including double-sided copying. In addition, purchasing and inventory practices that generate waste unnecessarily can be eliminated. For example, some companies might order large quantities of an item to receive a discounted unit price, only to have a portion of the order end up unused and discarded. Be cautious about over-ordering products with a limited shelf life.

- *Eliminating unnecessary items.* When reviewing your company's operations for opportunities to reduce waste, don't overlook the obvious. Your company may routinely use items that contribute little or nothing to your product or service. A number of effective waste reduction measures may involve simply eliminating the use of unnecessary materials and supplies.

While most of these waste prevention strategies involve daily facility operations, manufacturers can also consider lowering costs and preventing waste by altering the design of products or changing their manufacturing processes. Among the strategies to consider are:

- Using less raw material in a product that you manufacture.

- Avoiding or minimizing the use of hazardous substances in your manufacturing processes.

- Increasing the life span of your products by making them more durable and easier to repair.

- Cutting back on the amount of packaging associated with your products.

- Making your products packaging reusable.

Often, companies dispose of materials that other businesses, nonprofit organizations, or community groups could use in their operations. Instead of being thrown away, these materials can be traded, donated, or sold.

Materials exchanges are an effective waste reduction measure your company might want to consider to transfer items that could otherwise become waste. Materials involved in these exchanges include building supplies, manufacturing remnants, old equipment, and many other items. A company producing animal feed, for example, might use stale baked goods from a bakery as a feed supplement. The plastic wrapping used to protect paper shipments to printing companies typically gets tossed in the dumpster. In a materials exchange, instead of paying for its disposal, a company could give it to a local plastic bag manufacturer.

Your company also might participate in donations programs, giving away items such as office equipment and building supplies to charities and other nonprofit organizations. Some donations programs specialize in food donations.

Increasingly, businesses and state and local governments across the country are developing or participating in these kinds of programs. Contact your state or local solid waste or environmental agency for information on tapping into these networks.

Recycling

The next preferred alternative for waste reduction is recycling. Recycling offers businesses a way to avoid disposing of the waste that cannot be prevented. Many businesses are collecting bottles, cans, paper, corrugated cardboard, and other materials for recycling. If your company is interested in recycling, you will need to design a system to collect the recyclable materials. In many cases, these items also must be sorted and stored. Sometimes a company is responsible for transporting the collected materials to the recycling facility, too. It might be possible, however, to contract with your waste hauler or a local recycling company so that it is responsible for cleaning, transportation, and other steps in the recycling chain. Participation in existing municipal collection efforts also might be an option.

Composting

Composting and "grasscycling" are other effective ways to reduce the amount of waste materials that your company must dispose of. Simply by leaving grass clippings on the lawn, your company can significantly reduce waste that would require management and disposal, while conserving nutrients and reducing the need for fertilizer. Companies with other yard trimmings, such as leaves and branches may also consider collecting them into a pile and composting them on-site. Such programs typically require limited space and attention. In some cases, food scraps, mixed paper, and other organic matter may also be composted with yard trimmings.

Composting can make a significant contribution to achieving your company's waste reduction goals, especially if organic waste comprises a large proportion of its solid waste. In addition, the resulting compost can be used on company grounds as a soil amendment or mulch. If the quality of your compost is high enough, you might also be able to sell it to help cover expenses.

If your company is interested in composting, assistance can often be obtained from state solid waste or environmental agencies; contacting these agencies is particularly important in those states where standards for composting have been established. These agencies, as well as county extension services, can also be helpful in identifying new composting techniques and equipment. Businesses also might participate in municipal or county composting programs. Contact your local department of public works for more information.

Purchasing

Many waste prevention activities will invariably change the way you purchase supplies and equipment. For example, a switch to reusable plates in your cafeteria will eliminate the need to buy single-use plates. In addition, purchasing products with recycled content is another important element of waste reduction. An important complement to your recycling efforts, buying recycled products helps ensure that collected recyclables will actually be used in new products and kept out of disposal facilities. Furthermore, when your company buys a product or package that was manufactured with recycled material, natural resources and energy often are conserved. And many companies have found that recycled products are now priced competitively with products made from virgin materials. To be sure your company is purchasing recycled products whenever possible, it might be necessary to review purchasing specifications to ensure that they do not unnecessarily discriminate against products made from recycled materials.

PROGRAM PLANNING AND ORGANIZATION

Successful waste reduction programs hinge on careful planning and organization. This section discusses how to lay the groundwork for an effective program. The key steps to getting started are:

- Obtaining management support and involvement.

- Establishing your waste reduction team and team leader.

- Setting preliminary program objectives.

- Getting the whole company on board by announcing the program and its goals to all employees.

Management Support

The support of company management is essential for developing a lasting and successful waste reduction program. At the outset of a program, an endorsement from company management is needed to help establish your waste reduction team. Throughout the program, company management can support your team by endorsing program goals and implementation, communicating the importance of reducing waste within the company, guiding and sustaining the program, and encouraging and rewarding employee commitment and participation in the effort. Stressing the range of benefits that can come from waste reduction, such as cost savings and enhanced company image, will help sell the program to management.

Waste Reduction Team

The waste reduction team is a group of employees who are responsible for many of the tasks involved in planning, designing, implementing, and maintaining the program. A team approach allows these tasks to be distributed among several employees and enables employees from all over the company to directly contribute to reducing waste.

Typically, members of your waste reduction team are responsible for:

- Working with company management to set the preliminary and long-term goals of the waste reduction program.

- Gathering and analyzing information relevant to the design and implementation of the program. This is done through a waste assessment, which is discussed in the section on Conducting Waste Assessments.

- Promoting the program to employees and educating them about how they can participate in the effort.

- Monitoring the progress of the program.

- Periodically reporting to management about the status of the program.

The size of your team should relate to the size of your company and be representative of as many departments or operations as possible. For a modest waste reduction program, an effective team might consist of just one or two people. The ideal candidate for a one-person team would be an individual who wears many hats and is familiar with the overall operations of your company. A two-person team might consist of a company manager and an administrative or technical support person. Larger businesses might opt to create a team of employees from different departments to encourage widespread input and support. These individuals can include environmental managers, building supervisors, technical or operational staff, administrative staff, maintenance staff, and purchasing staff, or other employees interested in waste reduction.

Team members can be volunteers or appointed members. To increase the members' motivation and interest, it might be appropriate to make membership a sign of special recognition within the company. Whether volunteer or appointed, however, it is important that members be enthusiastic about the waste reduction program and able to commit time to the effort.

Company management or the team should appoint a knowledgeable and motivated team leader. Depending on the size of the company and the type of program being implemented, the position can require a significant amount of time and energy. The leader must be capable of directing team efforts; administering the planning, implementation, and operation of the waste reduction program; and acting as a liaison between management and the team. Likely candidates include a facilities manager, an environmental manager, or an employee who has championed waste reduction in your company. If possible, the task should be incorporated into the person's job description.

Once your team has been established, members should meet regularly to develop a plan and begin program implementation. The time needed to design and implement a waste reduction program will vary. Generally, large facilities incorporating many different options will need several months to start up a program. Department-specific or more modest programs might be implemented in less than a month. Some businesses might even be able to implement simple options within a matter of days. In any case, the investment of time and resources at this stage will likely be returned by the savings realized through a successful waste reduction program.

Program Objectives

While the general objective of any waste reduction program is to reduce the amount and/or toxicity of municipal solid waste being generated, your first task as a team will be to work with management to establish and record specific, preliminary goals for the program. These goals might include enhancing the company's corporate image or increasing operational efficiency. The goals should be based primarily on how much waste reduction is possible given the level of effort that the company is willing to dedicate to the task. The goals set by the team will provide a framework for specific waste reduction efforts to follow. Keep in mind, however, that the preliminary goals set by the team should be flexible, as they might need to be reexamined and adjusted as specific waste reduction options are considered later on.

Employee Involvement

Once the general direction of the waste reduction program has been established, present the program to the rest of the company. This is a good opportunity to get employees excited and generate some momentum behind the team's efforts. The first step is an announcement from the president or representative of the upper management of the company, demonstrating that the program has full management support and is a high priority for the company. The announcement should:

- Introduce employees to waste reduction.

- Explain how waste reduction can benefit both the company and the environment.

- Outline the design and implementation stages of the program.

- Offer the team leader's name and number and encourage employees to contact him or her with any ideas or suggestions.

Your program is more likely to succeed if you solicit suggestions from employees for reducing waste. To reduce paper, the announcement should be posted in a prominent place, circulated, or distributed through electronic or voice mail, if available.

Throughout the duration of the program, periodic communications (in the form of centrally posted memos or announcements, for example) can help maintain employee support. Employees are likely to appreciate being asked to join in your company's waste reduction efforts, and such offers will encourage consistent participation.

Conducting Waste Assessments

Having established the framework of your company's waste reduction program, the next step for the waste reduction team is to consider conducting a waste assessment. Some teams, especially those planning very limited programs or in companies where the waste stream is well understood, might opt to forego a waste assessment. In fact, many effective waste reduction measures can be adopted without the help of an assessment. The data generated in an assessment can, however, provide your team with a much greater understanding of the types and amounts of waste your company generates. These data can be invaluable in the design and implementation of a waste reduction program. The key steps to conducting a waste assessment are:

- Understanding the purpose of the waste assessment.
- Determining the approach.
- Examining company records.
- Taking a facility walk-through.
- Conducting a waste sort.
- Documenting the waste assessment.

Purpose

The waste assessment serves two basic purposes:

1. To establish a baseline of data by collecting background information on a facility and its current purchasing, waste generation, and management practices.

2. To identify potential waste reduction options for further evaluation.

The data you collect in the waste assessment can be used to identify and evaluate potential waste reduction options, including alternative purchasing practices, reuse, material exchanges, recycling, and composting. The waste assessment also will enable you to examine current waste reduction practices and to quantify their effectiveness. Furthermore, information generated by the assessment can act as baseline data against which the effectiveness of the waste reduction program can be evaluated.

If you do not have the time or resources to conduct a waste assessment, you might consider using industry averages of the amount of waste generated by companies in your field to approximate the amounts and types of waste your company generates. Often, waste generation estimates by general waste category can be obtained for a company's specific type of business and used as the basis for designing a waste reduction program. While this may be the easiest way to approximate your waste generation rate, these estimates are unable to account for specific conditions and may therefore result in inaccuracies. In addition, these potentially inaccurate data can hinder the evaluation process, since measuring waste reduction progress depends on comparing current waste generation data with information regarding the amounts and types of waste produced before program implementation.

Waste Assessment Approaches

Planning and executing an appropriate waste assessment involves determining its scope, scheduling the different assessment activities, communicating the necessary information to employees, and performing the actual assessment. Depending on the objective of your waste reduction program, a waste assessment can involve:

- Examining facility records.

- Conducting a facility walk-through.

- Performing a waste sort.

Your assessment may require just one of these activities, or a combination of approaches.

The team should determine what type of assessment is best for your company based on such factors as the type and size of the facility, the complexity of the waste stream, the resources (money, time, labor, equipment) available to implement the waste reduction program, and the goals of the program. For example, if your facility generates only a few types of waste materials, your team might only need to review company records and briefly inspect facility operations. On the other hand, if your company generates diverse types of waste and has established a goal to cut waste disposal by 50%, the team will need to thoroughly examine and quantify the wastes generated in most company operations by performing a waste sort. Each type of waste assessment activity is described below. Table 3.1 summarizes the strengths and limitations of each activity.

Table 3.1. Waste Assessment Approaches

Records Examination
(Hauler Records)

Strengths:
- May provide accurate data on the weight or volume of waste generated at the facility.
- Can require less time and effort than facility walk-throughs or waste sorts.

Limitations:
- Might not provide adequate data, if accurate waste hauling records do not exist.
- Not likely to provide information about specific waste components.
- Can be difficult to use if more than one business shares a dumpster.

Records Examination
(Purchasing Records)

Strengths:
- Can provide data on waste generation of specific materials or products.
- Tracks major potential waste from the point of origin.
- Can require less time and effort than facility walk-throughs or waste sorts.
- Can be more accurate than waste sorts for tracking small items, low-volume waste materials, and occasional or seasonal waste materials.
- Can help identify the most expensive components of a company's waste.

Limitations:
- Not likely to provide the full picture on waste generation.
- If company purchasing is not centralized, can be incomplete or require substantial effort to collect and analyze.

Table 3.1.—Continued

Facility Walk-Through

Strengths:
- Can require less time and effort than waste sorts.
- Allows first-hand examination of facility operations.
- Can provide qualitative information about major waste components and waste-generated processes.
- Allows interviews with workplace personnel, which can reveal waste prevention, recycling, composting, and purchasing opportunities that would not be found through records examinations or waste sorts.

Limitations:
- Might not identify all wastes generated.
- Might not be representative if only conducted once.
- Does not provide precise information about the quantity of waste generated.

Waste Sort
(Specific Functional Areas)

Strengths:
- Provides quantitative information about specific types of waste and functional areas; appropriate if targeting specific types of waste.
- Requires less time and effort than comprehensive waste sorts.

Limitations:
- Does not provide data on waste generated facility-wide.
- Might omit major components of the facility's waste.
- Might not be representative of the waste in the specific area if only conducted once.

Table 3.1. —*Continued*
Waste Sort (Facility-Wide)
Strengths: ■ Provides waste generation estimates for the entire facility. ■ Provides quantitative information on each waste component. *Limitations:* ■ Requires significant time and effort to conduct. ■ Might not be representative if only conducted once. ■ Does not provide qualitative information on how or why wastes are generated.

Records Examination

Examining company records can provide insight into your company's waste generation and removal patterns. The types of records you might find useful include:

- Purchasing, inventory, maintenance, and operating logs.

- Supply, equipment, and raw material invoices.

- Equipment service contracts.

- Repair invoices.

- Waste hauling and disposal records and contracts.

- Contracts with recycling facilities and records of earned revenues from recycling.

Facility Walk-Through

The walk-through involves touring the facility (and its grounds), observing the activities of the different departments, and talking with employees about waste-producing activities and equipment. A facility walk-through is a relatively quick way to examine a facility's waste-generating practices. Specifically, the walk-through will enable the team to:

- Observe the types and relative amounts of waste produced.

- Identify waste-producing activities and equipment.

- Detect inefficiencies in operations or in the way waste moves through the organization.

- Observe the layout and operations of various departments.

- Assess existing space and equipment that can be used for storage, processing recyclables, and other activities.

- Assess current waste reduction efforts.

- Collect additional information through interviews with supervisors and employees.

While a records examination provides the team with data (such as estimates of the types and amounts of waste generated by your company), the walk-through is an opportunity to observe the connection between the types of waste generated and the actual waste generating activities or processes. In addition, a facility walk-through that includes interviews with grounds keeping staff is a good way to assess the amount of yard trimmings generated by your company. The team should be careful during the walk-through not only to record the types of waste observed and the ways in which waste is generated, but also to consider the potential waste reduction opportunities that lie in increasing the efficiency of these operations.

Before conducting the walk-through, the team leader should check with the managers of the departments that will be toured to avoid disrupting special deliveries, rush orders, or other department functions. He or she also can request that the supervisor and employees of each department be available during the walk-through to answer questions or describe operations. These interviews can offer important additional detail on waste generation and removal practices. Moreover, interviews help keep employees informed and interested in the evolving waste reduction program, and offer an opportunity to ask questions. Employees can also be a valuable source of ideas for reducing waste.

The time necessary to complete a walk-through depends on the size and structure of your facility. For medium-sized businesses with few departments, the walk-through should be quick and relatively easy. It might take as little as an hour to tour facilities like small warehouses, office buildings, and restaurants. Teams in larger companies might need to devote more time and resources to this activity since more departments must be investigated and more employees interviewed. Large office buildings, complexes, or plants might require a day or more to tour.

Waste Sort

If you need more data than company records or a facility walk-through can provide, a waste sort can be conducted. A waste sort involves the physical collection, sorting, and weighing of a representative sample of the company's waste. The goal of the sort is to identify each waste component and calculate as precisely as possible its percentage of the waste your company generates. Depending on your needs, a waste sort can focus on the entire company's waste or target certain work areas. If the team believes one or more specific functional areas are responsible for much of the facility's waste, it may choose to concentrate its waste sort accordingly.

For some companies, it will be feasible to assemble and measure one day's worth of waste. In larger firms where this is impractical, team members might choose instead to assemble a portion of the waste from each department for measuring. However you choose to structure the waste sort, consider whether waste generation at your company varies significantly enough from one day to the next to distort results. Seasonal and periodic variations in waste generation are also common. If the potential for inaccuracy is large enough, the team might want to sort samples on more than one day. Multi-day sampling might provide a more accurate representation of the waste generated at your company. Since the data gathered in the waste sort will be used as the basis for key waste reduction program decisions, it is important that you obtain a truly representative sample of your company's waste. If a representative sample is not obtained, calculations on waste generation, waste composition, and waste removal costs can be skewed significantly.

In addition, waste reduction teams in companies with active recycling programs will need to decide whether their waste sort should measure all materials or target just the portion of the waste stream that is not currently being recycled. For a complete assessment of the types and amounts of waste being generated, the waste reduction team should locate all recycling collection areas and measure the contents to be sure that all waste components are included in the sort. If your company is more concerned with finding ways to reduce just the materials that are not being recycled, it can focus exclusively on the waste collected in company dumpsters. This might also help companies with existing recycling programs to identify the amount of materials that could have been recycled under the current program but ended up being thrown away.

To organize a waste sort, you will need to determine which waste categories to quantify. Typically, the major components of a business' waste include paper, plastic, glass, metal, and organic material such as yard trimmings and food scraps. A range of other types of waste also can be generated by a company depending on the nature of its operations. The team also should choose whether to limit its waste sort to identifying and measuring these major waste component categories or further sorting the waste into subcategories (for example, sorting paper into such subcategories as high-grade, low grade, newsprint, corrugated cardboard, magazines, and other). If possible, the team should strive to separate and measure the waste sample as completely as possible. These precise measurements will be useful later on when the team is determining which materials can be exchanged, reused, sold, or recycled.

Also, consider whether a particular waste component needs to be measured. For example, if you know that a market for recyclable, high-grade paper exists in the area, team members might want to design the sort to ensure that this waste type is quantified accurately.

Documenting the Waste Assessment

Once the team has determined the approach to use, it is time to perform the actual waste assessment. While examining the company's waste generation and management practices, team members also should search for opportunities to reduce waste and increase efficiency. Be sure to document all information gained through the waste assessment. Documenting your findings serves several purposes, including:

- Providing a record of the company's efforts to reduce waste.

- Developing a recordkeeping system so that costs, savings, and waste reduction quantities can be more easily tracked.

- Obtaining baseline data from which to investigate economic and technical feasibility of waste reduction options.

- Obtaining baseline data from which to evaluate the impact of these options once implemented.

Sample Waste Assessment

Waste assessments can be instrumental in gathering detailed information about your company's waste. For companies that plan to focus on a particular portion of their waste, an assessment can be indispensable. At one of Quaker Oats Company's largest food processing plants, in Newport, Tennessee, company officials suspected that a few types of recyclable materials—glass, aluminum, polyethylene terephthalate (PET) plastic, and corrugated cardboard comprised a large percentage of their waste. To find out, a waste assessment was conducted in 1990 focusing on recyclable materials.

Company Records Review. To conduct the assessment, the company set up a team of waste auditors. First, the team reviewed company records, a step that proved to be especially useful. In addition to indicating the total amount of waste generated annually, these records provided waste auditors with data on specific components of the company's waste stream. For example, shipments often arrive at the facility packaged in cardboard cartons. By researching how many cartons the facility received per year and estimating the weight of a single box, the waste auditors were able to calculate the total weight of the cardboard boxes discarded each year.

Facility Walk-Through. Waste auditors also spent a couple of days walking through the plant, observing waste generating processes and watching what was thrown into dumpsters. During the walk-through, the auditors solicited input from both employees and the plant's four department managers. In addition, the managers were asked to complete detailed questionnaires, recording every item thrown out in their department during one week. In place of a detailed waste sort, the auditors used data from the questionnaires and checked the contents of selected dumpsters.

Results. The waste assessment confirmed that glass, aluminum, PET plastic, and cardboard were the major components of the company's waste. Furthermore, the assessment provided highly specific data to help the company plan and set up its recycling program. With these data, the auditors were able to demonstrate to the company's management that, given the huge volume of recyclable waste the plant generated, certain pieces of equipment and capital improvements were justified. Management agreed to invest over $130,000 in a glass crusher, a plastic granulator, a horizontal baler for corrugated cardboard, and dozens of color-coordinated dumpsters (blue for bottles, yellow for plastic, and brown for cardboard).

The waste assessment data also proved useful when negotiating contracts with buyers of recyclable materials. For buyers with minimum shipment requirements, for example, plant managers were able to determine at a glance if the facility could meet the requirement. Assessment data also showed the volume of materials the company expected to process, enabling equipment vendors to recommend a machine of the correct size and capacity.

EVALUATING WASTE REDUCTION OPTIONS

The information collected in the waste assessment now can be used to list, analyze, and choose appropriate waste reduction measures for your company. This section will help the team consider both the operational and the economic feasibility of the options under consideration, as well as the extent to which they will help achieve the goals of your waste reduction program. In addition, the section on Implementing the Waste Reduction Program discusses the process of implementing your program and monitoring it over time to evaluate progress. The key steps of selecting, implementing, and monitoring waste reduction options are:

- Compiling and screening the options.

- Analyzing and selecting the options.

- Implementing the options.

- Educating and training employees.

- Monitoring and evaluating the program.

It is important that the waste reduction team throughly review the potential effects of each waste prevention, recycling, composting, and purchasing option. While a strong consideration is likely to be whether the option's costs are justified by potential savings, the waste reduction team also should consider:

- Effects on product or service quality and product marketing.

- Compatibility with existing operations.

- Equipment requirements.

- Space and storage requirements.

- Operation and maintenance requirements.

- Staffing, training, and education requirements.

- Implementation time.

- Effects on employee morale, environmental awareness, and community relations.

Compiling and Screening Options

Using the findings from the waste assessment, the team should list all the possible waste prevention, recycling, composting, materials exchange, and purchasing measures that it feels might be effective. This list should be compiled based on the goals of your waste reduction program. For example, if your company hopes to reduce waste removal costs as much as possible, and is charged for waste removal based on volume, the list of options should focus on high-volume waste materials. Brainstorming sessions among team members can provide additional options.

Managers and employees who participated in the walk-through also can be consulted for more ideas, if necessary.

After all potential options have been identified, the team should screen them based on criteria such as whether the options will substantially reduce waste removal costs, reduce purchase costs, have low start-up costs, or are likely to boost employee morale. Companies may weigh these criteria differently based on the particular goals of their program. This initial evaluation and screening will help you identify a subset of options that deserve further analysis and possible inclusion in your waste reduction program.

Analyzing and Selecting Options

Once a short list of waste reduction options has been identified, the team should begin the process of deciding which options are the most appropriate for your program. During this evaluation process, the team should be clear on the relative importance of the different criteria against which the options are being measured. Depending on your company's waste reduction goals, for example, cost-effectiveness may not always be the overriding criteria for selected options. Other criteria, such as improved environmental awareness, employee morale, and community relations, may be equally important. In addition, teams whose companies feel cost-effectiveness must be a key criteria should be sure to consider the long-term economic feasibility of an option. While the team may be inclined to disregard a particular option with large start-up costs, the measure may end up yielding impressive savings over several years.

In addition, after completing the evaluation of these options, review the long-term feasibility of the program as a whole. Successful programs can be designed around complementary options that take advantage of their different strengths. Before removing any items from consideration, for example, consider whether certain waste reduction efforts may, over time, save enough money to pay for other waste reduction activities that improve environmental awareness, employee morale, or community and customer relations.

Some options might not require extensive analysis. For example, if your facility already has a copy machine with the ability to make two-sided copies efficiently, then a policy mandating double-sided copying usually can be implemented easily. On the other hand, you will want to carefully analyze complex options that require a significant change in operations or large capital investments. For instance, a food service considering a switch from disposable to reusable service ware needs to assess factors ranging from the cost of new equipment and added labor expenses to the savings from reduced waste removal costs and the avoided purchase of disposable service ware. The health and sanitation aspects of such a switch also should be considered. For complex options, the team will want to contact suppliers, product refurbishers, packaging designers, and any other individuals who could help determine if the option is feasible. These individuals also can help pinpoint any unforeseen obstacles or complications that could hinder implementation.

Waste Prevention Options

When analyzing and selecting specific options, team members should focus first on waste prevention, which will enable your company to eliminate some of its waste. After studying your company's waste generation and management practices, you will likely have compiled a number of waste prevention options. Determine the capital and operating costs of these options and compare them against potential savings and revenues. Be sure to examine the potential operational effects, as well. For example, while modifying packaging can significantly reduce waste, you will want to consider carefully how these changes will affect storage, operations, and labor costs.

One waste prevention option may result in savings in several different areas, including avoided purchasing, storage, materials handling, and removal costs. For example, switching to double-sided copying can result in cost savings associated with reduced paper purchasing, reduced space necessary for paper storage, reduced employee time associated with handling paper and filling paper trays, and reduced paper packaging removal costs. Be sure to consider savings in each of these areas when evaluating waste reduction options.

Recycling Options

Next, evaluate the recycling options the team has identified to better manage waste that cannot be prevented. Before implementing any recycling option, the team needs to consider the marketability of the materials to be collected. To locate potential buyers, contact local recycling companies. Consult the Yellow Pages (under "recycling"), trade associations, chambers of commerce, and state or local government recycling offices for assistance. When conducting preliminary contract discussions with local buyers and haulers, there are a number of questions you should ask. The following list provides a good starting point.

Questions to Ask Recycling Companies

When meeting with recycling companies interested in purchasing your collected materials, there are a number of issues you should discuss, including:

1. What types of recylables will the company accept and how must they be prepared?

Recycling companies might request that the material be baled, compacted, shredded, granulated, or loose. Generally, recyclers will offer a better price for compacted or baled material. Compacting or densifying materials before transporting also can be a cost-effective method of lowering hauling costs for the buyer.

2. What contract terms will the buyer require?

Discuss the length of the potential contract with the buyer. Shorter contracts provide greater flexibility to take advantage of rising prices, while longer contracts provide more security in an unsteady market. Often, buyers favor long-term contracts to help ensure a consistent supply of materials. The terms of payment should be discussed as well, since some buyers pay after delivery of each load, while others set up a periodic schedule. Also, ask whether the buyer would be willing to allow changes to the contract over time. The buyer might want some flexibility as well; in many cases, the buyer will be willing to pay a higher rate in return for a stable supply of quality materials.

3. Who provides transportation?

If transportation services are not provided by the buyer, you will need to locate a hauler to transport materials to the buyer. The Yellow Pages, local waste haulers, and state or local waste management authorities can help provide this information.

4. What is the schedule of collections?

If the recycling company offers to provide transportation, check on the frequency of collections. Some businesses might prefer to have the hauler be on call, picking up recyclables when a certain weight or volume has been reached. Larger companies might generate enough recyclable material to warrant a set schedule of collections.

5. What are the maximum allowable contaminant levels and what is the procedure for dealing with rejected loads?

Inquire about what the buyer has established as maximum allowable contaminant levels for food, chemicals, or other contaminants. If these requirements are not met, the buyer might reject a contaminated load and send it back to your company. The buyer also might dispose of a contaminated load in a landfill or combustor, which can result in your company incurring additional costs.

6. Are there minimum quantity requirements?

Find out whether the buyer requires a minimum weight or volume before accepting delivery. If a buyer's minimum quantity requirements are difficult to meet, consider working with neighboring offices or retail spaces. By working together, it might be possible to collect recyclables in central storage containers and thereby meet the buyer's requirements.

7. Where will the waste be weighed?

Ask where the material will be weighed, and at what point copies of the weight slips will be available. Weighing the material before it is transported will eliminate the problem of lost weight slips and confirm the accuracy of the weight recorded by the buyer.

8. Who will provide containers for recyclables?

Buyers should be asked whether they will provide containers in which to collect, store, and transport the material, and whether there is a fee for this service.

9. Can "escape clauses" be included in the contract?

Such clauses establish the right of a company to be released from the terms of the contract under conditions of noncompliance by the buyer.

10. Be sure to check references.

Obtain and thoroughly check the buyer's references with existing contract holders, asking these companies specifically whether their buyer is fulfilling all contract specifications.

Be sure to carefully weigh the cost-effectiveness and potential operational effects of your recycling options. Recycling programs, especially more ambitious efforts, often require purchases of equipment like containers, compactors, and balers. Additional labor also might be required. Moreover, steps might be necessary to ensure that contamination of collected materials is minimized. Some companies also may have to pay a fee to have their collected recyclable material removed. In many cases, however, the savings and revenues (such as reduced removal costs and revenues from selling collected materials) will offset these costs. In addition, consider whether the new recycling program will affect current purchasing practices. For instance, your company might want to begin buying exclusively white legal pads instead of yellow ones to take advantage of the strong market for white office paper. Also examine the extent to which internal collection, transfer, and storage systems are needed and whether these new systems will be compatible with existing operations.

Composting Options

If the team discovers that yard trimmings or other organic matter make up a significant percentage of your company's waste, evaluate the feasibility of "grasscycling" or composting. Most companies can benefit by "grasscycling"—leaving cut grass on the lawn where it will decompose quickly and help add nutrients that improve the quality of the lawn. Though not necessary, a mulching mower can cut the grass clippings into smaller pieces, allowing them to decompose more quickly. Your company will save time and money by no longer bagging the clippings, and will reduce its disposal fees.

If your company has available outdoor space, on-site composting can be used. Companies with composting programs usually find them to be a cost-effective method for turning lawn trimmings into a product that may then be sold or used on company grounds. The team can design a program to collect all types of organic materials into piles for composting, or a simpler program designed to compost yard trimmings only might be used. If the local municipal government operates or participates in a composting project, off-site composting also may be an option. A program can be designed to collect and store organic materials and, if necessary, haul it to the composting facility. Even when hauling is necessary, however, these programs also tend to be cost-effective.

To determine if composting is appropriate for your company's waste reduction program, it is important to calculate likely startup and ongoing expenses against projected savings at the outset. Estimate likely costs, including training grounds keeping staff in compost management, educating company employees about participating in the program, equipment and operating costs, and compare these against projected savings in waste removal costs and the potential for selling the compost or using it on-site in place of commercial mulch.

In addition to grasscycling and composting, other practices can reduce yard trimmings at your facility and should be considered by the team. The team can investigate the possibility of chipping other ground debris, like branches, into mulch. The mulch can be used on company property to reduce weeds and conserve moisture around plantings. Other options include planting low maintenance plants. Slow growing species and evergreen trees generally do not create large amounts of debris.

Purchasing Options

During the waste assessment, the team may have noted purchasing changes that could help reduce waste, from buying supplies with reduced packaging to careful inventory control to avoid over ordering and possibly throwing away perishable items. In addition, during the team's exploration of local recycling markets, the need for favoring products made with recycled content also may have become evident. In any business, many opportunities exist to use the company's buying power to reduce waste and encourage the growth of recycling markets. To identify specific changes in purchasing that your company could adopt, the team might contact its suppliers and discuss alternative products that would meet the new purchasing criteria. Check with other suppliers, as well, to see what they may be able to offer. In addition, various industry groups, state solid waste agencies, and federal information services, such as EPA's RCRA Hotline can help identify ways to reduce waste through product purchasing and sources of products made from recycled materials.

After you have identified opportunities to purchase recycled products and products that can help you reduce waste, each item should be evaluated in terms of availability and cost. Reduced waste and recycled products do not necessarily cost more than other products. For example, while paper made from recycled fibers was once considerably more expensive than virgin paper, the price of paper with recovered content is now competitive with traditional paper. In addition, be sure to compare recycled or reduced-waste products to other products on the basis of long-term costs, rather than purchasing costs alone. For example, while benches and picnic tables made from recycled plastic may initially cost more than their wooden counterparts, they last up to four times longer and do not require maintenance. Similarly, while reusable products may cost more to purchase initially, they often save money over time by avoiding frequent purchases of single-use items.

IMPLEMENTING THE WASTE REDUCTION PROGRAM

Having determined the initial waste reduction measures to adopt, the team should now begin to implement the measures. Consider building your program slowly, implementing a few options at a time, so employees are not overwhelmed by changes in procedure. This is particularly important for more complex waste reduction programs. Building slowly also provides an opportunity to identify, assess, and solve any operational problems in the early stages. If, however, a program involves only a few simple measures, it might be possible to implement all options at once.

Employee Training and Education

As the team begins to implement the waste reduction program, it is essential that all employees be informed about the program and the importance of their cooperation and involvement. Be sure to update employees regarding the options being implemented, changes in work patterns or equipment, expected benefits, and their roles and responsibilities. These messages can be conveyed in a variety of ways, including:

- Staff meetings and training sessions.

- Employee newsletters.

- Posters, signs, or flyers.

- Notices on electronic mail.

- Special events, such as slogan contests, cash awards, or other recognition for waste reduction activities.

- New employee orientation.

- Job performance standards.

Some companies can effectively reach all their employees by circulating memos or holding informal meetings. Larger businesses might need to conduct a full-scale education or training campaign to be sure their entire company is aware of and involved in the program.

These outreach techniques also should be used to keep staff up-to-date on the program's successes and problems. Employees will feel a greater stake in the program if they receive frequent updates on the quantity of waste being reduced, reused, or recycled; the recycled products being purchased; and the cost savings that have resulted. These reports also might impress management, increasing their commitment to the program.

Another method of sustaining employee interest is to encourage them to submit new ideas for increasing the efficiency of company operations. You also might consider asking employees to help with program implementation. These employees could notify program coordinators or monitors when recycling containers are full or oversee waste reduction measures, such as double-sided copying in their department, to ensure that everyone understands and complies with the policy.

Employee Participation

In 1990, the independently owned and operated Boston Park Plaza Hotel & Towers expanded its modest white paper recycling initiative into a comprehensive environmental program that includes waste prevention, recycling, energy and water conservation, as well as education and communications. Recognizing that the effort's success depended on widespread employee support, management launched the program with a formal "Environmental Policy" statement signed by the president and the environmental program director. A copy of the statement was distributed in three languages—English, Spanish, and French—to all 600 employees, and was framed and posted in each department at the 977-room hotel.

The next step was to develop an ongoing effort to increase employee participation. The management at The Boston Park Plaza quickly learned that regular communication with individual employees—on the phone or one-on-one - kept the staff informed about the program and encouraged them to suggest improvements. The hotel adopted many other outreach methods, including:

- *Holding regular "Green Team" meetings.* Representatives from different departments meet monthly to discuss possible program changes. Green Team members also encourage fellow employees to participate in the program.

- *Holding monthly company-wide "Green Day" luncheons.* At lunchtime on the third Thursday of every month, the company celebrates Green Day. In the employee cafeteria, educational displays are set up and a raffle is held to give away prizes, such as canvas tote bags, plants, and water-saving devices.

- *Publicizing program changes and achievements in the company newsletter.* Articles highlight employees' efforts, press coverage of the program, and basic operations and changes.

- *Announcing special happenings in memos and paycheck stuffers.* When the hotel's program receives outstanding recognition, such as the President's Environment and Conservation Challenge Award (which it won in 1992), employees get a memo from the hotel president thanking them for their assistance and support.

- *Rewarding employees for program involvement.* Once a year, the most dedicated employees are awarded a small gift and a framed certificate praising them for their involvement.

In addition to attracting national attention, The Boston Park Plaza Hotel has proved that even a luxury hotel can "reduce, reuse, and recycle" without sacrificing quality. In fact, since the program was launched, hotel executives estimate having generated more than $750,000 in new convention business due to the clientele's desire to support such an environmentally-conscious establishment.

Program Evaluation

Waste reduction is a dynamic process. Once the program is underway, the team will need to evaluate its effectiveness to see if preliminary goals are being met. In addition, once the potential for reducing waste in the company becomes better understood, consider establishing long-term goals for the program. It is important to evaluate the program periodically to:

- Keep track of program success and to build on that success (e.g., waste reduced, recycling rates achieved, money saved).

- Identify new ideas for waste reduction.

- Identify areas needing improvement.

- Document compliance with state or local regulations.

- Determine the effect of any new additions to the program.

- Keep employees informed and motivated.

The best way to assess and monitor program operations is through continued documentation. Perform your first evaluation after the program has been in place long enough to have an effect on your company's waste generation rate, usually about one year. In addition, it might be worthwhile to conduct additional periodic waste assessments to determine further changes in your company's waste. If an assessment already has been performed, subsequent ones will be much easier to conduct. Also consider reviewing your company's waste removal receipts and purchasing records, or preparing a summary of recycling receipts and waste assessment worksheets.

Many companies are finding that waste reduction makes economic and environmental sense. By working with other employees in your company as a team, you can devise and implement a successful waste reduction program. Not only can such a program look good on the bottom line, but it also can reflect well on your company.

COMMON SOLID WASTE REDUCTION PRACTICES

This section summarized some common waste reduction practices that companies have used to reduce waste. You might be able to adapt some of these to your business after considering your particular needs and local conditions.

1. *Writing/Printing Paper*

    - Establish a company-wide double-sided copying policy, and be sure future copiers purchased by your company have double-sided capability.

    - Reuse envelopes or use two-way ("send-and-return") envelopes.

    - Keep mailing lists current to avoid duplication.

Solid Waste Reduction Programs / 85

- Make scratch pads from used paper.

- Circulate (rather than copy) memos, documents, periodicals, and reports.

- Reduce the amount of advertising mail you receive by writing to the Direct Marketing Association Mail Preference Service, P.O. Box 9008, Farmingdale, NY 11735-9008, and ask that your business be eliminated from mail lists.

- Use outdated letterhead for in-house memos.

- Put company bulletins on voice or electronic mail or post on a central bulletin board.

- Save documents on hard drives or floppy disks instead of making paper copies.

- Use central files to reduce the number of hard copies your company retains.

- Proof documents on the computer screen before printing.

- Eliminate unnecessary reports.

- Donate old magazines and journals to hospitals, clinics, or libraries.

2. *Packaging*

    - Order merchandise in bulk.

    - Purchase products with minimal packaging and/or in concentrated form.

    - Work with suppliers to minimize the packaging used to protect their products.

    - Establish a system for returning cardboard boxes and foam peanuts to suppliers for reuse.

- Request that deliveries be shipped in returnable and/or recyclable containers.

- Minimize the packaging used for your products.

- Use reusable and/or recyclable containers for shipping your products.

- Repair and reuse pallets or return them to your supplier.

- Reuse newspaper and shredded paper for packaging.

- Reuse foam packing peanuts, "bubble wrap," and cardboard boxes, or donate to another organization.

3. *Equipment*

    - Rent equipment that is used only occasionally.

    - Reuse worn out tires for landscaping, swings, etc.

    - Purchase remanufactured office equipment.

    - Establish a regular maintenance routine to prolong the life of equipment like copiers, computers, and heavy tools.

    - Use rechargeable batteries where practical.

    - Install reusable furnace and air conditioner filters.

    - Reclaim usable parts from old equipment.

    - Recharge fax and printer cartridges or return them to the supplier for remanufacture.

    - Sell or give old furniture and equipment to other businesses, local charitable organizations, or employees.

4. *Organic Waste*

    - Compost yard trimmings or ask your landscape contractor to compost them.

    - If unable to compost on-site, investigate participating in a municipal composting program.

    - Choose a landscape design that needs low maintenance.

    - Use a worm bin to convert non-fatty food wastes into potting soil (called vermicompost).

    - Use a mulching lawn mower and leave grass clippings on the lawn.

5. *Inventory/Purchasing*

    - Implement an improved inventory system (such as systems based on optical scanners) to provide more precise control over supplies.

    - Avoid ordering excess supplies that may never be used.

    - Advertise surplus and reusable waste items through a materials exchange.

    - Set up an area in your business for employees to exchange used items.

    - Donate surplus produce to food banks, if still edible.

    - Substitute less toxic or nontoxic products for products, such as inks, paints, and cleaning solvents.

    - Use products that promote waste reduction (products that are more durable, of higher quality, recyclable, reusable).

    - Where appropriate, order supplies in bulk to reduce excess packaging.

6. Consumer Education

- Teach your customers about the importance of reducing waste. Effective tools include:

  -Promotional campaigns
  -Brochures and newsletters
  -Banners
  -Newspaper advertisements
  -Product displays
  -Store signs
  -Information on labels

- Offer customers waste-reducing choices, such as:

  -Items in bulk or concentrate
  -Solar-powered items, such as watches, calculators, and flashlights
  -Rechargeable batteries
  -Durable, repairable merchandise
  -Returnable bottles

- Encourage reuse of shopping bags by offering customers the choice of buying their own bag, providing a financial incentive for reuse.

- Offer customers a rebate when they reuse grocery bags, containers, mugs, or cups for refilling.

COMMON RECYCLABLE MATERIALS

The following is a list of common recyclable materials:

A. Paper

1. *High-Grade Paper*

High-grade paper is usually generated in office environments and can earn recycling revenues when present in sufficient quantity. Types of high grade paper include:

-*Computer paper* (also known as Computer Print Out or CPO). Can be all white or have a white main fiber with bright green or blue bars.

-*White ledger*. Most white office paper, including white computer paper, copy machine paper, letterhead, white notebook paper, and white envelopes. Common contaminants include glossy paper, wax-coated paper, latex adhesive labels, envelopes with plastic windows, and carbon paper.

-*Tab cards*. Usually manilla-colored computer cards; may be other colors but must be separated by color to be valuable as a high grade paper.

2. *Other Papers*

These papers are less valuable than high-grade paper in terms of recycling, although they still can be cost-effective to recycle in many cases. Examples of other types of paper include:

-*Colored ledger*. Most non-white office paper, including carbonless paper, file folders, tablet paper, colored envelopes, and yellow legal paper.

-*Corrugated Cardboard* (also known as Old Corrugated Cardboard or OCC). Includes unbleached, unwaxed paper with a ruffled (corrugated) inner liner. It usually does not include liner board or press board, such as cereal boxes and shoe boxes. For most businesses, cardboard is a cost-effective material to recycle.

-*Newspaper* (also known as Old News Print or ONP). It is most valued when separated from other paper types, but can be recycled as mixed waste paper.

-*Miscellaneous waste paper*. Encompasses most types of clean and dry paper which do not fall into the categories mentioned above, including glossy papers, magazines, catalogs, telephone books, cards, laser-printed white ledger, windowed envelopes, paper with adhesive labels, paper bags, wrapping paper, packing paper, sticky-backed notes, and glossy advertising paper. This mixed paper has limited value in existing markets.

-*Mixed waste paper*. Paper that is unsegregated by color, quantity, or grade (e.g., combination of white ledger, newsprint, colored paper, envelopes without windows, computer paper, glossy paper, etc.). Mixed paper generally sells below the price of the least valuable paper in the mix.

B. Glass

1. *Color-Separated*

This includes all container glass that is separated into clear, green, and brown. When this glass is broken or crushed for recycling, it is called "flint," green," and "amber" cullet, respectively.

2. *Mixed Color*

This is the same as color-separated glass except clear, green, and brown glass are mixed together. It generally has very limited market value.

C. Plastic

There are 7 types of plastic which are identified by a Society of Plastics Industry (SPI) code number ranging from 1 to 7. These numbers are usually found on the bottom of plastic containers inside a three-arrow recycling symbol. A description of each kind of plastic is presented below. Also, you may check with the Society of the Plastics Industry at 1-800-2-HELP-90 for information about haulers/recyclers in your area. Some recyclers only accept a sub-category of the ones presented below. For example, a recycler may only accept HDPE milk jugs and not all HDPE products.

1. *PET (SPI=1)*

Polyethylene terephthalate (PET) is the most readily recyclable material at this time. It includes 1- and 2-liter clear soda bottles, as well as some bottles containing liquor, liquid cleaners, detergents, and antacids.

2. HDPE (SPI=2)

High-Density Polyethylene (HDPE) is currently recyclable in some areas. This class includes milk, juice, and water jugs, base cups for some plastic soda bottles, as well as bottles for laundry detergent, fabric softener, lotion, motor oil, and antifreeze.

3. PVC (SPI=3)

Polyvinyl Chloride (PVC, also referred to simply as "vinyl") includes bottles for cooking oil, salad dressing, floor polish, mouthwash, and liquor, as well as "blister packs" used for batteries and other hardware and toys.

4. LDPE (SPI=4)

Low-Density Polyethylene (LDPE) includes grocery bags, bread bags, trash bags, and a variety of other film products. LDPE is currently being recycled by some of the major retail chains.

5. Polypropylene (SPI=5)

Polypropylene includes a wide variety of packaging such as yogurt containers, shampoo bottles, and margarine tubs. Also cereal box liners, rope and strapping, combs, and battery cases.

6. Polystyrene (SPI=6)

Polystyrene includes Styrofoam™ coffee cups, food trays, and "clamshell" packaging, as well as some yogurt tubs, clear carry-out containers, and plastic cutlery. Foam applications are sometimes called EPS, or Expanded Polystyrene. Some recycling of polystyrene is taking place, but is limited by its low weight-to-volume ratio and its value as a commodity.

7. Other (SPI=7)

Can refer to applications which use some of the above six resins in combination or to the collection of the individual resins as mixed plastic (e.g., camera film can include several types of plastic resins). Technology exists to make useful items such as plastic "lumber" out of mixed plastic resins, but generally the materials are more useful and valuable if separated into the generic resin types described above.

D. Metals

1. *Aluminum*

Included in this category are aluminum beverage cans, as well as clean aluminum scrap and aluminum foil. Currently, aluminum is a highly valued material for recycling.

2. *Tin-Coated Steel Containers*

Includes cans used for food packaging (i.e., canned foods). Some local recyclers may require cans to be cleaned and crushed with labels removed.

3. *Bimetal Containers*

A typical example includes tin-plated steel cans with an aluminum "pop top" (e.g., peanut cans). These containers can be separated from aluminum cans by using a magnet. [*Note:* Technically, tin cans are bimetal, but we do not consider them when referring to bimetal cans.] Many recyclers accept bimetal containers with tin-coated steel cans.

4. *Non-Ferrous Metals*

Includes most types of scrap metal which do not contain iron (such as copper and brass). This scrap can be a relatively valuable commodity, depending on quantity. It is often recycled through scrap metal dealers, although some general recyclers will handle it with other materials.

5. *Ferrous Metal*

Includes iron and iron-containing metal scrap. Ferrous metal is handled in the same manner as non-ferrous metal but generally has lower market value.

E. Compostables

Usually, compostable materials include food scraps and yard trimmings. Paper that cannot be recycled also can be composted.

1. *Food Scraps*

Includes grease-free organic scraps from restaurants, cafeterias, motels, and other places producing food waste. It is technically possible to compost food waste in a manner similar to yard trimmings, although additional problems with rodents and other scavengers need to be addressed. Some states allow farmers to sterilize food waste and use it as animal feed.

2. *Yard Trimmings*

Includes landscaping debris, grass clippings, branches, and leaves. There are large-scale facilities which compost yard trimmings, producing a product which can be used for mulch, potting soil, landfill cover, and soil amendment. Also, composting can be performed directly on-site or in backyards.

F. Miscellaneous Recyclables

1. *Lead-Acid Batteries*

Lead-acid batteries are used in automobiles, back-up lighting systems, lawn mowers, and computers. Lead-acid batteries contain lead, a toxic metal, and sulfuric acid. Many states prohibit disposal of lead-acid batteries as municipal solid waste, and many require either retailers, wholesalers, or distributors to take back batteries. Currently, about 90% of lead-acid batteries used in automobiles are recycled.

2. *Household Batteries*

Household batteries come in a variety of types, including alkaline, carbon-zinc, mercuric-oxide, silver-oxide, zinc-air, and nickel-cadmium. Currently, only button batteries containing mercury and silver or nickel-cadmium batteries can be recycled, often at a net cost.

Chapter 4

Hazardous Waste Minimization Programs

INTRODUCTION

On May 18, 1993, the EPA issued guidance to hazardous waste generators on how to develop and implement a RCRA waste minimization plan.[84] This chapter discusses the benefits of waste minimization and explains the necessary components of an effective hazardous waste minimization program.

EPA considers waste minimization, the term employed by Congress in the RCRA statute, to include:

1. Source reduction; and

2. Environmentally sound recycling.

The first category, source reduction, is defined in Section 6603(5)(A) of the Pollution Prevention Act,[85] as any practice which:

1. Reduces the amount of any hazardous substance, pollutant, or contaminant entering any waste stream or otherwise released into the environment (including fugitive emissions) prior to recycling, treatment, or disposal; and

2. Reduces the hazards to public health and the environment associated with the release of such substances, pollutants, or contaminants.

---

[84] "Guidance to Hazardous Waste Generators on the Elements of a Waste Minimization Program," 58 Fed. Reg. 31114 (May 28, 1993).
[85] 42 USC 13102(5)(a).

The term includes equipment or technology modifications, process or procedure modifications, reformulation or redesign of products, substitution of raw materials, and improvements in housekeeping, maintenance, training, or inventory control. EPA relies on this definition for use in identifying opportunities for source reduction under RCRA.

The second category, environmentally sound recycling, is the next preferred alternative for managing those pollutants that cannot be reduced at the source. In the context of hazardous waste management, there are certain practices or activities that the RCRA regulations define as "recycling." The definitions for materials that are "recycled" are found in 40 C.F.R. Section 261.1(c).

EPA considers recycling activities that closely resemble conventional waste management activities not to constitute waste minimization. Unfortunately, it is not always easy to distinguish recycling from conventional treatment.[86] Treatment for the purposes of destruction or disposal is not part of waste minimization, but is, rather, an activity that occurs after the opportunities for waste minimization have been pursued.

Transfer of hazardous constituents from one environmental medium to another also does not constitute waste minimization. For example, the use of an air stripper to evaporate volatile organic constituents from an aqueous waste only shifts the contaminant from water to air. Furthermore, concentration activities conducted solely for reducing volume do not constitute waste minimization unless, for example, concentration of the waste is an integral setup in the recovery of useful constituents prior to treatment and disposal. Similarly, dilution as a means of toxicity reduction would not be considered waste minimization, unless dilution is a necessary step in a recovery or a recycling operation.

BENEFITS OF WASTE MINIMIZATION

Waste minimization provides additional environmental improvements over "end of pipe" control practices, often with the added benefit of cost savings to generators of hazardous waste and reduced levels of treatment, storage, and disposal. Waste minimization has already been shown to result in significant benefits for industry, as evidenced in numerous success stories documented in available literature.

---

[86] See 56 Fed. Reg. 7143 (Feb. 21, 1991); 53 Fed. Reg. 522 (Jan. 8, 1988).

The benefits to companies that pursue waste minimization often include:

1. Minimizing quantities of hazardous waste generated, thereby reducing waste management and compliance costs and improving the protection of human health and the environment;

2. Reducing or eliminating inventories and possible releases of "hazardous chemicals";

3. Possible decrease in future Superfund and RCRA liabilities, as well as future toxic tort liabilities;

4. Improving facility mass/energy efficiency and product yields;

5. Reducing worker exposure; and

6. Enhancing organizational reputation and image.

Waste minimization programs are being implemented by a wide array of organizations. Numerous state governments have also enacted legislation requiring facility specific waste minimization programs, and other states have legislation pending that may mandate some type of facility-specific waste minimization program.

ELEMENTS OF A HAZARDOUS WASTE MINIMIZATION PROGRAM

EPA's guidance on the elements of a waste minimization program is intended to assist companies and individuals to properly certify that they have implemented a program to reduce the volume and toxicity of hazardous waste to the extent "economically practicable." The guidance is directly applicable to generators who generate 1000 or more kilograms per month of hazardous waste ("large quantity" generators) or to owners and operators of hazardous waste treatment, storage, or disposal facilities who manage their own hazardous waste on-site.

Small quantity generators who generate greater than 100 kilograms but less than 1000 kilograms of hazardous waste per month are not subject to the same "program in place" certification requirement as large quantity generators. Instead, they must certify on their hazardous waste manifests that they have "made a good faith effort to minimize" their waste generation. Nevertheless, EPA encourages small quantity generators to develop their own waste minimization programs to show good faith efforts.

According to the EPA guidance (which is not a formal regulation, and therefore, not enforceable) the following basic elements should be part of most waste minimization programs:

1. Top management support;

2. Characterization of waste generation and waste management costs;

3. Periodic waste minimization assessments;

4. Appropriate cost allocation;

5. Encouragement of technology transfer; and

6. Program implementation and evaluation.

Thus, generators should consider these elements when designing multimedia pollution prevention programs directed at preventing or reducing wastes, substances, discharges, and/or emissions to all environmental media—air, land, surface water, and groundwater. Each of these elements is discussed below.

Management Support

Top management should support a company-wide effort. There are many ways to accomplish this goal. Some of the methods described below may be suitable for some companies, while not for others. However, some combination of these techniques or similar ones will demonstrate top management support:

1. Make waste minimization a part of the company policy. Put this policy in writing and distribute it to all departments and individuals. Each individual, regardless of status or rank, should be encouraged to identify opportunities to reduce waste generation. Encourage workers to adopt the policy in day-to-day operations and encourage new ideas at meetings and other organizational functions. Waste minimization, especially when incorporated into company policy, should be a process of continuous

improvement. Ideally, a waste minimization program should become an integral part of the company's strategic plan to increase productivity and quality.

2. Set explicit goals for reducing the volume and toxicity of waste streams that are achievable within a reasonable timeframe. These goals may be quantitative or qualitative. Both can be successful.

3. Commit to implementing recommendations identified through assessments, evaluations, and waste minimization teams.

4. Designate a waste minimization coordinator who is responsible for facilitating effective implementation, monitoring, and evaluation of the program. In some cases (particularly in large multifacility organizations), an organizational waste minimization coordinator may be needed in addition to facility coordinators. In other cases, a single coordinator may have responsibility for more than one facility. In these cases, the coordinator should be involved in or be aware of operations and should be capable of facilitating new ideas at each facility. It is also useful to set up self-managing waste minimization teams chosen from a broad spectrum of operations: engineering, management, research and development, sales and marketing, accounting, purchasing, maintenance, and environmental staff personnel. These teams can be used to identify, evaluate, and implement waste minimization opportunities.

5. Publicize success stories. Set up an environment and select a forum where creative ideas can be heard and tried. These techniques can inspire additional ideas.

6. Recognize individual and collective accomplishments. Reward employees who identify cost-effective waste minimization opportunities. These rewards can take the form of collective and/or individual monetary or other incentives for improved productivity/waste minimization.

7. Train employees on the waste-generating impacts that result from the way they conduct their work procedures. For example, purchasing and operations departments could develop a plan to purchase raw materials with less toxic impurities or return leftover materials to vendors. This approach can include all departments, such as those in research and development,

capital planning, purchasing, production operations, process engineering, sales and marketing, and maintenance.

Waste Generation and Management Costs

Maintain a waste accounting system to track the types and amounts of wastes as well as the types and amounts of the hazardous constituents in wastes, including the rates and dates they are generated. Each organization must decide the best method to obtain the necessary information to characterize waste generation. Many organizations track their waste production by a variety of means and then normalize the results to account for variations in production rates.

In addition, a waste generator should determine the true costs associated with waste management and cleanup, including the costs of regulatory oversight compliance, paperwork and reporting requirements, loss of production potential, costs of materials found in the waste stream (perhaps based on the purchase price of those materials), transportation/ treatment/storage/disposal costs, employee exposure and health care, liability insurance, and possible future RCRA or Superfund corrective action costs. Both volume and toxicities of generated hazardous waste should be taken into account. Substantial uncertainty in calculating many of these costs, especially future liability, may exist. Therefore, each organization should find the best method to account for the true costs of waste management and cleanup.

Waste Minimization Assessments

Different and equally valid methods exist by which a waste minimization assessment can be performed. Some organizations identify sources of waste by tracking materials that eventually wind up as waste, from point of receipt to the point at which they become a waste. Other organizations perform mass balance calculations to determine inputs and outputs from processes and/or facilities. Larger organizations may find it useful to establish a team of independent experts outside the organization structure, while some organizations may choose teams comprised of in-house experts. Most successful waste minimization assessments have common elements that identify sources of waste and calculate the true costs of waste generation and management. Each organization should decide the best method to use in performing a waste minimization assessment that addresses these two general elements:

1. Identify opportunities at all points in a process where materials can be prevented from becoming a waste (for example, by using less material, recycling materials in the process, finding substitutes that are less toxic and/or more easily biodegraded, or making equipment/process changes). Individual processes or facilities should be reviewed periodically. In some cases, performing complete facility material balances can be helpful.

2. Analyze waste minimization opportunities based on the true costs associated with waste management and cleanup. Analyzing the cost effectiveness of each option is an important factor to consider, especially when the true costs of treatment, storage, and disposal are considered.

Cost Allocation

If practical and implementable, organizations should appropriately allocate the true costs of waste management to the activities responsible for generating the waste in the first place (e.g., identifying specific operations that generate the waste, rather than charging the waste management costs to "overhead"). Cost allocation can properly highlight the parts of the organization where the greatest opportunities for waste minimization exist; without allocating costs, waste minimization opportunities can be obscured by accounting practices that do not clearly identify the activities generating the hazardous wastes.

Technology Transfer

Many useful and equally valid techniques have been evaluated and documented that are useful in a waste minimization program. It is important to seek or exchange technical information on waste minimization from other parts of the organization, from other companies, trade associations, professional consultants, and university or government technical assistance programs. EPA- and/or State-funded technical assistance programs (e.g., Minnesota Technical Assistance Program—MnTAP, California Waste Minimization Clearinghouse, EPA Pollution Prevention Information Clearinghouse) are becoming increasingly available to assist in finding waste minimization options and technologies.

Program Implementation and Evaluation

Implement recommendations identified by the assessment process, evaluations, and waste minimization teams. Conduct a periodic review of program effectiveness. Use these reviews to provide feedback and identify potential areas for improvement.

Some examples of general documents to assist organizations with more detailed guidance on conducting waste minimization assessments and developing pollution prevention programs are:

1. *Facility Pollution Prevention Guide,* EPA/600/R-92/088, October 1992 (Rockville, MD: Government Institutes).

2. *Waste Minimization: Environmental Quality with Economic Benefits,* EPA/530-SW-90-044, April 1990, available by calling the RCRA Information Center at (202) 260-9327.

EPA has also developed numerous waste minimization and pollution prevention documents that are tailored to specific manufacturing and other types of processes, and periodically sponsors pollution prevention workshops and conferences.

HAZARDOUS WASTE MINIMIZATION CASE STUDY: AGENT REGENERATION

In the following case study, IBM evaluated waste minimization opportunities in its process for manufacturing printbands, a component part of most impact printers manufactured by IBM.[87] In this case, RCRA solid waste regulations at 40 CFR 258 (Criteria for Municipal Solid Waste Landfills), applied and could not be met, given the amount of chromium contained in the waste sludge produced by the problem manufacturing process. Further, the company's waste sludge failed to meet the standards of the RCRA hazardous waste regulations at 40 CFR 261 (Regulation for

---

[87] See "Case Study: Agent Regeneration and Hazardous Waste Minimization, The IBM Corporation, Endicott, New York Facility," National Pollution Prevention Center for Higher Education, University of Michigan, Document # 93-1 (April 1993).

Identifying Hazardous Wastes) because of excessive chromium leaching potential.

Problem Identification

During the evolution of this case, and as a matter of stated corporate policy, IBM has become fully committed to the minimization of environmental impacts from its various manufacturing operations. Hazardous waste reduction programs are in place throughout the firm to:

1. Effect waste reduction at the source.

2. Encourage recycling.

3. Develop and implement waste treatment technologies.

IBM's Endicott, New York facility has expanded its operations to include science centers, research laboratories, product development activities, and manufacturing. It currently employs approximately 8,500 people and its primary mission is to:

1. Develop and manufacture technology packaging.

2. Develop systems software.

3. Manufacture bank systems.

Composed of five million square feet of floor space, Endicott is one of IBM's largest plantlab sites. Its product line consists of a variety of computer processors, banking systems, circuit packaging, and printers. It is this latter product series which serves as the focus of the case. More specifically, a component part—print bands—of most of the impact printers manufactured at this location.

Impact printers have over two hundred component parts. Print bands are stainless steel belts which are used in most such printers. Panels of raw stainless steel, measuring one foot by five feet by ½ the thickness of a dime, from which ten print bands can be produced, are used in the basic manufacturing process described below. The resultant print bands are approximately 3/4 of an inch wide with raised characters for printing. The character sets on the bands are subject to change, based upon specific customer requirements.

The manufacturing process used to produce print bands at the Endicott facility involves a photolithographic process in which the stainless steel panels are chemically machined using a strong ferric chloride etching solution containing hydrochloric acid. The waste material which results from this chemical reaction is a solution composed primarily of ferrous chloride.

The chemical process which takes place involves an oxidation/reduction reaction in which the iron (Fe^{+3}) ions are reduced, while iron (Fe) is oxidized. The net product of this reaction is the formation of, or dissolution of iron in the solution in the form of iron ions in the +2 oxidation state, called ferrous ions (Fe^{+2}). This reaction may be chemically represented as:

$$Fe^0 \;+\; 2\,Fe^{+3}Cl_3 \;=\; 3Fe^{+2}Cl_2$$

| Stainless Steel | Etchant | Dissolved Steel Spent Etchant |

Typically, approximately 50,000 print bands are manufactured annually through this process at the Endicott plant. In 1986, this chemical process consumed 158,300 gallons of ferric chloride etching agent, and generated 1,890 tons of hazardous waste sludge, along with 34,200 gallons of concentrated waste etchant which was trucked off-site for treatment prior to its disposal in a landfill.

For each print band panel, up to that period, 1.48 pounds of material had to be etched away from the surface to form the final product. This waste material included 1.24 pounds of iron, .2 pounds of chromium, and .04 pounds of molybdenum. On the average, 32 gallons of the ferric chloride solution were required for the etching of one print band panel. This solution weighed 400.5 pounds, including 60.3 pounds of iron.

The manufacturing process was a batch operation. Etching machines were filled with fresh etchant solution once a shift. The etchant was heated to 130 degrees F, 15 pints of hydrochloric acid was added, and the etchant was balanced by the addition of water. The operator started the print band panels through the machine and, because the etching solution was new, the bath's oxidation-reduction potential (ORP) was high. Thus, the etch rate was swift for the first group of panels to be etched.

As additional panels entered the solution, the bath lost some of its etching power as the chemical reaction took place. The operator had to keep adjusting the conveyor speed to compensate for this decrease in etching power. After 12 to 14 panels had been etched, the conveyor speed would have dropped to less than half of the original starting speed. Thus, the efficiency of the process was dependent, in large part, on the skill of the operator to make a high quality product in an environment of constantly changing etching parameters.

At the end of each of the three daily work shifts, the etching solution bath was dumped, the solution-spray nozzles were removed, the etched machine sump was cleaned, clean nozzles were installed, fresh etchant added, and the bath heated up for the next shift. This set-up operation required about two hours per shift to complete, making it a fairly expensive operation in terms of both downtime and labor.

The spent etching solution, plus the waste flow from the panel rinsing operations, was dumped into the site waste collection system for treatment. This waste was pumped to the facility's waste treatment plant where the soluble iron ions were converted to a precipitate by combination with hydroxide ions, the usual practice used in treating heavy metal wastes. This treatment resulted in clean water, which could be safely discharged, and a sludge which was hauled away from the plant in large containers by a vendor and disposed in a landfill at some considerable expense. At that time, the plant shipped its waste to a specific location in Quebec, where it was formulated into a non-leachable, stabilized matrix prior to burial.

The waste etch solution contained a large quantity of hydrochloric acid, which made it extremely corrosive. This acid was neutralized via the addition of a hydrated lime slurry solution. All (99.9%) of the metals end up in the precipitate. As noted, the chemical composition of the sludge is ferrous chloride. It also includes a residual of the chromium deposits contained in the stainless steel, which was etched away in the chemical reaction. This metal is the component which classified the sludge as "hazardous" and added to its disposal cost.

During the second quarter of 1986, the production of 80 print band panels per day created excessive demands on the waste treatment plant equipment. The sludge processing equipment simply could not keep up with the waste etchant output. As an expedient alternative, the etchant that could not be processed on-site was hauled in tank trucks to outside vendors for treatment. About ten large tank trucks of such material were removed from the plant that year—again, at considerable expense.

Environmental engineers at the Endicott plant were aware at the time that ferric chloride solutions had a market that was readily available. At many sewage treatment plants in the region, such a solution is injected into the treatment process as an aid to accelerate flocculation. They pursued this alternative option as an attractive means of disposal. Unfortunately, the chromium content of the solution caused it to be unacceptable for any alternative use application.

As noted, the above manufacturing/treatment/disposal process was utilized in 1986 and for at least 15 years prior. Demand projections for print bands from 1986 to 1991 showed a sharp increase. Such an increase would normally require either:

1. Additional sludge handling equipment to be installed at the plant's waste treatment site; or

2. An increase in off-site waste treatment/disposal dependency. (1991 print band production plan volumes would necessitate a 300% increase.)

In late 1986, this complex problem situation was brought to a head. A chemical outage occurred on the ferric chloride supply system which resulted in a print band production shut-down for several days. This shut-down occurred at a time when some critical print band orders were being processed. This shortage served to heighten management's sensitivity to the chemical supply/waste generation problem associated with this manufacturing process.

Management's response to the above critical situation was to immediately appoint a task force to directly address this problem. The task force was charged to:

1. Ensure adequate virgin ferric chloride storage capacity to prevent manufacturing interruptions.

2. Increase the conservation practices of ferric chloride in the manufacturing cycle.

3. Enlist engineering groups throughout the company to invent a process to regenerate the ferric chloride solution used to manufacture print bands (in line with the basic IBM site strategy of waste minimization).

Strategic Alternatives

Since the volume of ferric chloride etchant was so great, and the waste formed in the print band manufacturing process was so voluminous, the incentive to investigate the various methods of affecting ferric chloride etchant regeneration was considered to be not only justifiable, but imperative.

All chemical reactions are reversible. However, the ease and cost of reversibility varies widely, depending on the chemicals involved. In the case of the ferric chloride etching solution, it was recognized that various methods existed which could be utilized to reverse the chemical process and reform the iron (Fe^{+3}) ions from the spent etchant. However, it was also clearly understood that regeneration of the etchant was the *only* alternative which could result in a significant reduction in the volumes of etchant used in the manufacturing process and the waste created by that same process.

Inherent in all of the regeneration processes examined up to that time was the use of extremely reactive, hazardous, and toxic chemicals, or electrical energies which were extremely high. Chemicals such as chlorine gas, hydrogen peroxide, sodium chlorate, and other strong oxidants were often employed. Although effective in regenerating ferric chloride etchant, personnel safety would almost certainly be at risk when chemicals of this nature were employed.

Rather than place its employees in a hazardous manufacturing environment which employed such strong chemicals (specifically, the three cited in the above paragraph), IBM scientists and engineers decided to investigate three strategic, problem-solving alternatives:

1. Creation of an electrochemical process of regeneration via equipment not yet developed.

2. The injection of ozone gas into the ferric chloride solution during the manufacturing process.

3. The use of air oxidation as the regeneration agent.

Electrochemical Option

This option was initiated at the behest of the Environmental Engineering Department at the IBM Endicott plant. The management of Printband Manufacturing was provided with an in-depth briefing of the significant impact of their etching process on the site's waste treatment systems. In order to show good faith in addressing this concern, Printband Manufacturing management sponsored and funded the electrolytic process as part of the Endicott Technical Development (ETD) organization's annual budget for development projects.

A high-ranking member of the ETD organization had prior experience with an electrolytic process, and he assigned the investigation of this option to an employee with whom he shared the responsibility for its development. An aggressive approach was taken, which projected enormous waste volume reductions. Progress was presented in highly technical terms at monthly development status meetings.

In presenting the results of this option, many negative references were made to the Ozone option which was not being developed at the Endicott plant and, as a result, not represented at these meetings. When rebuttals of arguments against the Ozone option were presented, Endicott ETD management countered that the sort of competition presented by this (the Ozone) option was not beneficial to the Electrolytic option and should not be pursued. Consequently, only viable test results were reported at these meetings, with development project assessors being led to believe that *every* test was successful.

Ozone Gas Option

Like the Electrolytic Option, the Ozone approach to agent regeneration was sponsored by Printband Manufacturing management. It was initiated via a paper presented at an IBM technology-based meeting, authored by a development chemist at the East Fishkill, New York, plant. His etchant regeneration system was in the process of being implemented at that location after months of testing and approvals. Like the Endicott plant, the Fishkill process involved a ferric etchant.

Printband Manufacturing management only became aware of the proposed Ozone regeneration system after they had already authorized the Endicott Technology Development organization to investigate the Electrolytic option. ETD had already been provided the funding for its activities and no funds were available to support the Ozone option development efforts without a considerable loss of "face."

The Ozone option development team at the East Fishkill location wanted very badly to sell its system elsewhere, and so had agreed to do the development testing at no cost to Printband Manufacturing management. Some minimally valued equipment was made available to the Ozone process developers and it was refurbished for use at that location.

Infrequent trips were made to visit the Ozone project by Printband Manufacturing engineering personnel for feasibility and review purposes. Two or three development status memorandums were weekly transmitted to the Endicott facility. All testing and results continued to be favorable. No unsuccessful test results, of which there were a number, were ever reported directly to Endicott, nor was the development chemist free to travel to that location to directly report progress and status.

Air Oxidation Option

Early in 1987, two Printband Manufacturing engineers and an Environmental Engineering acquaintance began work at the Endicott facility on an idea being used in the copper etching process areas in the plant. They investigated the injection of compressed air into the piping loop through which the etchant was circulated during etching. This method worked with some degree of success with cupric chloride etch systems, but chemical thermodynamics did not appear to favor a similar promise for ferric chloride.

Etchant conservation practices were immediately implemented to:

1. Monitor the etchant bath thru-put to better define bath life and ensure complete bath exhaustion prior to dumping to the treatment/disposal process.

2. Modify a preliminary step in printband manufacturing called "flash etch" to decrease ferric chloride consumption.

However, with a minimal investment of both time and capital, the engineers were authorized by Printband Manufacturing management to set up a pilot test system using compressed air on one of three ferric chloride etchers in the manufacturing area. The net result of the preliminary tests for this simple $12,000 system was a repeatable 17% decrease in the volume of the etchant required in the etch process. Most of the components of this air oxidation system were either used, donated from inactive systems, or fabricated on-site to expedite installation.

These six month test results and the consequent savings were significant enough to warrant implementation of similar systems on two new etchers being installed at that time (at a total cost for all etchers of less than $50,000). This investment was readily recouped through chemical and waste volume reductions. The return was over 130%, and the payback period was achieved in about *nine* months. As a result, no further off-site treatment of the ferric chloride waste would be necessary.

Additionally, planned follow-up testing, employing oxygen enriched air and pure oxygen, was conducted in 1988. The net result of the oxygen testing was a system which would regenerate etchant at a rate which matched normal production requirements. Etchant regeneration could now be carried out in a working etcher at such a rate that no fresh etchant was required to maintain the etch rate. The print band panel conveyor could be set at a standard speed, thus reducing operator time and expense, and simplifying the etching process.

These tests came at a very critical moment. An equipment problem on the bulk ferric chloride system made it impossible to pump virgin etchant from either of the two 5,000 gallon bulk supply tanks to the etchers. This would have shut down production for several days had it not been for the (then) pilot oxygen regeneration system in place. Manufacturing continued on an uninterrupted basis during the period when this pumping failure was being corrected.

These test results were presented to management and capital funds were made available to enhance and upgrade the initial air-oxidizing systems to the more effective oxygen systems. Oxygen generators were installed and the regeneration of ferric chloride etchant was fully operational in October 1989, three months ahead of schedule. Further testing was being conducted on this process at the very same time that the Ozone and Electrolytic options were still being developed. Additionally, a patent application was filed and remains active today.

No toxic or hazardous chemicals were employed, no exotic, sophisticated equipment requiring a trained staff of maintenance personnel was necessary, and operational costs were minimal.

Option Selection; Results

The Oxygen Option was ultimately approved on economic grounds. Since it was not developed by the Endicott Technical Development organization, some of the ties to that unit were severed as a result of this decision.

The "winning" system was easily implemented by making simple hookups to replace plant compressed air with oxygen generators. While some bugs had yet to be fully worked out of the system, Printband Manufacturing management was satisfied with its overall effectiveness. The approval for purchase, installation and testing of the full system was *not* shared with the Ozone and the Electrolytic development teams at the time the "go-ahead" decision was made—on the grounds that the "politics" were not right.

The air oxidation option was a unique method of ferric chloride regeneration, developed and implemented utilizing a simple combination of readily available hardware and a non-toxic gas (Oxygen). The final return on investment (ROI) was over 185%, with a four month payback period. The entire system was actually paid for in the savings associated with virgin chemical and waste costs experienced during the engineering pilot testing periods for this project. Because of the reduced virgin ferric chloride demand, the additional bulk holding capacity was never required. Additionally, this system is now being installed to replace regeneration methods on other etch systems at the Endicott Plant.

This new regeneration system requires minimal maintenance and repair activity, as well as very infrequent operator attention. The cost savings and environmental benefits, in terms of reduced volumes of solid and hazardous waste for land filling, continue with every print band produced today.

In 1991, only 4,122 gallons of ferric chloride etchant were consumed in the printband manufacturing process. This resulted in only 78 tons of hazardous waste sludge being shipped to a landfill, a production normalized reduction of 94.2% in etchant use and a 90.8% reduction in the hazardous waste generated by this manufacturing process.

A total system to support the oxygen regeneration option for all print band manufacturing etchers could be purchased and installed for less than one quarter the cost of either the ozone or the electrochemical alternative methods still in the development stage. The inherent risks to personnel associated with those two options were thus avoided.

It is understood that without management's commitment to continuing development efforts, a project such as this would never have realized its full potential. It is the never-ending quest to continually improve on existing processes that can sometimes lead to dramatic advancements, as evidenced in the case presented.

Chapter 5

Water Pollution Prevention Programs

INTRODUCTION

Best management practices (BMPs) are recognized as an important part of the Clean Water Act's National Pollutant Discharge Elimination System (NPDES) permitting process to prevent the release of toxic and hazardous chemicals.[88] Over the years, as BMPs for many different types of facilities have been developed, case studies have demonstrated not only the success but the flexibility of the BMP approach in controlling releases of pollutants to receiving waters. More recently, pollution prevention practices have become part of the NPDES program, working in conjunction with BMPs to reduce potential pollutant releases. Pollution prevention methods have been shown to reduce costs as well as pollution risks through source reduction and recycling/reuse techniques. The purpose of this chapter is to provide guidance for NPDES permittees in the development of BMPs to control and reduce water pollution discharges at their facilities.

POLLUTION PREVENTION AND BEST MANAGEMENT PRACTICES

Best management practices are inherently pollution prevention practices. Traditionally, BMPs have focused on good housekeeping measures and good management techniques intending to avoid contact between pollutants and water media as a result of leaks, spills, and improper waste disposal. However, based on the authority granted under Clean Water Act regulations, BMPs may include the universe of pollution prevention

---

[88] See discussion beginning on p. 33.

encompassing production modifications, operational changes, materials substitution, materials and water conservation, and other such measures.

EPA recognizes that significant opportunities exist for industry to reduce or prevent pollution through cost-effective changes in production, operation, and raw materials use. In addition, such changes may offer industry substantial savings in reduced raw materials, pollution control, and liability costs, as well as protect the environment and reduce health and safety risks to workers. Where pollution prevention practices can be both environmentally beneficial and economically feasible, EPA finds their implementation to be prudent.

EPA believes that the intent of pollution prevention practices and BMPs are similar and that they can be concurrently developed in a technologically sound and cost-effective manner. Thus, although this chapter primarily focuses on best management practices, which pertain to the NPDES program, the reader may be compelled to also consider pollution prevention for all media in order to maximize the benefits achieved.

Types of BMPs

BMPs may be divided into general BMPs, applicable to a wide range of industrial operations, and facility-specific (or process-specific) BMPs, tailored to the requirements of an individual site. General BMPs are widely practiced measures that are independent of chemical compound, source of pollutant, or industrial category. General BMPs are also referred to as baseline practices, and are typically low in cost and easily implemented. General BMPs are practiced to some extent at almost all facilities. Common general BMPs include good housekeeping, preventive maintenance, inspections, security, employee training, and recordkeeping and reporting.

Facility-specific BMPs are measures used to control releases associated with individually identified toxic and hazardous substances and/or one or more particular ancillary source. Facility-specific BMPs are often developed when a facility notes a history of problem releases of toxic or hazardous chemicals, or when facility personnel believe that actual or potential pollutant discharge problems should be addressed. Facility-specific BMPs may include many different practices such as source reduction and on-site recycle/reuse.

Facility-specific BMPs will vary from site to site depending upon site characteristics, industrial processes, and pollutants. For example, a site-specific BMP in the form of area dikes may be adopted due to the location of the facility: facilities in flat areas or on slopes are likely to utilize dikes to control spills whereas there may be no need for dikes for facilities located in basins. Additionally, plants handling and storing large amounts of liquid chemicals would be more likely to utilize dikes than facilities storing and using dry chemicals. Facilities experiencing erosion and sediment control problems may consider establishing vegetative buffer strips or indigenous ground cover for purposes of soil stabilization and infiltration of run off. Facility sites located adjacent to other industrial areas may consider run-on controls to prevent extraneous spills and contaminated run-on from entering the facility site. Other site-specific considerations, such as endangered species, may motivate facilities to store materials in an alternate location so as to prevent exposure.

Processes also drive the determination of appropriate specific BMPs. Materials handling procedures that expose employees to toxic chemicals (e.g., hand drawing) may prompt the consideration of procedures that reduce the potential for exposure (e.g., automated pneumatic pumping in enclosed conduits). Some examples of process-specific BMPs include the following:

- Using splash plates designed to prevent spills at a metal finishing facility.

- Installing solvent recovery equipment to control benzene releases at a petroleum refinery.

- Purchasing solvents in reusable containers rather than 55-gallon drums to ease storage concerns and reduce wastewater resulting from requirements to triple rinse drums.

Pollutant characteristics, such as volatility and toxicity, also affect BMP selection. More and more, harmful chemicals are being considered for replacement with less toxic alternatives, or for elimination from the process. Ozone layer-depleting solvents which are used for cleaning at many facilities are being replaced with detergent-based cleaning agents. Additionally, facilities using materials with toxic properties have been inspired to take more proactive control measures (e.g., double walled containment).

Factors Affecting BMP Selection

The following general factors should be considered when selecting specific BMPs for the facility:

Chemical nature. The need to control materials based on toxicity and fate and transport.

Proximity to waterbodies. The need to control liquid spills prior to their release to media, such as water, from which materials may not later be separated.

Receiving waters. The need to protect sensitive receiving waters which are more severely impacted by releases of toxic or hazardous materials. The need to protect the water uses including recreational waters, drinking water supplies, and fragile aquatic and biota communities.

Proximity to populace. The need to control hazardous materials with potential to be released near populated areas.

Climate. The need to prevent volatilization and ignitability in warmer climates. The need to reduce wear on moving parts in freezing climates. The need to avoid spills in climates and under circumstances where mitigation cannot occur.

Age of the facility/equipment. The need to prevent releases caused by older equipment with greater capacity for failure. The need to address obsolete and outdated instruments and processes which are not environmentally protective.

Process complexity. The need to address problems of materials incompatibility.

Engineering design. The need to address design flaws and deficiencies.

Employee safety. The need to prevent unnecessary exposure between employees and chemicals.

Environmental release record. The need to control releases from specific areas demonstrating previous problems.

COMPONENTS OF BMP PLANS

Suggested components of BMP plans are defined and described in this section. The suggested elements of a good BMP plan can be separated into three phases: planning, development and implementation, and evaluation/reevaluation. Generally, the planning phase, discussed in the section on Planning Phase, includes demonstrating management support for the BMP plan and identifying and evaluating areas of the facility to be addressed by BMPs. The goal of plan development should be to ensure that its implementation will prevent or minimize the generation and the potential for release of pollutants from the facility to the U.S. waters. The development phase consists of determining, developing, and implementing general and facility-specific BMPs and is described in the section on Planning Phase. The evaluation/reevaluation phase described in the section on Plan Evaluation and Reevaluation consists of an assessment of the components of a BMP plan and reevaluation of plan components periodically, or as a result of factors such as environmental releases and/or changes at the facility. The suggested elements of a baseline BMP plan are as follows:

1. *Planning Phase Considerations*

   - BMP committee
   - BMP policy statement
   - Release identification and assessment

2. *Development Phase Considerations*

   - Good housekeeping
   - Preventive maintenance
   - Inspections
   - Security
   - Employee training

- Recordkeeping and reporting

3. Evaluation and Reevaluation Phase Considerations

- Evaluate plan implementation benefits

PLANNING PHASE

In the planning phase, a facility must decide who will take the responsibility for establishing and carrying out the BMP plan. The plan should be initiated with clear support and input from facility management and employees. The facility must also identify and evaluate areas of the facility that, because of the substances involved and their management, will be addressed in the BMP plan. Each of these elements is discussed in detail in the next several subsections.

BMP Committee

A BMP committee is comprised of interested staff within the facility's organization. The committee will represent the company's interests in all phases of BMP plan development, implementation, oversight, and plan evaluation.

It should be noted that a BMP committee may function similarly to other committees that might exist at an industrial facility (e.g., pollution prevention committee) and may include the same employees.

The BMP committee is developed to assist a facility in managing all aspects of the BMP plan. The committee functions to conduct activities and shoulder the responsibilities for the following:

- Develop the scope of the BMP plan.

- Make recommendations to management in support of company BMP policy.

- Review any existing accidental spill control plans to evaluate existing BMPs.

- Identify toxic and hazardous substances.

- Identify areas with potential for release to the environment.

- Conduct assessments to prioritize substances and areas of concern.

- Determine and select appropriate BMPs.

- Set forth standard operating procedures for implementation of BMPs.

- Oversee the implementation of the BMPs.

- Establish procedures for recordkeeping and reporting.

- Coordinate facility environmental release response, cleanup, and regulatory agency notification procedures.

- Establish BMP training for plant and contractor personnel.

- Evaluate the effectiveness of the BMP plan in preventing and mitigating releases of pollutants.

- Periodically review the BMP plan to evaluate the need to update and/or modify the BMP plan.

To be most effective, the committee must perform tasks efficiently and smoothly. In large part, the personnel selected to act as committee members will determine the committee's success. Some of the considerations for personnel selection include the following:

- A lead committee member must be determined.

- Committee members must include persons knowledgeable of the plant areas involved (e.g., process areas, tank farms) and utilization of chemicals and generation of pollutants (e.g., solvents, products, chemical reactants) at the facility.

- Committee members should have the authority to make decisions effecting BMP plan development and implementation.

- The size of the committee must be appropriate to the facility.

- The committee must represent affected areas of the plant and employees.

An example of the effectiveness of the formation of a committee is provided in the following case study:[89]

The 3M company, a manufacturer of diverse products such as coated abrasives, pressure sensitive tape, photographic film, electrical insulation materials, and reposition notes, has developed a corporate philosophy that Pollution Prevention Pays (referred to as the 3P program). As part of the 3P program, 3M has created a 3P Coordinating Committee which includes employee representatives from the engineering manufacturing, laboratory, and corporate environmental sectors.

The 3P Coordinating Committee provides support and coordination for nationwide teams establishing 3P programs. These 3P Teams are organized by employees that have identified pollution problems and recognize potential solutions.

The 3P Coordinating Committee and the 3P Teams have been instrumental in source reduction of hydrocarbons, odor, water, dissolved solids, sulfur, zinc, alcohol, and incinerated scrap. In the first year of the 3P programs, air pollutants have been reduced by 123,000 tons, water pollutants by 16,400 tons, wastewater by 1,600 million gallons, and solid waste pollutants by 409,000 tons. This has resulted in savings of more than $500 million.

The BMP committee is responsible for developing the BMP plan and assisting the facility management in its implementation, periodic evaluation, and updating. While the BMP committee is responsible for developing the plan and overseeing its implementation, all activities need not be limited to committee members. Rather, appropriate company personnel who are knowledgeable in the areas of concern can carry out certain activities associated with BMP plan development. With this in mind, the selection of the committee members can be limited to a select set of individuals, while the resources of interested and knowledgeable employees can still be utilized.

---

[89] See T. Zeal, "Case Study: How 3M Makes Pollution Prevention Pay Big Dividends," *Pollution Prevention Review* (Winter 1990-91).

In order to ensure a properly run organization, one person should be designated as the lead committee member. Thus, the first step in developing a BMP committee is to determine the appropriate committee chairperson. The determination of a single leader will assist in the smooth conduct of meetings and the designation of tasks, and will aid in the decision making process. Generally, the chairperson should be highly motivated to develop and implement the BMP plan, familiar with all committee members and their areas of expertise, and experienced in managing tasks of this magnitude. The chairperson will be responsible for ensuring that all tasks are assigned to appropriate personnel, keeping facility management and employees informed, and cohesively developing the BMP plan. Potential candidates for this role are plant managers, environmental coordinators, or other distinctly knowledgeable technical and management personnel.

The next step is to select the appropriate personnel to comprise the committee. Personnel selected should represent all affected facility areas. Members might also be selected based on their areas of expertise (e.g., industrial processes). Personnel might be selected who have a full understanding of the manufacture processes from raw materials to final products, as well as of the recycling, treatment, and disposal of wastes. Possible candidates include foremen in manufacturing, production, or waste treatment and disposal; maintenance engineers; environmental and safety coordinators; and materials storage and transfer managers. Not only must committee members understand the activities conducted throughout the entire facility, members of the BMP committee must also include individuals who are in the decision making positions within the company structure. Some committee members must represent company management and have the authority to implement measures adopted by the committee.

While the BMP committee should reflect the lines of authority within the company, it should also be sensitive to general employee interests. It is crucial to ensure that employees are aware of and in support of the BMP plan and the responsible committee, as it is primarily the employees who will implement the changes resulting from committee decisions. Forming a committee comprised solely of upper level management and administrative personnel would exclude general personnel whose input is critical for the development and implementation of the plan. Selecting employee-chosen representatives, such as union stewards, may be an appropriate means to ensure employee involvement.

The size of a BMP committee should reflect the size and complexity of the facility, as well as the quantity and toxicity of the materials at the facility. The committee must be small enough to communicate in a open and interactive manner, yet large enough to allow for input from all necessary parties.

Where needed, committee members should call upon the expertise of others through the establishment of project-specific task forces. For example, personnel involved in research and development may be asked to research the effectiveness of product substitution and process changes that are being considered as part of BMP plan development. This method of calling upon specialists, when the need arises, should allow the committee to remain a manageable size. Generally, the size selection process outlined below presents a good rule of thumb:

- For small facilities, a single committee member is acceptable as long as that person has the requisite expertise and authority.

- For larger facilities, selection of six to eight people as permanent members of the committee should be ideal.

BMP Committee—What to Do

- Develop a roster of BMP committee members which includes area of specialization and projected responsibilities. This list helps identify any holes in the planned BMP development activities and any missing expertise.

- Include a list of alternate BMP committee members where transfers are expected to occur during the life of the BMP plan.

- Post BMP committee member names and including them in the plan to allow any interested parties the opportunity to contact BMP committee members.

- While developing and updating the BMP plan, include input from interested employees not on the committee. Employee input sessions and suggestions boxes can be used to meet this goal.

- Extend technical reviews to personnel not on the BMP committee, where specialized expertise is necessary or where interest is expressed.

- Follow up with all responsible parties on a periodic basis to ensure they are aware of their BMP-related responsibilities.

- Encourage BMP committee members to spend time on-the-line in order to communicate with other potentially interested parties.

- Set schedules with milestone dates for the performance of important activities. This avoids possible procrastination and allows the BMP plan development to remain on schedule.

BMP Committee—What Not to Do

- The committee should enable, not impede, the decision making process for preventing or mitigating spills or otherwise responding to events addressed by the BMP plan.

- Remember that personnel contributing to the design of a BMP need not be member of the BMP committee. This is of particular importance where a technical specialist or manager simply would not have the time to contribute on a regular basis.

BMP Policy Statement

A BMP policy statement describes the objectives of the BMP program in clear, concise language and establishes the company policies related to BMPs. The following case study provides an practical example of the successful use of a policy.[90]

Dow Chemical has observed a significant impact as a result of their company's environmental policy. As part of their "Environmental Policy and Guidelines," Dow has set forth a hierarchy similar to that developed as part of the Pollution Prevention Act of 1990. Dow's policy sets forth

---

[90] See D. Sarokin, W. Muir, C. Miller, S. Sperber, *Cutting Chemical Wastes: What 29 Organic Chemical Plants are Doing to Reduce Hazardous Wastes*, INFORM, Inc., New York, New York (1985).

preferences to handle materials by reducing pollutants at the source, followed by recycle and use of materials whenever possible. Where disposal is necessary, Dow has specified that incineration be considered first, followed by land disposal on Dow-owned property, and finally land disposal on property not owned by Dow.

Dow's decision to follow its disposal hierarchy was based in large part on the liability associated with disposal. Dow reasoned that incineration was the most appropriate disposal method since it resulted in the pollutants in the ash materials being in elemental form. In many cases, the company has identified opportunities for recycle of materials found in the incinerator ash. Dow also imposed a $215 per drum surcharge for hazardous wastes going to a landfill to provide incentives for finding alternatives to landfilling. Dow also believes that they can better exercise control of on-site disposal, thus influencing the preferences for on-site rather than off-site disposal.

Dow's policy has resulted in an impact on the environmental releases. For example, at the Dow Pittsburg, California facility, wastewater discharges have been reduced by 95% over the past 10 years. Additionally, the approximately 10.2 million pounds of chlorinated organics wastes generated are either incinerated or recycled.

The policy statement provides two major functions: (1) it demonstrates and reinforces management's support of the BMP plan; and (2) it describes the intent and goals of the BMP plan. It is very important that the BMP policy represent both the company's goals and general employee concerns. Several steps to take in developing an effective BMP policy statement include the following:

- Determine the appropriate author.

- Develop tone and content that are positive, but that establish realistic and achievable goals.

- Distribute the policy statement effectively.

The first step in creating a BMP policy statement is determining the appropriate author. To indicate management's commitment, the policy statement should be signed by a responsible corporate officer. A responsible corporate officer can be the president, vice president, or the principal manager of manufacturing, production, or operations. Generally, the policy statement author should be a person who performs policy- or decision making functions for the corporation/facility.

The next step in developing the BMP policy statement is to craft the specific language. The policy statement may include references to the company's commitment to being a good environmental citizen, expected improvements in-plant safety, and potential cost savings. Regardless of personal style, in all cases the policy should: (1) indicate the company's support of BMPs to improve overall facility management and (2) introduce the intent of the BMP plan.

The length and level of detail of the policy statement will vary depending on the writer's personal style. The following variations may be included in a BMP policy statement:

- An outline of steps that will be taken.

- A discussion of the timeframes for development and implementation.

- An indication of the areas and pollutants of focus.

- A projection of the end result of the BMP plan.

- Create enthusiasm and support for the BMP plan by all employees.

The tone of the BMP policy statement is also important. The projected positive impacts of BMP implementation should be discussed in general terms. If specific goals are outlined, the level of information and the expectations presented should be reasonable, to avoid overwhelming the reader. Ultimately, the policy should provide an upbeat message of the improved working environment that will result from BMP implementation. Since gaining employee support is so important, it may be appropriate to solicit employee concerns prior to the development of the BMP policy. These concerns can be highlighted as areas which will be evaluated during BMP plan development.

Finally, to ensure that all employees are aware of the impending BMP plan, the policy statement should be printed on company letterhead (for an official appearance) and distributed to all employees. Complete distribution can be best ensured if the statement is both delivered to each employee and posted in common areas.

BMP Policy Statement—What to Do

- Utilize meetings and open sessions to solicit employee participation in the development of the BMP policy.

- Demonstrate that employee ideas are welcome by immediate follow-up on suggestions. Discuss possible implementation opportunities or reasons that implementation is feasible.

- Keep the statement clear and concise.

- Post the BMP policy statement in key locations where employees congregate so that employees will discuss it.

- Use the policy statement to promote an emblem/motto that represents the BMP plan and its benefits.

- Emboss the objectives of the BMP policy on a plaque.

BMP Policy Statement—What Not to Do

- Do not include details of the BMP plan in the policy statement.

- The BMP policy statement should not be issued solely by the BMP committee. It should be issued by the company.

Release Identification and Assessment

Release identification is the systematic cataloging of areas at a facility with ongoing or potential releases to the environment. A release assessment is used to determine the impacts on human health and the environment of any ongoing or potential releases identified. The identification and assessment process involves the evaluation of both current discharges and potential discharges.

The release identification and assessment process can provide a focus for the range of BMPs being considered on those activities and areas of a facility where the risks (considering the potential for release and the hazard posed) are the greatest. In some cases, the assessment may be performed based on experience and knowledge of the substances and circumstances involved. In other cases, more detailed analyses may be necessary to provide the correct focus, and release assessments may then rely on some of the techniques of risk assessment (e.g., pathway analysis, toxicity, relative risk). Understanding the dangers of releases involves both an understanding of the hazards each potential pollutant poses to human health and the environment, as well as the probability of release due to the facility's methods of storage, handling, and/or transportation.

Some facilities may identify a number of situations or circumstances representing actual or potential hazards that should all be addressed in some detail through the BMP plan. However, in some instances prioritizing potential hazards is the most sensible and cost effective approach. The following example illustrates the need for BMP prioritization:

> ACME Concrete is a concrete and supply facility with a designated area used to house maintenance vehicles and materials, as well as to stockpile construction materials and equipment. Among other things, this facility contains a large stockpile of building sand used to prepare concrete, a vehicle maintenance area where oil is drained from company vehicles, and a shed where drums of solvents used in cleaning operations are stored. Although each of the three materials mentioned at the site (sand, used oil, and solvents) can cause environmental or health damage unless they are controlled, it would not be feasible or reasonable to control losses of small amounts of clean building sand with the same careful attention given to the release of toxic solvents.

As this basic example shows, priorities for BMPs should reflect a basic understanding of the loss potential and hazards posed by these potential losses. A prudent manager of the ACME Concrete maintenance yard could first limit the potential for escape of solvents through careful training and periodic preventative maintenance and inspection of drums and storage facilities, then prevent run off of used oil to surface waters or groundwater by collecting and recycling used oil, and finally control major losses of sand through constructions of filter fences or sediment ponds.

Identifying and assessing the risk of pollutant releases for purposes of a BMP plan can best be accomplished in accordance with a five-step procedure:

1. Reviewing existing materials and plans, as a source of information, to ensure consistency, and to eliminate duplication.

2. Characterizing actual and potential pollutant sources that might be subject to release.

3. Evaluating potential pollutants based on the hazards they present to human health and the environment.

4. Identifying pathways through which pollutants identified at the site might reach environmental and human receptors.

5. Prioritizing potential releases.

Once established, these priorities may be used in developing a BMP plan that places the greatest emphasis on the sources with the greatest overall risk to human health and the environment, considering the likelihood of release and the potential hazards if a release should occur, while still implementing low cost BMPs that might contribute to safety or other worker driven needs.

The first step in the conduct of a release identification and assessment involves the review of existing materials and plans to gather needed information. Many industrial facilities are already subject to regulatory requirements to collect and provide information that may be useful in the identification and assessment of releases. In some cases, these plans may have been developed by persons in-plant safety or process engineering who do not normally consider themselves part of the environmental staff. In particular, the following plans should be identified and reviewed:

- *Preparedness, prevention, and contingency plans (see 40 CFR Parts 264 and 265)* require the identification of hazardous wastes handled at a facility.

- *Spill control and countermeasures (SPCC) plans (see 40 CFR Part 112)* require the prediction of direction, rate of flow, and total quantity of oil that could be discharged.

- *Storm water pollution prevention plans (see 40 CFR 122.44)* require the identification of potential pollutant sources which may reasonably be expected to affect the quality of storm water discharges.[91]

- *Toxic organic management plans (see 40 CFR Parts 413, 433, and 469)* may require the identification of toxic organic compounds.

- *Occupational Safety and Heallh Administration (OSHA) emergency action plans (see 29 CFR Part 1910)* require the development of a list of major workplace fire and emergency hazards.

Other sources of information that might be pertinent to the release identification and assessment process include the facility's NPDES permit application and, where applicable, information collected for SARA Title III, Section 313 Form R. SARA Title III requires facilities with certain chemicals to annually submit toxic release data annually as part of community right-to-know requirements.

The second step of conducting a release identification and assessment is to characterize current and potential pollutant sources. This step may be conducted through assembling a description of facility operations and chemical usage and then verifying information through inspections. This process allows facility personnel to confirm the accuracy of information on hand (e.g., the amount of chemicals used in a specific location) while also tracking changes that might have evolved over time (e.g., changing the staging of lubricants in a particular part of the plant).

Generally, the preparation of a site map or maps covering the entire facility is very useful in this evaluation. Maps should cover the entire property and illustrate plant features including material storage areas for raw materials, by-products, and products; loading and unloading areas; manufacturing areas; and waste/wastewater management areas. The map should also indicate site topography, including facility drainage patterns. Any existing structural control measures already used to reduce pollutant releases should be highlighted, and conveyance mechanisms or pathways to surface water bodies should be noted. The facility site map should also indicate property boundaries, buildings, and operation or process areas.

---

[91] See also M. Dennison, *Storm Water Discharges: Regulatory Compliance and Best Management Practices* (Lewis Publishers, 1995).

Any neighboring properties that have potential sources of contaminants that might migrate onto the facility (because of drainage patterns) should also be noted on the map.

Following preparation of a site map, a materials inventory should be prepared. Generally, purchasing records should be helpful in determining the raw materials that are part of the inventory. However, the products manufactured and the by-products resulting during the manufacturing process should also be considered.

The materials inventory should include descriptions of the amounts of pollutants released or with the potential to be released based on methods of storage or on-site disposal, loading and access methods, and management and control practices (including structural measures or treatment). The inventory should refer to the location of the material keyed to the site map. Materials inventories will vary with the size and the complexity of the facility. It may be helpful to conduct separate inventories for different areas (e.g., manufacturing areas 1, 2, and 3; water drainage areas 1 and 2).

The site map and materials inventory developed to this point have relied solely on plant records. The next part of this process requires a field evaluation/inspection that verifies the facts compiled to this point, and determines the reasons for any discrepancies. Determining the cause of discrepancies is an important step as it may result in the identification of new locations of concern (e.g., storage areas or process lines have been moved to a different part of the plant). This process may also add/delete chemicals or other materials to/from the list being evaluated (e.g., where a chemical is no longer in use, or where a chemical substitution has been made).

The field evaluation also provides an opportunity to look for evidence of past releases or situations that represent potential releases to the environment. Notes should be assembled indicating the substances that might be released and the migration pathway that would be followed by any such release. This information should be correlated with the facility map. Where evidence of past leaks is found, further study should be undertaken to determine if the evidence correlates with the release information already obtained.

The third step in the release identification and assessment process involves evaluating potential pollutants based on the hazards they present to human health and the environment. No single measure of toxicity or hazardous characteristics exists because chemicals may have a variety of effects (both direct and indirect) that are characterized by a range of physical/chemical properties and associated effects. Some chemicals, for example, may be hazardous because of flammability and therefore represent fire hazards. Other products may be toxic and represent a threat to waterways and their associated flora and fauna, contaminate groundwaters, and/or threaten workers cleaning up spills who are not provided with the proper protective equipment (e.g., respirators). Potential releases of pollutants to the environment might be subject to regulation under environmental permits, and represent threats to the facility in the form of noncompliance.

Detailed information on material properties should be available from plant safety personnel. When evaluating the threats posed by chemicals, facility personnel should consult available technical literature, manufacturer's representatives, and technical experts, such as safety coordinators within the plant. A variety of technical resources can provide information of chemical properties, including the following:

- Material safety data sheets.

- American Conference of Government and Industrial Hygienist publications.

- N. Sax, *Dangerous Properties of Industrial Materials,* Eighth Edition, Volumes 1-3, Van Nostrand Reinhold, New York, New York (1994).

- *National Institute of Occupational Safety and Health (NIOSH) Pocket Guide to Chemical Hazards, U.S.* Department of Health and Human Services (1990).

- M. Dennison, *Understanding Solid and Hazardous Waste Identification and Classification*, John Wiley & Sons, Inc, New York, New York (1993).

- EPA guidance documents. (Call EPA Public Information Center at (202) 260-7751.)

These references can provide information on specific physical/chemical properties that should be considered in evaluating hazards, including toxicity, ignitability, explosivity, reactivity and corrosivity. Careful evaluation of these data will provide a basis for determining the intrinsic threat posed by materials at the facility. Armed with such understanding and subsequent identification of exposure pathways and potential receptors (the next step in the process), the need for developing BMPs comes into focus.

The fourth step in the release identification and assessment process involves identifying pathways by which pollutants identified at the site might reach environmental and human receptors. Identifying the pathways of current releases can easily be accomplished based on visual observations.

However, identifying the pathways of potential releases requires the use of sound engineering judgement in determining the point of release, estimating the direction and rate of flow of potential releases toward receptors of concern, and identification and technical evaluation of any existing means of controlling chemical releases or discharges (such as dikes or diversion ditches).

Information from the site map and observations made during the visual inspection (e.g., location of materials, potential release points, drainage patterns) should prove useful in this analysis. Of primary concern will frequently be exposures to workers in the immediate area of a release where concentrations will be highest. Migration pathways for other exposures will often be of secondary concern.

When identifying pathways and receptors, all logical alternative pathways should be considered. Contaminations may be released through a number of methods (e.g., volatilization, leaching, run off) to a number of media (e.g., air, groundwater, soil), all which may result in release to water. The analyst should consider all pathways carefully in combination with the materials inventory to identify possible release mechanisms and receptor media.

During the site-assessment, each area should be evaluated for potential problems. These problems might include equipment failure, evidence of wear or corrosion, improper operation (e.g., a tank overflow or leakage or exposure of raw material to run off), problems caused by natural conditions (e.g., cracks or joint separation due to extremes in temperature), and materials incompatibility. The adequacy of control and planned remedial measures should also be examined. For example, the volume of oil in a storage tank holding liquid petroleum fuel oil might exceed the amount that could be controlled by a dike or a berm in the case of a tank failure. Increasing the size of containment can remedy such a problem. The availability and location of absorbent materials and/or booms would be of interest in case of spill or tank failure and should be evaluated to determine sufficiency.

The fifth and final step in the release identification and assessment process requires the application of best professional judgment in prioritizing potential releases. Priorities should be established for both known and potential releases. A combination of information identified in the previous steps about releases (the probability of release, the toxicity or hazards associated with each pollutant, and descriptions of the potential pathways for releases) should be evaluated. Using this information, a facility can rank actual and potential sources as high, medium, or low priority. These priorities can then be used in developing a BMP plan that places the greatest emphasis on BMPs for the sources that present the greatest risk to human health and environment.

Release Identification and Assessment—What to Do

- Consider using other resources when conducting the release identification and assessment. Corporate or brother/sister company personnel may be available for consultation and assistance. Also, non-regulatory on-site assistance may be available.

- Utilize worksheets and boilerplate formats to ensure that information is organized, easily evaluated, and easily understood.

- Utilize videotapes and photos to capture a visual picture of the facility site for use in later assessment evaluations. These representations may also be useful in BMP plan evaluation/reevaluation.

- Consider conducting monitoring to identify pollutants, pollutant loadings, and sources.

- Conduct brainstorming sessions to gather creative solutions for prioritized problems, followed by screening to eliminate impractical resolution.

- Evaluate technical merits and economic benefits of alternatives in an organized fashion. Consider ranking alternatives based on effects to product quality, costs, environmental benefits, ease of implementation, and success in other applications.

Release Identification and Assessment — What Not to Do

■ Do not make the site map so busy that information cannot be discerned. Enlarge the site map, or separate information on transparencies to later superimpose on the base map.

■ At large facilities, be cognizant of not overloading BMP committee members with release identification and assessment responsibilities. Consider the establishment of several evaluation teams, each assigned to assess a specific area.

■ Do not make changes in processes prior to allowing for an update in the release identification and assessment. Allow for the determination as to whether alternate methodologies or materials can be identified which are more environmentally protective or cost-effective.

■ Do not take on more than the company can handle at one time. Consider implementing changes in stages. Simple, procedural changes can be implemented immediately, while evaluations may need to be performed prior to the adoption of other measures.

DEVELOPMENT PHASE

After the BMP policy statement and committee have been established and the release potential identification and assessment has defined those areas of the facility that will be targeted for BMPs, the committee can begin determining the most appropriate BMPs to control environmental releases. The BMP plan should consist of both facility-specific BMPs and general BMPs.

General BMPs are relatively simple to evaluate and adopt. As previously indicated, general BMPs are practiced to some extent at all facilities. It is EPA's belief that all BMP plans should consist of six basic components:

1. *Good housekeeping.* A program by which the facility is kept in a clean and orderly fashion.

2. *Preventive maintenance.* A program focused on preventing releases caused by equipment problems, rather than repair of equipment after problems occur.

3. *Inspections.* A program established to oversee facility operations and identify actual or potential problems.

4. *Security.* A program designed to avoid releases due to accidental or intentional entry.

5. *Employee training.* A program developed to instill an understanding of the BMP plan in employees.

6. *Recordkeeping and reporting.* A program designed to maintain relevant information and foster communication.

A discussion of each of these basic components follows.

Good Housekeeping

Good housekeeping is essentially the maintenance of a clean, orderly work environment. Maintaining an orderly facility means that materials and equipment are neat and well-kept to prevent releases to the environment. Maintaining a clean facility involves the expeditious remediation of releases to the environment. Together, these terms, clean and orderly, define a good housekeeping program.

Maintaining good housekeeping is the heart of a facility's overall pollution control effort. Good housekeeping cultivates a positive employee attitude and contributes to the appearance of sound management principles at a facility. Some of the benefits that may result from a good housekeeping program include ease in locating materials and equipment; improved employee morale; improved manufacturing and production efficiency; lessened raw, intermediate, and final product losses due to spills, waste or releases; fewer health and safety problems arising from poor materials and equipment management; environmental benefits resulting from reduced releases of pollution; and overall cost savings.

Good housekeeping measures can be easily and simply implemented. Some examples of commonly implemented good housekeeping measures include the orderly storage of bags, drums, and piles of chemicals; prompt cleanup of spilled liquids to prevent significant run off to receiving waters; expeditious sweeping, vacuuming, or other cleanup of accumulations of dry chemicals to prevent them from reaching receiving waters; and proper disposal of toxic and hazardous wastes to prevent contact with and contamination of storm water run off.

The primary impediment to a good housekeeping program is a lack of thorough organization. To overcome this obstacle, a three-step process can be used, as follows:

1. Determine and designate an appropriate storage area for every material and every piece of equipment.

2. Establish procedures requiring that materials and equipment be placed in or returned to their designated areas.

3. Establish a schedule to check areas to detect releases and ensure that any releases are being mitigated.

The first two steps act to prevent releases that would be caused by poor housekeeping. The third step acts to detect releases that have occurred as a result of poor housekeeping.

As with any new or modifed program, the initial stages will be the largest hurdles; ultimately, though, good housekeeping should result in savings that far outweigh the efforts associated with initiation and implementation. Generally, a good housekeeping plan should be developed in a manner that creates employee enthusiasm and thus ensures its continuing implementation.

The first step in creating a good housekeeping plan is to evaluate the facility site organization. In most cases, a thorough release identification and assessment has already generated the needed inventory of materials and equipment and has determined their current storage, handling, and use locations. This information together with that from further assessments can then be used to determine if the existing location of materials and equipment are adequate in terms of space and arrangement.

Cramped spaces, and those with poorly placed materials, increase the potential for accidental releases due to constricted and awkward movement in these areas. A determination should be made as to whether materials can be stored in a more organized and safer manner (e.g., stacked, stored in bulk as opposed to individual containers, etc). The proximity of materials to their place of use should also be evaluated. Equipment and materials used in a particular area should be stored nearby for convenience, but should not hinder the movement of workers or equipment. This is especially important for waste products. Where waste conveyance is not automatic (e.g., through chutes or pipes) waste receptacles should be located as close as possible to the waste generation areas, thereby preventing inappropriate disposal leading to environmental releases.

Appropriately designated areas (e.g., equipment corridors, worker passageways, dry chemical storage areas) should be established throughout the facility. The effective use of labeling is an integral part of this step. Signs and adhesive labels are the primary methods used to assign areas. Many facilities have developed innovative labeling approaches, such as color coding the equipment and materials used in each particular process. Other facilities have stenciled outlines to assist in the proper positioning of equipment and materials.

Once a facility site has been organized in this manner, the next step is to ensure that employees maintain this organization. This can be accomplished through explaining organizational procedures to employees during training sessions, distributing written instructions, and most importantly, demonstrating by example.

Support of the program must be demonstrated, particularly by responsible facility personnel. Shift supervisors and others in positions of authority should act quickly to initiate activities to rectify poor housekeeping. Generally, employees will note this dedication to the good housekeeping program and will typically begin to initiate good housekeeping activities without prompting. Although initial implementation of good housekeeping procedures may be challenging, these instructions will soon be followed by employees as standard operating procedures.

Despite good housekeeping measures, the potential for environmental releases remains. Thus, the final step in developing a good housekeeping program involves the prompt identification and mitigation of actual or potential releases. Where potential releases are noted, measures designed to prevent release can be implemented. Where actual releases are occurring, mitigation measures, such as those described below, may be required.

Mitigation practices are simple in theory: the immediate cleanup of an environmental release lessens chances of spreading contamination and

136 / Pollution Prevention Strategies and Technologies

lessens impacts due to contamination. When considering choices for mitigation methods, a facility must consider the physical state of the material released and the media to which the release occurs. Generally, the ease of implementing mitigation actions should also be considered. For example, crushed stone, asphalt, concrete, or other covering may top a particular area. Consideration as to which substance would be easier to clean in the event of a release should be evaluated.

Good Housekeeping—What to Do

- Integrate a recycling/reuse and conservation program in conjunction with good housekeeping. Include recycle/reuse opportunities for common industry wastes such as paper, plastic, glass, aluminum, and motor oil, as well as facility-specific substances such as chemicals, used oil, dilapidated equipment, etc. into the good housekeeping program. Provide reminders of the need for conservation measures including turning off lights and equipment when not in use, moderating heating/cooling, and conserving water.

- When reorganizing, keep pathways and walkways clear with no protruding containers.

- Create environmental awareness by celebrating Earth Day (April 22) and/or developing a regular (e.g., monthly) good housekeeping day.

- Develop slogans and posters for publicity. Involve employees and their families by inviting suggestions for slogans and allowing children to develop the facility's good housekeeping posters.

- Provide suggestion boxes for good housekeeping measures.

- Develop a competitive program that may include company-wide competition or facility-wide competition. Implement an incentive program to spark employee interest (i.e., 1/2 day off for the shift which best follows the good housekeeping program).

- Conduct inspections to determine the implementation of good housekeeping. These may need to be conducted more frequently in areas of most concern.

- Pursue an ongoing information exchange throughout the facility, the company, and other companies to identify beneficial good housekeeping measures.

- Maintain necessary cleanup supplies (i.e., gloves, mops, brooms, etc.).

- Set job performance standards which include aspects of good housekeeping.

Good Housekeeping—What Not to Do

- Do not allow rubbish or other waste to accumulate. Properly dispose of waste, or arrange to have it removed in a timely fashion.

- Do not limit good housekeeping measures to industrial locations. Office areas should also be involved in the good housekeeping program.

Preventive Maintenance

Preventive maintenance (PM) is a method of periodically inspecting, maintaining, and testing plant equipment and systems to uncover conditions which could cause breakdowns or failures. As part of a BMP plan, PM focuses on preventing environmental releases. Most facilities have existing PM programs. It is not the intent of the BMP plan to require development of a redundant PM program. Instead, the objective is to expand the current PM program to address concerns raised as part of the release potential identification and assessment. Ulitimately, this will result in the focus of preventive maintenance on the areas and pollutants determined to be of most concern. Where no refocusing is necessary, the PM program suggested as part of the BMP plan and the existing PM program can be identical.

A PM program accomplishes its goals by shifting the emphasis from a repair maintenance system to a preventive maintenance system. It should be noted that in some cases, existing PM programs are limited to machinery and other moving equipment. The PM program prescribed to meet the goals of the BMP plan includes all other items (man-made and natural) used to contain and prevent releases of toxic and hazardous materials. Ultimately, the well-operated PM program devised to support the BMP plan should produce environmental benefits of decreased releases to the environment, as well as reducing total maintenance costs and increasing the efficiency and longevity of equipment, systems, and structures.

In terms of BMP plans, the PM program should prevent breakdowns and failures of equipment, containers, systems, structures, or other devices used to handle the toxic or hazardous chemicals or wastes. To meet this goal, a PM program should include a suitable system for evaluating equipment, systems, and structures; recording results; and facilitating corrective actions. A PM program should, at a minimum, include the following activities:

- Identification of equipment, systems, and structures to which the PM program should apply.

- Determination of appropriate PM activities and the schedule for such maintenance.

- Performance of PM activities in accordance with the established schedule.

- Maintenance of complete PM records on the applicable equipment and systems, and structures.

Generally, the PM program is designed to prevent and/or anticipate problems resulting from equipment and structural failures. However, it is unrealistic to expect that the PM program will avert the need for repair maintenance as a result of unanticipated problems. Adjustments and repair of equipment will still be necessary where problems occur, and replacement of equipment will be necessary when adjustment and repair are insufficient.

Generally, all good PM programs will consist of the four components noted above. However, it is of particular importance that the PM program address those areas and pollutants identified during the release identification and assessment step.

Although creating and implementing PM programs sounds easy, it is often impeded by lack of funding and organization. Lack of funding must be overcome by a facility's commitment to its PM program based on the simple truth that PM is less costly than replacement. Lack of organization can be overcome by better planning, which can be achieved by following the steps to developing an effective PM plan discussed below.

At the outset of a PM program, an inventory should be devised. This inventory should provide a central record of all equipment and structures including: location; identifying information such as serial numbers and facility equipment numbers/names; size, type, and model; age; electrical and mechanical data; the condition of the equipment/structure; and the manufacturer's address, phone number, and person to contact. In addition to the equipment inventory, an inventory of the structures and other non-moving parts to which the PM program is to apply should also be determined.

Inventories can be developed through inspections and/or reviews of facility specifications and operations and maintenance manuals. In some cases, it is effective to label equipment and structure with assigned numbers/names and some of the identifying information. This information may be useful to maintenance personnel in the event of emergency situation or unscheduled maintenance where maintenance information is not readily available. Several different methods are effective for recording inventory information including the use of index cards, prepared forms and checklists, or a computer database.

Since the PM program involves the use of maintenance materials (i.e., spare parts, lubricants, etc.), some additional considerations may apply. First, good housekeeping measures, as discussed in the section on Good Housekeeping, are particularly important for organizing maintenance materials and keeping areas clean. A tracking system may also be necessary for organizing maintenance materials. The inventory should include information such as materials/parts description, number, item specifications, ordering information, vendor addresses and phone numbers, storage locations, order quantities, order schedules and costs. A large facility may require a parts catalog to coordinate such information. Large facilities may also find it necessary to develop a purchase order system which maintains the stock in adequate number and in the proper order by keeping track of the minimum and maximum number of items required to make timely repairs, parts that are vulnerable to breakage, and parts that have a long delivery time or are difficult to obtain.

Once the inventory is completed, the facility should determine the PM requirements including schedules and specifications for lubrication, parts replacement, equipment and structural testing, maintenance of spare parts, and general observations. The selected PM activities should be based on the facility-specific conditions but should be at least as stringent as the manufacturer's recommendations. Manufacturer's specifications can generally be found in brochures and pamphlets accompanying equipment. An operations and maintenance manual also may contain this information. If these sources are not available, the suggested manufacturer's recommendation can be obtained directly from the manufacturer. In cases of structures or non-moving parts, the facility will need to determine appropriate maintenance activities (e.g., integrity testing). As with inventory information, PM information should be recorded in an easily accessible format.

After establishment of the materials inventory and the development of PM requirements, a facility should schedule and carry out PM on a regular basis. Personnel with expertise in maintenance should be available to conduct maintenance activities. In a small facility where one person may conduct regular maintenance activities, specialized contractors may supplement the maintenance program for more complex activities. An up-to-date list of outside firms available for contract work beyond the capability of the facility staff should be readily available. Additionally, procedures explaining how to obtain such support should be provided in the pollution prevention plan. Larger facilities should have sufficient PM expertise within the staff, including a PM manager, an electrical supervisor, a mechanical supervisor, electricians, technicians, specialists, and clerks to order and acquire parts and maintain records. Ongoing training and continuing education programs may be used to establish expertise in deficient areas.

Maintenance activities should be coordinated with normal plant operations so that any shutdowns do not interfere with production schedules or environmental protection. The maintenance supervisory staff should also consider other timing constraints such as the availability of the PM staff for both regularly scheduled PM and unanticipated corrective repairs.

The final step in the development of a PM program involves the organization and maintenance of complete records. A PM tracking system which includes detailed upkeep, cost, and staffing information should be utilized. A PM tracking system assists facilities in identifying potential equipment or structural problems resulting from defects, general old age, inappropriate maintenance, or poor engineering design; preparation of a maintenance department budget; and deciding whether a piece of equipment or a structure should continue to be repaired or replaced.

There are many commercial software systems that enable facilities to track maintenance. Computer systems allow for input of inventory and PM information and generate daily, weekly, monthly, and/or yearly maintenance sheets which include the required the item to be maintained, the maintenance duties, and materials to be used (e.g., oil, spare parts, etc.). The system can be continually updated to add information gathered during maintenance activities. Some of the maintenance information that proves useful includes the work hours spent, materials used, frequency of downtime for repairs, and costs involved with maintenance activities. This information in turn can generate budgets and determinations of the cost effectiveness of repair versus replacement, etc. Computerized systems for maintenance tracking are usually most effective at larger facilities.

Useful manual systems may involve index cards, maintenance logs, and a maintenance schedule. Initially, inventory and PM information can be recorded on index cards. This information can be consulted during maintenance activities. Maintenance logs should also be developed for each piece of equipment and each structure, and should contain information such as the maintenance specifications, and data associated with the completion of maintenance activities. Maintenance personnel should complete relevant information including the date maintenance was conducted, hours spent on duties, materials used, worker identification, and the nature of the problem.

Preventive Maintenance—What to Do

- When attaching information to equipment and structures, use bold and bright colors consistent with the approaches described for good housekeeping.

- Consider and discuss additional PM procedures beyond those normally recommended by the manufacturer.

- Conduct extensive safety training for PM personnel.

- Coordinate scheduling of PM activities with facility or unit downtime.

- Keep track of how long materials have been stored. This will support an evaluation of the integrity of storage containers.

- Develop a PM staff team approach including team names (i.e., the A-team) to create enthusiasm.

- Utilize blackboards and charts to assist in organizing and conveying an annual PM schedule.

Preventive Maintenance—What Not to Do

- Do not forget to stock important replacement parts and any specialized tools required to repair equipment.

- Do not create a paperwork nightmare. Develop the minimum number of well-organized logs necessary to maintain information.

- Do not let untrained, unskilled personnel conduct PM activities. Employees taking part in the PM program must be familiar with equipment and maintenance procedures.

- Do not forget to determine the availability and time needed to obtain vital parts or contractor assistance.

Inspections

Inspections provide an ongoing method to detect and identify sources of actual or potential environmental releases. Inspections also act as oversight mechanisms to ensure that selected BMPs are being implemented. Inspections are particularly effective in evaluating the good housekeeping and PM programs previously discussed.

Many facilities may be currently conducting inspections, but in a less formalized manner. Security scans, site reviews, and facility walk throughs conducted by plant managers and other such personnel qualify as inspections. These types of reviews, however, are often limited in scope and detail. To ensure the objectives of the BMP plan are met, these types of reviews should be conducted concurrently with periodic, in-depth inspections as part of a comprehensive inspection program.

Inspections implemented as part of the BMP plan should cover those equipment and facility areas identified during the release identification and assessment as having the highest potential for environmental releases. Since inspections may vary in scope and detail, an inspection program should be developed to prevent redundancy while still ensuring adequate oversight and evaluation.

A BMP inspection program should set out guidelines for each of the following:

- Scope of each inspection.

- Personnel assigned to conduct each inspection.

- Inspection frequency.

- Format for reporting inspection findings

- Remedial actions to be taken as a result of inspection findings.

Despite the different requirements of each type of inspection, the focus of inspections conducted as part of the BMP plan should not vary. Some of the areas within the facility that may be the focus of the BMP plan include solid and liquid materials storage areas, in-plant transfer and materials handling areas, activities with potential to contaminate storm water run off, and sludge and hazardous waste disposal sites.

An inspection program's goal will be to ensure thoroughness, while preventing redundancy. Ultimately, this will ensure that the use of resources is optimized. In addition, it should be clear that the inspection team's efforts are directed to support the operating groups in carrying out their responsibilities for equipment and personnel safety, and work quality, and to ensure that all standards are met. In achieving these goals, written procedures discussing the scope, frequency and scheduling, personnel, format, and remediation procedures should be provided.

The scope of each inspection type should be discussed in the written procedures. Many different types of inspections are conducted as part of the inspection program. Guidelines for the scope of these inspections include:

- *Security scan.* Search for leaks and spills which may be occurring. Specifically examine problems areas which have been identified by the plant manager or equivalent persons.

- *Walk through.* Conduct oversight of the duties associated with a security scan. In addition, ensure that equipment and materials are located in their appropriate positions.

- *Site review.* Conduct oversight of duties associated with a walk through. Additionally, evaluate the effectiveness of the PM, good housekeeping, and security programs by visual oversight of their implementation.

- *BMP plan oversight inspection.* Conduct oversight of duties associated with a site review. Evaluate the implementation of all aspects of the written BMP plan including the review of the records generated as part of these programs (e.g., inspection reports, PM activity logs).

- *BMP plan evaluation/reevaluation inspection:* Conduct an evaluation/reevaluation of the facility and determine the most appropriate BMPs to control environmental releases.

An appropriate mix of these types of inspections should be developed based on facility specific considerations. The proper frequency for conducting inspections will vary based on the type of the inspection and other facility-specific factors. Some general guidelines for establishing frequency follow:

- Security scans can be conducted various times daily.

- Walk through inspections can be conducted once per shift to once per week.

- Site reviews can be conducted once per week to once per six months.

- BMP plan oversight inspections can be conducted once per month to once per year.

- BMP plan reevaluation inspections can be conducted once per year to once every five years.

There are no hard and fast rules for conducting inspections as part of the BMP plan. Inspection frequencies should be based on a facility's needs. Two points should be considered when establishing an inspection program: (1) As would be expected, more frequent inspections should be conducted in the areas of highest concern; and (2) inspections must be conducted more frequently during the initial BMP implementation until the BMP plan procedures become part of standard operating procedures.

It may be useful to set up a schedule to ensure a comprehensive inspection program. Varying the dates and times of inspection conduct is also good practice in that it ensures all stages of production and all situations are reviewed.

Individuals qualified to assess the potential for environmental releases should be assigned to conduct formal inspections. Members of the BMP committee can generally fulfill this requirement, but they may not be available to conduct all inspections. Thus, it may be appropriate to identify and train personnel to conduct specific types of inspections. For example, shift foremen and other equivalent supervisory personnel may appropriately conduct walk throughs and site reviews as a result of their position of authority and ability to require prompt correction if problems are observed. Personnel with immediate responsibility for an area should not be asked to conduct inspections of that area as they may be tempted to overlook problems. Additionally, plant security and other personnel who routinely conduct walk throughs should not be assigned to conduct BMP plan inspections since their familiarity with the facility may result in their not being suited to best identify opportunities for improvement.

Different perspectives are useful when conducting inspections. By developing a team inspection approach or by alternating inspectors, facilities can receive a more thorough review. One inspector may observe something that another will overlook, and an inspector tends to focus on the areas with which he/she is most familiar.

An inspection checklist of areas to inspect with space for a narrative report is a helpful tool when conducting inspections. A standard form helps ensure inspection consistency and comprehensiveness. Checklists may, however, not be necessary for each inspection performed. This may be particularly true for facilities conducting frequent inspections (once per hour, once per shift, etc.); procedures for using inspection checklists should be reasonable to prevent excessive paperwork.

The findings of inspections will be useless unless they are brought to the attention of appropriate personnel and subsequently acted upon. To ensure that reports are acted upon in an expeditious and appropriate manner, procedures for routing and review of reports should be developed and followed.

Despite the usefulness of written reports, in no way should a written report replace verbal communication. Where a problem is noted, particularly environmental releases currently occurring or about to occur, it should be verbally communicated by the inspector to the responsible personnel as soon as possible.

Inspections—What to Do

- Encourage workers to conduct visual inspections and report any actual or potential problems to the appropriate personnel.

- Develop inspection checklists for each type of inspection. Vary them where necessary for each part of the facility subject to BMPs.

- Consider utilizing non-regulatory support from EPA, states, or university-supported resources when conducting site assessments.

Inspections—What Not to Do

- Do not rely solely on the use of a checklist for inspections. Narrative descriptions should be included in the reports to ensure that problems are identified and discussed.

- Do not conduct inspections and then fail to provide feedback of findings of concern to the person responsible for the area inspected.

Security

A security plan describes the system installed to prevent accidental or intentional entry to a facility that might result in vandalism, theft, sabotage, or other improper or illegal use of the facility. In relation to a

BMP plan, a security system should prevent environmental releases caused by any of these improper or illegal acts.

Most facilities already have a program for security in place; this security program can be integrated into the BMP plan with minor modifications. Facilities developing a program for security as part of the BMP plan may be hesitant to describe their security measures in detail due to concerns of compromising the facility. The intent of including a security program as part of the BMP plan is not to divulge facility or company secrets; the specific security practices for the facility may be kept as part of a separate confidential system. The security program as part of the BMP plan should cover security in a general fashion, and discuss in detail only the practices which focus on preventing environmental releases.

The security program as part of the BMP plan should be designed to meet two goals. First, the security plan should prevent security breaches that result in the release of hazardous or toxic chemicals to the environment. The second goal is to effectively utilize the observation capabilities of the security plan to identify actual or potential releases to the environment. Some typical components of a security plan include the following:

- Routine patrol of the facility property by security guards in vehicles or on foot.

- Fencing to prevent intruders from entering the facility site.

- Good lighting to facilitate visual inspections at night, and of confined spaces.

- Vehicular traffic control (i.e., signs).

- Access control using guardhouse or main entrance gate, where all visitors and vehicles are required to sign in and obtain a visitor's pass.

- Secure or locked entrances to the facility.

- Locks on certain valves or pump starters.

- Camera surveillance of appropriate sites, such as facility entrance, and loading / unloading areas.

- Electronic sensing devices supplemented with audible or covert alarms.

- Telephone or other forms of communication.

Typically, security systems focus on the areas with the greatest potential for damage as a result of security breaches. As part of the BMP plan, the security program will focus on the areas that result in environmental releases. Typically, these areas have been identified in the release identification and assessment step. In many cases, the findings of this step may indicate a need to change the focus or broaden the scope of the security program to include areas of the facility addressed by the BMP plan. Since the security program may not be common knowledge, general BMP committee members may not be able to recommend changes. As a result, security personnel should be involved in the decisions made by the committee, with one person possibly serving as a member.

While performing their duties, security personnel can actively participate in the BMP plan by checking the facility site for indications of releases to the environment. This may be accomplished by checking that equipment is operating properly; ensuring no leaks or spills are occurring at materials storage areas; and checking on problem areas (i.e., leaky valves, etc).

The advantages of integrating security measures into the BMP plan are considerable. Security personnel are in positions that enable them to conduct periodic walk throughs and scans of the facility, as well as covertly view facility operations. They are in an excellent position to identify and prevent actual or potential releases to the environment.

Where security personnel are utilized as part of the oversight program, two obstacles generally must be overcome: (1) support must be gained from the security staff; and (2) security personnel must be knowledgeable about what may and may not be a problem, and to whom to report when there is a problem. Involving the security staff in the BMP plan development at an early stage should assist in gaining their support. Integration of the security staff into the employee training, and recordkeeping and reporting programs discussed in the sections on

Employee Traning and Recordkeeping and Reporting, respectively, can also be used to overcome these barriers.

Security—What to Do

- File detailed documentation of the security system separately from the BMP plan to prevent unauthorized individuals from gaining access to confidential information.

- Make certain that all security personnel are aware of their assigned responsibilities under the BMP plan.

- Post security and informational signs and distribute security and direction information to visitors. This may be particularly useful for frequently visited buildings.

Security—What Not to Do

- Do not assume that isolation is adequate security.

- Do not locate alarms or indicator lights where they cannot be readily seen or heard.

Employee Training

Employee training conducted as part of the BMP plan is a method used to instill in personnel, at all levels of responsibility, a complete understanding of the BMP plan, including the reasons for developing the plan, the positive impacts of the plan, and employee and managerial responsibilities under the BMP plan. The employee training program should also educate employees about the general importance of preventing the release of pollutants to water, air, and land.

Training programs are a routine part of facility life. Most facilities conduct regular employee training in areas, including fire drills, safety, and miscellaneous technical subject areas. Thus, the training program developed as a result of the BMP plan should be easily integrated into the existing training program.

Employee training conducted as part of the BMP plan should focus on those employees with direct impact on plan implementation. This may include personnel involved with manufacturing, production, waste treatment and disposal, shipping/receiving, or materials storage; areas where processes and materials have been identified as being of concern; and PM, security, and inspection programs. Training programs, which include all appropriate personnel, should include instruction on spill response, containment, and cleanup. Generally, the employee training program should serve to improve and update technical, managerial, or administrative skills; increase motivation; and introduce incentives for BMP plan implementation.

Employee training programs function through the following four step process:

1. Analyzing training needs

2. Developing appropriate training materials

3. Conducting training; and

4. Repeating training at appropriate intervals in accordance with steps 1 through 3.

The first stage in developing a training program is analyzing training needs. Generally, training needs to be conducted during the planning and development phases of the BMP plan, and as follow-up to BMP implementation for selected areas of concern. In all three cases, it is important to analyze training needs and develop appropriate training tools to use during conduct of the training.

The initial BMP development session educates employees of the need for, objectives of, and projected impact of the BMP plan. As would be expected, this initial training should be conducted at the onset of the BMP development. The message portrayed at this session should be the positive impacts of the BMP plan, including ease in locating materials and equipment; improved employee morale; improved manufacturing and production efficiency; lessened raw, intermediate and final product losses due to releases; fewer health and safety problems arising from unmitigated releases and/or poor placement of materials and equipment; environmental benefits resulting from reduced releases of pollution; and overall cost savings. When providing this message, it is essential that the benefits for employees, as well as the company itself be stressed.

While it is important to point out the reasons that lead to the decision to implement a BMP plan, it is also important to provide a realistic picture of the changes and impacts which will result. These modifications should be discussed in terms of their positive impact to help maintain a high level of enthusiasm.

After the BMP plan is developed, the BMP implementation training sessions should be developed. The training sessions should review the BMP plan and associated procedures, such as the following:

- The good housekeeping program, including the use of labeling (signs, color coding, stenciling, etc.) to assign areas and procedures to return materials to assigned areas.

- The PM program, including new PM schedules and procedures.

- Integration of the security plan with the BMP plan.

- Inspection programs

- Responsibilities under the recordkeeping and reporting system.

In some cases, it may be appropriate to provide a general session explaining BMP plan implementation followed by specialized training for each area. For example, since all employees should be aware of the good housekeeping program, this program should be discussed at the general session. Training for selected facility-specific BMPs may be necessary only for employees in the production and manufacturing areas. PM information could be presented only to the personnel conducting maintenance, while security personnel need only be briefed of security-related responsibilities under the BMP plan.

Training sessions are only as effective as the level of preparation. It is vital that workshop materials are technically accurate, easily read, and well-organized. More importantly, training materials must leave a strong impression, such that their message is remembered and any distributed training materials are consulted in the future. The use of audiovisual aids supplemented with informational handouts is one of the best methods of conveying information. Including copies of any slide or overhead helps avoid distractions during presentation caused by employees' copying contents of overheads. Other techniques which assist in effectively conveying information include the following:

- Providing aesthetically pleasing covers and professional looking handouts.

- Developing detailed tables of contents with well numbered pages.

- Frequently assimilating graphics into presentations.

- Integrating break-out sections and exercises.

- Incorporating team play during exercises.

- Allowing for liberal question/answer sessions and discussions during or after presentations.

- Providing frequent breaks.

- Integrating field activities with classroom training.

The use of qualified personnel to conduct training presentations also supports the facility's commitment to BMP plan implementation. Speakers should be identified in the initial training preparation stages based on their expertise in the topics to be presented. However, expertise is not the only consideration. Expertise must be supplemented with a well executed, interesting, enthusiastic presentation. Preparation prior to the training event will allow speakers to organize presentations, establish timing, and develop tone and content. Speakers should consider undergoing a dry run during which the speaker provides the full presentation, including use of audio/visual aids.

Proper planning should ensure the execution of an effective training event. Once the training event has been conducted, some follow-up activities should be conducted. For example, evaluation forms requesting feedback on the training should be distributed to employees. These evaluation forms can be used to identify presentation areas needing improvement, ideas needing clarification, and future training activities. Ultimately, information gathered from these forms can help direct the employee training program in the future.

Once BMP plan implementation is underway, training should be conducted both routinely and on an as-needed basis. Special training sessions may also be prompted when new employees are hired, environmental release incidents occur, recurring problems are noted during inspections, or changes in the BMP plan are necessary.

Employee Training—What to Do

- Show strong commitment and periodic input from top management to the employee training program to create the necessary interest for a successful program.

- Ensure that announcements of training events are posted well in advance and include the times and dates of the sessions, the names and positions of the instructors, the lesson plans, and the subject material covered.

- Make the training sessions interesting. Use film and slide presentations. Bring in speakers to demonstrate the use of cleanup materials or equipment. Contact the state health and environmental agencies or the EPA Regional office for films, volunteer speakers, or other training aids.

- Use employee incentive programs or environmental excellence awards to reinforce training programs.

- Conduct demonstrative hands-on field training to show the effectiveness of good housekeeping, PM, or inspection programs.

- Give frequent refresher courses and consider pop quizzes to keep employees sharp.

Employee Training—What Not to Do

- Do not provide training to permanent facility employees only. Overlooking temporary and contractor personnel can increase the possibility of environmental releases.

154 / Pollution Prevention Strategies and Technologies

- Do not allow training session attendance to be optional. Employees in the positions that incur the most stress in terms of meeting schedules should be reminded to avoid taking shortcuts when handling toxic or hazardous chemicals.

- Do not become too standardized. Reusing an annual employee training session will be tedious to employees. Integrate new information and improve on old information.

Recordkeeping and Reporting

As part of a BMP plan, recordkeeping focuses on maintaining records that are pertinent to actual or potential environmental releases. These records may include the background information gathered as part of the BMP plan, the BMP plan itself, inspection reports, PM records, employee training materials, and other pertinent information.

Maintenance of records is ineffective unless a program for the review of records is set forth. In particular, a system of reporting actual or potential problems to appropriate personnel must be included. Reporting, as it relates to the BMP plan, is a method by which appropriate personnel are kept informed of BMP plan implementation, such that appropriate actions may be determined and expeditiously taken. Reporting may be verbal or follow a more formal notification procedure. Some examples of reporting include the following:

- Informational memos distributed to upper management or employees to keep them updated on the BMP plan.

- Verbal notification by BMP inspectors to supervisors concerning areas of concern noted during inspections.

- Corrective action reports from the BMP committee to the plant manager which cite deficiencies with BMP plan implementation.

- Verbal and written notification to regulatory agencies of releases to the environment.

An effective recordkeeping and reporting program functions through the following three step procedure:

1. Developing records in a useful format.

2. Routing records to appropriate personnel for review and determination of actions to address deficiencies.

3. Maintaining records for use in future decision making processes.

Recordkeeping and reporting play an overlapping role with the programs previously discussed. In general, these programs will involve the development, review, maintenance, and reporting of information to some degree. For example, an inspection program may include the development and use of an inspection checklist, submittal of the completed checklist to relevant personnel, evaluation of the inspection information, and determination of appropriate corrective actions. This may, in some cases, involve the development of a corrective action report to submit to appropriate persons (which may include regulatory agencies where necessary/required). The checklist and the corrective action reports should be maintained in organized files.

As part of the BMP plan, a recordkeeping and reporting program will primarily be developed for the PM and inspection programs. However, effective communication methods can also be useful in the development of the release identification and assessment portion of the BMP plan.

The first step to ensuring an effective recordkeeping and reporting program is the development of records in a useful format. The use of standard formats (i.e., checklists) can help to ensure the completion of necessary information, thoroughness in reviews, and understanding of the supplied data. For example, a standard inspection format may specify a summary of findings, recommendations, and requirements on the first page; then, detailed information by geographical area (e.g., materials storage area A, materials storage area B, the north loading and unloading zone) may be discussed. With a standard format, an inspection report reviewer may quickly review the findings summary to determine where problems exist, then refer to the detailed discussion of areas of concern. Ultimately, the use of a standard format minimizes the review time, expedites decision making concerning corrective actions, and simplifies reporting.

Despite the recommended use of standard formats, inspectors should not feel constrained by the format. Sufficient detail must be provided in order for the report to be useful. Narratives should accompany checklists where necessary to provide detailed information on materials that have been released or have the potential to be released; nature of the materials involved; duration of the release or potential release; potential or actual volume; cause; environmental results of potential or actual releases; recommended countermeasures; people and agencies notified; and possible modifications to the BMP plan, operating procedures, and/or equipment.

The second step to ensuring an effective recordkeeping and reporting system involves routing information to appropriate personnel for review and determination of actions to address deficiencies. Regardless of whether the system for recordkeeping and reporting is structured or informal, the BMP plan should clearly indicate: (1) How information is to be transferred (i.e., by checklist, report, or simply by verbal notification); and (2) to whom the information is to be transferred (i.e., the plant manager, the supervisor in charge, or the BMP committee leader).

Customarily, formal means to transfer information would be more appropriate in larger, more structured companies. For example, reviews of findings and conclusions as part of inspection reports may be conducted by supervisory personnel and the information may be routed through the chain of command to the responsible personnel, such as shift supervisors or foremen. Less formal communication methods, such as verbal notification may be appropriate for smaller facilities.

The key to ensuring a useful communication system is identifying one person (or, at larger facilities, several persons) to receive and dispense records and information. This person will be responsible for ensuring that designated individuals review records where appropriate, that corrective actions are identified, and that appropriate personnel are notified of the need to make corrections. Additionally, this person will ensure that information is maintained on file for use in later evaluations of the BMP plan effectiveness.

It should be noted that the recordkeeping and reporting system is designed to help, not hinder, the communications process. Verbal communications of impending or actual releases should be made regardless of whether a formal communications process has been set forth.

A communications system for notification of potential or actual release should be designated. Such a system could include telephone or radio contact between transfer operations, and alarm systems that would signal the location of a chemical release. Provisions to maintain communication in the event of a power failure should be addressed. Reliable communications are essential to expedite immediate action and countermeasures to prevent incidents or to contain and mitigate chemicals released.

A reporting system should include procedures for notifying regulatory agencies. A number of federal and state agencies may require reporting of environmental releases. It is outside the scope of this chapter to provide a summary of all necessary reporting requirements.[92] However, reporting requirements specified under the NPDES permitting program include, at a minimum, the following:

- Releases in excess of reportable quantities which are not authorized by an NPDES permit.

- Planned changes which:

    -subject the facility to new source requirements
    -significantly change the nature or quantity of pollutants discharged
    -change a facility's sludge use or disposal practices
    -may result in noncompliance

- Notification within 24 hours of any unanticipated discharges (including bypasses and upsets) which may endanger human health or the environment, and the submission of a written report within five days.

- The discharge of any toxic or hazardous pollutant above notification levels.

- Any other special notification procedure or reporting requirement specified in the NPDES permit.

Reports maintained in the recordkeeping system can be used in evaluating the effectiveness of the BMP plans, as well as when revising

---

[92] See M. Dennison, *Environmental Reporting, Recordkeeping, and Inspections: A Compliance Guide for Business and Industry* (Van Nostrand Reinhold, 1995).

the BMP plan. Additionally, these records provide an oversight mechanism which allows the BMP committee to ensure that any detected problem has been adequately resolved. As such, the final step in developing a recordkeeping and reporting program involves the development and maintenance of an organized recordkeeping system.

In general, an organized filing system involves selecting an area for maintaining files, labeling files appropriately, and filing information in an organized manner. A single location should be designated for receiving the data generated for and related to the BMP plan. At larger facilities, several locations may be appropriate (e.g., maintenance records in one location, other BMP related documentation in another). A centralized location will help to consolidate materials for later review and consideration. Without a designated location, materials may become dispersed throughout a facility and subsequently lost.

Filing information by subject and date is a practice followed by most facilities. The most effective filing system usually includes hard copies of the information on a file. Additionally, keeping inventory lists of documents maintained in file folders assists in quick reviews of file contents. Small facilities may be able to file all BMP-related information in the same folder in chronological order; larger facilities may have to file information by subject. For example, PM information may be filed by equipment type in separate folders, while good housekeeping information and related oversight and evaluation information may be filed based on facility area. In some cases, larger facilities may find it convenient to develop an automated tracking system (e.g., a database system) for efficiently maintaining records.

Recordkeeping and Reporting—What to Do

- Clearly designate review and filing responsibilities for BMP related materials.

- Designate a file copy of any BMP correspondence.

- Set up procedures for materials release notification that include those plant personnel to be immediately notified, in order of priority, including backups, and then the appropriate governmental regulating agencies (Federal, State, and local). Include the fire department, police, public water supply agency, fish and wildlife commission, and municipal sewage treatment plant, where appropriate.

- Develop a standard form for submitting a report for and the internal review of a release or near release.

- Share knowledge gained through BMP implementation with others. Report successes of BMP plan implementation in the Pollution Prevention Information Clearinghouse, magazines, or corporate newsletters.

Recordkeeping and Reporting—What Not to Do

- Do not keep the details of a materials release a secret known only to the facility management. Share the information learned from incidents so that all employees may benefit from the experience.

- Do not forget to keep employees informed. Continually provide updates (e.g., quarterly memo, newsletter) of BMP committee initiatives and progress. Lack of communication with employees may be interpreted as lack of continuing interest in the BMP plan's implementation.

PLAN EVALUATION AND REEVALUATION

Planning, development, and implementation of the BMP plan require the dedication of important resources by company management. The benefits derived, however, serve to justify the costs and commitments made to the BMP plan. To illustrate the plan's benefits, it may be appropriate and even necessary in some cases to measure the plan's effectiveness.

An evaluation can be performed by considering a number of variables, including:

- Benefits to the employees.

- Benefits to the environment.

- Reduced expenditures.

Benefits to the employees can be assessed in terms of health and safety, productivity, and other factors such as morale. Comparisons before and after plan implementation can be made to determine trends that show BMP plan effectiveness. The following information can be utilized in this determination:

- Time off due to on-the-job injury or illness resulting from exposure to chemicals.

- Production records which track worker productivity.

Benefits to the environment can be measured by several factors. First, pollutant monitoring prior to the inception of the BMP plan may show significant quantities of pollutants and or wastes that are minimized or eliminated after plan implementation. Discharge monitoring report records may show reductions in the quantity or variability of pollutants in the discharges. In addition, the reductions in volumes of and/or hazards posed by solid waste generation and air emissions may demonstrate the success of the BMP plan.

Other derived environmental benefits may include reduced releases to the environment resulting from spills, volatilization, and losses to storm water run off. These benefits may be measured through reductions in the number and severity of releases and of lessened losses of materials.

Reduced expenditures are the bottom line in substantiating the need for the BMP plan. Cost considerations can be easily tracked through expense records including chemicals usage, energy usage, water usage, and employee records. The development of production records on product per unit cost before and after BMP plan implementation may show a significant drop, thereby demonstrating the effectiveness of the plan.

The operations at an industrial facility are expected to be dynamic and therefore subject to periodic change. As such, the BMP plan can not remain effective without modifications to reflect facility changes. At a minimum, the BMP plan should be revisited annually to ensure that it fulfills its stated objectives and remains applicable. This time-dated approach allows for the consideration of new perspectives gained through the implementation of the BMP plan, as well as the reflection of new directives, emerging technologies, and other such factors. However, plan revisions should not be limited to periodic alterations. In some cases, it may be appropriate to evaluate the plan due to changed conditions such as the following:

- Restructuring of facility management.

- Substantial growth.

- Significant changes in the nature or quantity of pollutants discharged.

- Process or treatment modifications.

- New permit requirements.

- New legislation related to BMPs.

- Releases to the environment.

Many changes at a facility may warrant modifications to the BMP plan. Growth may require more frequent employee training or a redesign of the good housekeeping program to ensure the site is maintained in a clean and orderly fashion. An evaluation of or modifications to existing process, treatment, and chemical handling methods may substantiate the need for additional facility-specific BMPs.

Where new permit requirements or legislation focus on a specific pollutant, process, or industrial technology, it may be appropriate to consider establishing additional controls. These permit requirements or legislative changes do not necessarily have to be directly related to environmental issues. For example, new OSHA standards may result in modification of the BMP plan to include procedures that address the protection of worker health and safety.

If there has been a spill or other unexpected chemical release, the reasons for the release and corrective actions taken should be investigated. This investigation should include evaluation of all control programs including good housekeeping, PM, inspections, security, employee training, and recordkeeping and reporting. Additionally, facility-specific BMPs should be evaluated at that time to determine their effectiveness.

Ultimately, the BMP plan reevaluation may pinpoint areas of the facility not addressed by the plan, or activities that would benefit from further development of facility-specific BMPs or revision of the general programs contained in the BMP plan. It is useful to bear in mind that as the BMP plan improves, costs can continue to be minimized as a result of reduced waste generation, less hazardous or toxic materials use, and prevented environmental releases.

Chapter 6

Industry-Specific Pollution Prevention Strategies

INTRODUCTION

This chapter summarizes various waste minimization methods that have been identified, evaluated, and implemented by different types of facilities. While the industry-specific waste minimization strategies discussed in this chapter are surely representative of the many waste minimization options currently available and in use, it is not possible to provide an exhaustive treatment of every available option for every conceivable type of facility. Thus, the reader may wish to consult other sources for additional guidance on pollution prevention strategies for industries and operations not specifically covered here.

PAINT MANUFACTURING INDUSTRY

This section discusses recommended waste minimization methods for paint manufacturers. The primary waste streams associated with paint manufacturing are listed in Table 6-1, along with recommended control methods. In order of occurrence at a facility, the waste streams are: equipment cleaning wastes; spills and off-specification paint; leftover inorganic pigment in bags and packages; pigment dust from bag houses; filter cartridges; and obsolete products/customer returns.

The waste minimization methods listed in Table 6-1 can be classified generally as source reduction, which can be achieved through material substitution, process or equipment modification, or better operating practices; or as recycling. An example of a source reduction method in the table is the use of countercurrent rinsing to reduce the volume of cleaning waste, while an example of recycling is the working of spilled product back into the process.

Table 6.1. Waste Minimization Options for the Paint Manufacturing Industry

Waste Stream	Waste Minimization Options
Equipment cleaning wastes (rinse water, solvent, and sludge)	-Use mechanical wipers on mix tanks. -Use high pressure wash systems. -Install Teflon liners on mix tanks. -Use foam/plastic pigs to clean lines. -Reuse equipment cleaning wastes. -Schedule production to minimize need for cleaning. -Clean equipment immediately. -Use countercurrent rinse methods. -Use alternative cleaning agents. -Increase spent rinse settling time. -Use de-emulsifiers on spent rinses.
Spills and off-spec paint	-Increase use of automation. -Use appropriate clean-up methods. -Recycle back into process. -Implement better operating practices.
Leftover inorganic pigment in bags and packages	-Use water soluble bags and liners. -Use recyclable/lined/dedicated containers.
Air emissions, including pigment dust	-Modify bulk storage tanks. -Use paste pigments. -Install dedicated bag house systems.
Filter cartridges	-Improve pigment dispersion. -Use bag or metal mesh filters.
Obsolete products/customers returns	-Blend into new products.

Better operating practices are procedural or institutional policies that result in a reduction of waste. They include:

- Waste stream segregation

- Personnel practices

    -Management initiatives
    -Employee training
    -Employee incentives

- Procedural measures

    -Documentation
    -Material handling and storage
    -Material tracking and inventory control
    -Scheduling

- Loss prevention practices

    -Spill prevention
    -Preventive maintenance
    -Emergency preparedness

- Accounting practices

    -Apportion waste management costs to departments that generate the waste

Better operating practices apply to all waste streams. In addition, specific better operating practices that apply to certain waste streams are identified in the appropriate sections that follow.

Equipment Cleaning Wastes

Equipment cleaning generates most of the waste associated with paint manufacturing. Following production of either solvent or water-based paints, considerable waste or "clingage" remains affixed to the sides of the preparation tanks. The three methods of tank cleaning used in the paint industry are solvent washing for solvent-based paint, caustic washing for either solvent or water-based paint, and water washing for water-based paint.

Equipment used for preparation of solvent-based paint is rinsed with solvent, which is then generally reused in the following ways:

- Collected and used in the next compatible batch of paint as part of the formulation.

- Collected and re-distilled either on or off-site.

- Collected and used with or without settling for equipment cleaning, until spent. When the solvent is finally spent, it is then drummed for disposal.

In 1985, a survey conducted by the National Paint & Coatings Association's Manufacturing Management Committee showed that over 82% of the respondents recycled all of their solvent waste either on-site or off-site. With current costs of disposal, on-site distillation of solvent can be economically justified for as little as eight gallons of solvent waste generated per day. Of all the solvent that is recycled, 75% is recovered with the remaining portion disposed of as sludge.

Caustic rinse is used for equipment cleaning of both solvent and water-based paints, but more often with water-based paints. Water rinsing is usually insufficient in removing paint that has dried in the mix tanks. Since solvent rinsing can usually remove solvent-based paint that has dried, the need for caustic is less.

There are two major types of caustic systems commonly used by the paint industry. In one type of system, caustic is maintained in a holding tank (usually heated) and is pumped into the tank to be cleaned. The caustic drains to a floor drain or sump from which it is returned to the holding tank. In the second type of system, a caustic solution is prepared in the tank to be cleaned, and the tank is soaked until it is clean. Most plants reuse the caustic solution until it loses most of its cleaning ability. At that time, the caustic is disposed of either as a solid waste or wastewater, with or without neutralization.

Water wash of equipment used in the production of water-based paint is the source of considerable wastewater volume, which is usually handled as follows:

- Collected and used in the next compatible batch of paint as part of the formulation.

- Collected and used with or without treatment for cleaning until spent.

- Disposed with or without treatment as wastewater or as a solid waste in drums.

Sludges from settling tanks are drummed and disposed of as solid waste. Spent recycle rinse water is drummed and disposed of as solid waste after the soluble content prohibits further use.

The percentage of solvent-base and water-base paints produced is the most important factor that affects the volume of process wastewater generated and discharged at paint plants. Due to their greater use of water-wash, plants producing 90% or more water-base paint discharge more wastewater than plants producing 90% or more solvent-base paint. Additional factors influencing the amount of wastewater produced include the pressure of the rinse water, spray head design, and the existence or absence of floor drains. Where no troughs or floor drains exist, equipment is often cleaned externally by hand with rags; when wastewater drains are present, there is a greater tendency to use hoses. Several plants have closed their floor drains to force the use of dry clean-up methods and discourage excessive water use.

Waste associated with equipment cleaning represents the largest source of waste in a paint facility. Methods that reduce the need or frequency of tank cleaning or allow for reuse of the cleaning solutions are the most effective. Waste minimization methods considered include:

- *Use of mechanical devices such as rubber wipers.* In order to reduce the amount of paint left clinging to the walls of a mix tank, rubber wipers are used to scrape the sides of the tank. This operation requires manual labor and hence the percentage of waste reduction is a function of the operator. Since the benefits will be offset by increased labor, mechanization/automation should be considered. Many new mixers are available that are designed with automatic wall scrapers. These mixers can be used with any cylindrical mix tank (flat or conical bottom).

- *Use of high pressure spray heads and limiting wash/rinse time.* After scraping the tank walls, high pressure spray hoses can be used in place of regular hoses to clean water-based paint tanks. Based on studies, high pressure wash systems can reduce water use by 80 to 90%.[93] In addition, high pressure sprays can remove partially dried-on paint so that the need for

---

[93] USEPA. 1979. Development document for proposed effluent limitation guidelines, new source performance standards, and pretreatment standards for the paint formulating point source category, EPA-440-1-79-049b, Washington, D.C..

caustic is reduced. Tanks used for making solvent-based paints normally employ a built-in high pressure cleaning system. At Lilly, in High Point, N.C., a high pressure cleaning system was installed in several mix tanks. By continuously pumping a fixed amount of solvent into a tank until it was clean, the overall volume of solvent required for cleaning was reduced.[94]

- *Use of Teflon™-lined tanks to reduce adhesion and improve drainage.* The reduced amount of "clingage" will make dry cleaning more attractive. This method is probably applicable only to small batch tanks amenable to manual cleaning.

- *Use of a plastic or foam "pig" to clean pipes.* It was reported that much of the industry is currently using plastic or foam "pigs" (slugs) to clean paint from pipes. The "pig" is forced through the pipe from the mixing tank to the filling machine hopper. The "pig" pushes ahead paint left clinging to the walls of the pipe. This, in turn, increases yield and reduces the subsequent degree of pipe cleaning required. Inert gas is used to propel the "pig" and minimize drying of paint inside the pipe. The equipment (launcher and catcher) must be carefully designed so as to prevent spills, sprays, and potential injuries, and the piping runs must be free of obstructions so that the "pig" does not become stuck or lost in the system.

- *Better operating practices.* At Desoto, in Greensboro, N.C., wash solvent from each solvent-based paint batch is separately collected and stored. When the same type of paint is going to be produced, waste solvent from the previous batch is used in place of virgin solvent. In 1981, Desoto produced 25,000 gallons of waste mineral spirits. In 1982, when the system was implemented, waste solvent production amounted to 400 gallons. This same technique is currently being applied to their latex paint production operation.[95]

Other waste minimization measures based on good operating practices would be to schedule paint production for long runs or to cycle from light to dark colors so that the need for equipment cleaning would be reduced. For facilities using small portable mix tanks for water-based paints, immediate

---

[94] Kohl, J., P. Moses, and B. Triplett. 1984. Managing and Recycling Solvents, North Carolina Practices, Facilities, and Regulations, North Carolina State University, Raleigh, N.C.
[95] Id.

cleaning after use would reduce the amount of paint drying in the tank and, hence, reduce the need for caustic. Many times, dirty equipment is sent to a central cleaning operation where it waits until a given shift (usually night) to be cleaned. While tanks wait to be cleaned, the residual paint dries up, often necessitating the use of caustic solution for cleaning. By designing and operating the cleaning operation to handle any peak load continuously, all need for caustic should be eliminated or drastically reduced.

For plants employing CIP (clean-in-place) and recycle systems for wash/rinse operations, the inventory replacement frequency and waste volume can be minimized by using these following waste reduction methods:

- *A countercurrent rinsing sequence.* For facilities that have additional storage space available, countercurrent rinsing can be employed. This technique uses recycled "dirty" solution to initially clean the tank. Following this step, recycled "clean" solution is used to rinse the "dirty" solution from the tank. Since the level of contamination builds up more slowly in the recycled "clean" solution than with a simple reuse system, solution life is greatly increased. Countercurrent rinsing is more common with CIP systems, but can be used with all systems.

- *Alternative cleaning agent.* Many facilities use caustic to clean their mixing equipment. When the buildup of solids and dissolved organics reaches a given concentration, the cleaning efficiency decreases and the solution must be replaced. As reported by one of the audited facilities, substituting a proprietary alkaline cleaning solution for their caustic solution cut the solution replacement frequency in half and thereby reduced the volume of cleaning solution requiring disposal.

- *Sludge dewatering by filtration or centrifugation.* The above three methods are useful in reducing the amount of waste entering the environment provided they allow the continued use of the cleaning solution. Dewatering only to reduce sludge disposal volumes should not be viewed as waste minimization.

- *Provision for adequate solid settling time in spent rinse solution.*

- *Use of de-emulsifiers in rinse water to promote emulsion breakdown and organic phase separation.*

Off-Site Specification Paint

Most off-specification paint is produced by small shops that deal in specialty paints. Since these paints cost more to produce and, therefore, sell at a premium price, most off-spec paint is reworked into a saleable product. Since elimination of off-spec paint production has built-in economic incentives, the following techniques are widely used:

- *Increased automation.*

- *Better operating practices.* Unless the sludge from wet cleanup can be recycled into a marketable product, the use of dry cleanup methods should be maximized wherever possible. By closing floor drains and discouraging employees from routinely and needlessly washing down areas, some facilities have been able to achieve a large decrease in wastewater volume. Other effective ways to reduce water use include employing volume-limiting hose nozzles, using recycled water for cleanups, and actively involved supervision.

Bags and Packages

Inorganic pigments, which may contain heavy metals and, therefore, may be classified as hazardous, are usually shipped in 50 pound bags. After emptying the bag, an ounce or two of pigment usually remains inside. Empty containers of liquid raw materials that constitute hazardous waste (e.g. solvents and resins) are typically cleaned or recycled to the original raw material manufacturer or to a local drum recycler. Empty liquid containers are excluded from the following discussion. The following waste reduction techniques for bags and packages were noted:

- *Use of water soluble bags for toxic pigments and compounds used in water-based paints.* When empty, the bags could be dissolved or mixed in with the paint. Such a method is commonly used for handling mercury compounds and other paint fungicides. This method could not be used, however, when producing high quality, smooth finish paint since the presence of this material could affect the paint's film-forming property or could increase the load on the filters, which would increase filter waste.

- *Use of rinseable/recyclable drums with plastic liners instead of paper bags.*

170 / Pollution Prevention Strategies and Technologies

- *Better operating practices.* The most effective way of reducing hazardous waste associated with bags and packages (or any other waste stream) was to segregate the hazardous materials from the nonhazardous materials. As an example, empty packages that contained hazardous materials should be placed into plastic bags (so as to reduce or eliminate dusting leading to nonhazardous material contamination) and should be stored in a special container to await collection.

Air Emissions

The two major types of air emissions that occur in the paint manufacturing process are volatile organic compounds and pigment dusts. Volatile organics may be emitted from the bulk storage of resins and solvents and from their use in open processing equipment such as mix tanks. Since most existing equipment is of open design, reducing or controlling organic emissions from process equipment could require substantial expenditures in retrofit costs. Additional work on control methods appears to be warranted in this area, and as a result, the following measures only address bulk storage and pigment handling.

- *Control bulk storage air emissions.* Many methods are available for reducing the amount of emissions resulting from fixed roof storage tanks. Some of these methods include use of conservation vents, conversion to floating roof, use of nitrogen blanketing to suppress emissions and reduce material oxidation, use of refrigerated condensers, use of lean-oil or carbon absorbers, or use of vapor compressors. When dealing with volatile materials, employment of one or more of these methods can result in cost savings to the facility by reducing raw material losses.

Some of the dusts generated during the handling, grinding, and mixing of pigments can be hazardous. Therefore, dust collection equipment (hoods, exhaust fans, and bag houses) are provided to minimize a worker's exposure to localized dusting and to filter ventilation air exhaust.

- *Use of pigments in paste form instead of dry powders.* Pigments in paste form are dry pigments that have been wetted or mixed with resins. Since these pigments are wet, less dust or no dust is generated when the package is opened. In addition, most pigments in paste form are supplied in drums (which can be recycled) and, therefore, would eliminate the waste due to empty bags. While this method would increase the amount of pigment handling occurring at the supplier's facility, it can be argued that the overall number of handling/transfer points for dry powder will be greatly reduced along with the probability of spills and dust generation.

Spills

Spills are due to accidental or inadvertent discharges usually occurring during transfer operations or equipment failures (leaks). Spilled paint and the resulting clean up wastes are usually discharged to the wastewater treatment system or are directly drummed for disposal. If the plant has floor drains, large quantities of water may be used to clean up water-based paint spills. Dry cleaning methods are employed for cleaning of solvent-containing spills or for water-based spills where floor drains are not available. Waste reduction methods similar to those for off-spec paint include:

- *Increased automation.*

- *Better operating practices.* Unless the sludge from wet cleanup can be recycled into a marketable product, the use of dry cleanup methods should be maximized whenever possible. By closing floor drains and discouraging employees from routinely and needlessly washing down areas, facilities may achieve a large decrease in wastewater volume. Other effective ways to reduce water use include employing volume-limiting hose nozzles, using recycled water for cleanups, and actively involved supervision.

Filter Cartridges

Spent filter cartridges are produced during the paint loading operation. These cartridges are designed to remove undispersed pigment from the paint during loading and are saturated with paint when removed. Hence, waste minimization and economy both call for as small a cartridge as possible so as to reduce the amount of paint lost and the capital spent for the filters. If frequent filter plugging is a problem, then it should be first addressed from the standpoint of improving pigment dispersion, and not from the standpoint of increasing filter area.

Viable alternatives to cartridge filters include bag filters and metal mesh filters. Metal mesh filters are available in very fine micron sizes and they can be cleaned and reused. Since it is very important to minimize all wastes, the issue of mesh filter cleaning waste reuse or recyling would need to be addressed before switching to these filters.

Obsolete Products/Customer Returns

Obsolete products and customer returns can be blended into new batches of paint. Obsolete products result from changes in customer demand, new superior products, and expired shelf life. Marketing policies, such as discounting older paints, can reduce the amount of obsolete products requiring disposal.

AUTOMOTIVE REPAIR INDUSTRY

This section discusses recommended waste minimization methods for automotive repair shops. The primary waste streams associated with automotive repair are listed in Table 6-2 along with recommended waste minimization options. Typical waste streams are shop clean up wastes, parts cleaning wastes, and automotive maintenance wastes.

The waste minimization options listed in Table 6-2 can be classified generally as source reduction (which can be achieved through material substitution, process or equipment modification, or better operating practices) or as recycling. Better operating practices are procedural or institutional policies that result in a reduction in waste. They include:

- Waste stream segregation.

- Personnel practices.

    -Management initiatives.
    -Employee training.
    -Employee incentives.

- Procedural measures.

    -Documentation.
    -Material handling and storage.
    -Material tracking and inventory control.
    -Scheduling.

- Loss prevention practices.

    -Spill prevention.
    -Preventive maintenance.

-Emergency preparedness.

- Accounting practices.

    -Apportion waste management costs to departments that generate the waste.

Better operating practices apply to all waste streams. In addition, specific better operating practices that apply to certain waste streams are identified in the appropriate sections that follow.

Table 6.2. Waste Minimization Options for the Automotive Repair Industry

Waste Stream	Waste Minimization Options
Shop Clean-up	
Out-dated supplies	-Computerize inventory control. -Use first-in, first-out policy. -Minimize storage quantities. -Perform routine storage area inspections.
Dirty rags and sawdust	-Use good housekeeping measures to reduce spills and leaks. -Lease rags from a laundry cleaningservice
Alkaline floor cleaner	-Use good housekeeping measures to reduce spills and leaks, such as an award program for worker with cleanest bay. -Use of drip trays under leaking cars and removed parts. -Use of proper storage of waste materials, such as use of pallet/containment systems and installation of self-closing, non-leak safety faucets on portable waste oil drums

Table 6.2—Continued

Waste Stream	Waste Minimization Options
Clarifer sludges	-Use good housekeeping measures to reduce spills and leaks. -Don't flush dust or floor sweepings to the clarifer unit.
Parts Cleaning	
Solvent baths/ air emissions	-Use less hazardous or safer solvents. -Determine how clean parts need to be. -Use solvents properly; don't use to clean floors. -Increase cleaning efficiency. -Moniter solvent composition. -Operate solvent sinks properly. -Use drip trays and allow more drainage time. Keep lids closed when not in use. -Contract with a service company to maintain solvent sinks. -Install on-site solvent recovery equipment
Aqueous baths	-Switch from caustic to detergent-based cleaning solutions. -Use dry pre-cleaning methods, such as wire brushing. -Maintain solution quality by monitoring composition. -Maintain equipment in proper working order. -Filter solids from the bath on a continuous basis. -Screen solids before they reach the waste sump. -Employ two-stage parts cleaning sequence. -Install or convert free running rinses to still rinse. -Use a hot tank or jet spray washer lease service. -Switch to bake-off ovens.

Table 6.2—*Continued*

Waste Stream	Waste Minimization Options
Automative Maintenance	
Spent fluid (oil coolant, and transmission fluids)	-Store all wastes properly and keep segregated to promote the potential for recycling.
Rebuildable parts	-Give or sell to a parts remanufacturer.
Batteries	-If unbroken, sell batteries to an off-site recycling.
CFC-12	-Purchase recycling system to recover refrigeration.

Shop Cleanup

The human aspects of industrial activity can be very important in waste reduction. Often termed "good operating practices" or "good housekeeping," these methods can be very effective in reducing the amount of shop cleanup wastes generated. Typical wastes include outdated supplies, dirty rags, sawdust, area wash downs, and clarifier sludges.

Good housekeeping methods include improved employee training, management initiatives to increase employee awareness of the need for and benefits of waste minimization, and requiring increased use of preventive maintenance in an effort to reduce the number of leaks and spills that occur. Additional ways to reduce or minimize waste include:

- Improve inventory control.

- Use first-in, first-out (FIFO) policy.

- Minimize storage quantities.

- Increase storage area inspections.

- Conduct employee training.

- Employ spill containment techniques.

In one survey of automotive repair businesses,[96] all shops assessed used computerized inventory control; however, none reported the use of rigid control to maximize the use of supplies. This suggests that workers were free to obtain supplies at will. Workers should be made to return empty containers of materials before they are issued new supplies. This type of policy has been reported to be effective in reducing the solvent use at several automotive refinishing businesses.[97]

Cleanup wastes can be minimized by improving spill containment techniques and by implementing policies to reduce spillage. When a spill of raw material or hazardous waste occurs, sawdust (or some other adsorbent) may be used to adsorb it. Depending on the nature of the spilled material, the sawdust may become a hazardous waste and be subject to all hazardous waste regulations. If floors are heavily soiled with oil and other hazardous materials when washed, then large quantities of washwater may acquire hazardous waste classification. Ways to reduce spillage include:

- *Award program for worker with cleanest bay.* Awards should be based on the care a mechanic takes in preventing spills, as well as the worker's efficiency of cleanup after a spill. At some facilities, work bays out of sight of the customer tend to be dirtier than bays in sight. Special attention should be given to inspection of areas where the general feeling might be "out of sight, out of mind".

- *Use of drip trays.* To assist workers in keeping their assigned bays clean, drip pans should be provided and used. Most shops do not use drip pans and the occurrence of fluids leaking from automobiles and parts placed on the floor is common. By using drip pans, shop floors will remain cleaner and hence less frequent cleaning of the floors should be required. Added benefits would be a reduction in use of rags and adsorbent to clean the floors, and a safer work environment.

---

[96] City of Santa Monica, California. 1989. Hazardous waste minimization audits of automotive repair and refinishing facilities. Prepared by Jacobs Engineering Group Inc., Pasadena, Calif. for the City of Santa Monica Department of General Services. September 1989.

[97] For more information, refer to discussion beginning on p. 250 concerning pollution prevention options for automotive refinishers.

- *Proper storage of waste materials.* Waste materials should always be kept segregated and stored in proper containers. Storage areas should be bermed or diked so that accidental spills can be contained. This is especially important for spent battery storage. Batteries may be stored in the parts supply room awaiting pick up. If these were to leak, the resulting acid spill could be dangerous. For shops with limited space, combination pallet/containment systems are available.

Another option for reducing leaks and spills is the use of self-closing non-leak safety faucets on the portable waste oil collection tanks. Leaking valves should be replaced as soon as possible whenever leaking is noticed. If immediate replacement of the valve is not feasible, then a small collection cup should be hung under the valve to catch drippings. Accidental opening of the valve, which might occur if the valve handle were kicked or hit, can be prevented by using padlockable valves and inserting lock pins.

Parts Cleaning

The recommended strategy for developing effective waste minimization options for parts cleaning operations relies on systematic exploration of the following sequence of steps:

1. Avoid the need to clean.

2. Select the least hazardous cleaner.

3. Maximize cleaning efficiency.

4. Segregate cleaning wastes.

5. Maximize recycling and reuse.

This strategy is consistent with the multi-media approach and general emphasis of reducing the waste at the source. The following sections discuss waste minimization options for users of solvents and aqueous cleaners.

Solvent wastes were among the first to be banned from land disposal by the EPA. The 1984 amendments to the Resource Conservation and Recovery Act specify five categories of solvent waste (F-001 to F-005) which were banned from land disposal effective November 1986.[98] Due to the diverse problems associated with solvent use, solvents should always be used only when no other cleaner is suitable for the job. The major ways to avoid or reduce the generation of solvent waste include eliminating the need to use solvent; finding adequate substitutes for solvents; minimizing losses associated with solvent use; and segregation, recycle, recovery, and reuse of waste solvents.

- *Product reformulation or substitution.* The auto repair industry has reformulated carburetor cleaner compound to exclude the use of 1,1,1 trichloroethane (TCA), which had been used typically in 5% concentrations with methylene chloride and cresylic acid. TCA is a known toxic substance and irritant which can be absorbed through the skin.

Another potential product substitution is the increased use of terpene cleaners in place of Stoddard solvent. The terpene cleaners are available commercially in neat form or as water solutions with surfactants, emulsifiers, rust inhibitors, and other additives. Terpenes have tested favorably as substitutes for halogenated solvents for removal of heavy greases, oily deposits, and carbonized oils. Reported disadvantages of terpenes include inability to separate long chain aliphatic oils for recycling of the cleaning solution both in neat form and in aqueous emulsions. Ultrafiltration to remove oil is not viable for recycling and is only useful for treating dilute emulsions prior to wastewater treatment. Recovery by distillation is impractical since terpenes boil at around 340°F, which means that many light oils would be carried over with the solvent. Energy cost for distillation recovery, even with vacuum assist, would be high.

- *Determine how clean parts need to be.* Before using a solvent or aqueous cleaner, one should determine whether cleaning is necessary and just how clean a part needs to be. Rigorous chemical cleaning should only be performed when parts require it (e.g., bearings, engine internals, etc.). Stationary structural members typically require cleaning only for inspection.

- *Use solvents properly.* Solvent should never be used for the general cleaning of shop floors, and should only be used in a well-maintained self-contained cleaning system. When not in use, all solvent cleaning tanks

---

[98] RCRA 3004(e)(1), 42 USC 6904(e)(1).

must be covered and/or drain plugs closed. Solvent losses due to inappropriate usage, equipment leaks or spills, and evaporation can range from 25 to 40% of total solvent usage. Cans of spray cleaner should only be used when parts cannot be removed from the car and the placement of a cleaning sink or a pan under the part to catch drip page is not feasible.

- *Increase cleaning efficiency.* The need to dispose of or replace dirty solvent can often be reduced by increasing the degree of cleaning efficiency. While cold cleaning operations can be successfully performed at up to 10% soil solids content, solvent baths are often replaced when the contamination level reaches two to three percent, due to slow cleaning action. A simple way to increase cleaning efficiency is to employ manual brushing. Manual brushing is extremely effective at removing caked-on solids and is a common pre-cleaning technique. Use of ultrasonic or mechanical agitation also increases the cleaning efficiency.

- *Monitor solvent composition.* Because decisions to replace dirty solvent are made arbitrarily, much solvent is disposed of prematurely. Solvent monitoring may be performed to ensure that solvent is replaced only when it is truly dirty. In the dry cleaning industry, the level of solvent contamination is monitored by measuring the transmittance of light through a sample of dirty solvent. Work performed by the military on monitoring the quality of Stoddard solvent used for cleaning showed that light transmittance, as measured by visible absorbance at 500 nanometers (nm), was a reliable indicator of contamination. Solvent replacement was required when light transmittance dropped below 25%.

- *Operate solvent sinks properly.* Improper use of solvent sinks can lead to excessive solvent losses and increased waste generation. To reduce solvent losses and waste generation, solvent sinks should be operated properly. Ways to reduce losses include using a solvent sink with recirculating base tank, as opposed to a rinse tank or open bucket; placing of sinks in a convenient location; removing parts slowly after immersion to reduce drippage; installing drip trays or racks to drain cleaned parts; allowing more drainage time over the sink after withdrawal; and turning off the solvent stream, and covering or plugging the sink when not in use.

- *Contract with a solvent service company.* For a monthly fee, solvent service companies will pick up dirty solvent, clean and maintain the solvent sink, and refill the sink with clean solvent. Depending on the arrangement, solvent sinks may be owned by the shop or leased from the solvent service company. The cost for contracting with a solvent service company is often less than the combined cost of solvent purchase, tank maintenance, and waste disposal. Safety-Kleen Corporation, Safe-Way Chemical Company, and others offer this service. Over 95% of automotive repair operations have some type of solvent sink. The use of solvent sinks for parts washing either on an owned or leased basis is being accepted as general good practice.

Safety-Kleen, a nationally-franchised organization, recovers for reuse approximately 10,000 gallons of solvent per month in the San Francisco Bay area alone. Dirty solvent is taken to the recycling facility, where it is distilled and returned to the users. Safe-Way Chemical Company sells waste solvent to solvent recovery operations, such as Solvent Services of San Jose, where a waste fuel is produced by distillation.

In addition to cleaning solvent, both companies offer similar services for carburetor cleaner. Safety-Kleen recovers carburetor cleaner for reuse by distillation. Product fees charged include pickup and disposal of spent solutions. Safe-Way Chemical sends waste carburetor cleaner for solvent recovery to Solvent Services, where a lacquer wash is produced. Lacquer wash is used in paint stripping, among other uses.

- *Install on-site solvent recovery equipment.* Purchase of an on-site solvent recovery system is often viewed as a viable waste minimization option for solvent wastes. Recent prices for 5 and 15 gallon batch stills designed to process Stoddard solvent were $3,600 and $12,500, respectively. These stills utilize a bag liner (for ease of cleaning) and microprocessor control. Based on the results of California Department of Health Services (DHS) assessments, the low volume of solvent normally used at most small to medium repair operations does not justify the added expense of on-site solvent recovery equipment and maintenance costs.[99] For large operations that do generate significant volumes of solvent, labor costs to operate the equipment and additional costs for disposal of waste residues are not competitive with current solvent sink lease and maintenance service operations. Given the poor economics of this option, not to mention the increased liabilities and regulatory requirements that may be associated with

---

[99] California DHS. 1988. Hazardous waste reduction assessment handbook, automotive repair shops. California Department of Health Services, Toxic Substances Control Division, Alternative Technology Section. October 1988.

on-site recycling, this option would be viable only for a few automotive repair facilities.

Aqueous cleaning comprises a wide range of methods that use water, detergents, acids, and alkaline compounds to displace soil rather than dissolving it in organic solvent. Aqueous cleaning has been found to be a viable substitute for many parts cleaning operations currently using solvents. Its principle disadvantage is that the parts are wet after cleaning and carbon steel parts rust easily in this environment. Techniques for reducing wastes from aqueous cleaning include:

- *Switch to bake-off ovens.* Small bake-off ovens are being adopted for use in this industry to replace caustic cleaners. Bake-off ovens are designed to pyrolize the dirt and grease, leaving a dry residue that can be brushed off. In most cases, abrasive blasting of the parts is required to remove all of the residue. The advantage of a bake-off oven is that it produces a small volume of dry solid wastes compared to a large volume of liquid waste. Disadvantages of bake-off ovens include potential for increased air emissions, need for abrasive blasting equipment, and potential distortion or alteration of the part shape.

- *Switch to detergent-based cleaners.* Many shops are switching from solvent or caustic-based cleaners to less hazardous detergent-based cleaners. Operators should check that the type of cleaner used consists of surfactants that are good detergents but are poor emulsifiers (stable oil emulsions limit reuse of the cleaner and hasten its disposal). Agitation of the bath during use keeps the solids in suspension. Following prolonged periods of inactivity, however, the oily solids separate via flotation or settle to form a bottom sludge. Solution strength is maintained and bath life prolonged by removing these solids frequently.

- *Use dry pre-cleaning (wire brushing).* To reduce the loading of dirt and grime on chemical cleaners and reduce the generation of chemical-laden sludge, the use of dry wipes and wire brushing to pre-clean the part prior to soaking should be considered. While these methods would not be appropriate for precision cleaning, they can be used to remove the bulk of the dirt and grime from external surfaces.

- *Maintain solution quality.* In addition to dirt loading, excessive consumption of alkaline cleaner can also be caused by using air for agitation and hard water for make-up. Air agitation introduces carbon dioxide which reacts with alkali, and use of hard water can result in the formation of particulate solid sludge. In some applications, the decrease in cleaner effectiveness due to carbon dioxide and hard water salts can equal the decrease due to soil loading. Mechanical agitation by means of jet sprays and use of demineralized water for make-up is preferred. Analytical checks of solution strengths, performed by the operator using simple titration techniques, should be made routinely. The correction of solution strength by making small and frequent additions is more effective than making a few large additions.

- *Maintain equipment in good working order.* Rack systems should be maintained in good condition, free from cracks, rust, and corrosion which can flake off and contaminate the bath. Metal tanks should be properly coated with protective finishes both inside and out. Spray nozzles should be inspected regularly to avoid clogging. A still rinse following the cleaning tank is a good way to avoid the loss of cleaner and reduce the discharge of contaminants to the sewer.

Another important item that should be maintained regularly is the float valve that supplies make-up water to tanks of heated cleaning solutions. While maintaining an adequate level is necessary, it is also imperative that the valve does not leak and result in dilution of the cleaner. In addition to maintenance, routine analytical checks of solution strength is a good way to detect slow leaks. Decreases in solution strength during a time when the tank has not been used is a sure indication of a leaking valve (provided that the tank is not leaking).

- *Screen solids before they reach the waste sump.* The majority of the heavy metal residue, oil and grease removed from the hot tank operations occurs after the actual hot tank use. The heavier concentrations of solid residues are found in the waste sump. Standard practice currently is to use a high velocity spray wand to dislodge these solids into the sump. Proper capture and disposal of these wastes is necessary. This can be done by use of a solids collection tray with overflow to the sump or periodic clean out of the sump by a waste hauler for disposal at a Class I landfill.

- *Two-stage parts cleaning.* Use of a two-stage parts cleaning arrangement can help to reduce the amount of spent cleaning solution requiring disposal. In a single-stage washer, cleaning solution must be replaced when it can no longer remove or dissolve all of the surface contamination on the part. In a two-stage washer, dirty solution is used to mechanically dislodge bulk contamination from the part followed by the use of clean solution. When the clean solution can no longer be used in the second stage, it is used to replace the dirty solution in the first stage.

- *Install or convert free running rinses to still rinse.* Installing a still rinsing tank immediately after an aqueous cleaning tank allows for cleaner recovery and lowered rinse water discharges. In such a system, the workpiece is immersed in the still rinse tank following the cleaning operation. Since the still rinse has no free running inflow or outflow of water, cleaner concentration builds up in it. As water evaporates from the heated cleaning system, water from the still rinse is used as make-up. Fresh water is then added to the still rinse. In the case of radiator cleaning, use of a still rinse following the boilout tank is an effective way of reducing heavy metal discharge (most notably lead) to the sewer.

- *Use a leasing service.* Similar to solvent lease arrangements, some companies offer a leasing service for hot tanks and jet spray washers. Hot tank arrangements include monthly leasing of a hot tank and monthly general maintenance service with removal of 10 gallons of solution and sludge and recharge of solution with caustic or alkaline detergent and make-up water. Jet spray washer arrangements include monthly leasing of a jet spray and monthly general maintenance service with removal of 10 gallons of solution and sludge and recharge of solution with caustic or alkaline detergent and make-up water.

Automotive Maintenance

To minimize problems associated with disposal of maintenance wastes, automotive repair shops should manage wastes properly. Proper management includes keeping all incompatible wastes segregated and contracting with appropriate recyclers and waste handlers. Viable options include:

- *Solvent segregation.* Proper manifesting and recovery of spent solvent solutions can only occur if small- and medium-size repair operations segregate solvent wastes in suitable storage containers. Current practice at many operations is to commingle the wastes with the waste oil. This practice should be discouraged.

- *Carburetor cleaner segregation.* Similarly, carburetor cleaner is often mixed with waste oils. Carburetor cleaner is a corrosive liquid and contains chlorinated compounds. This waste should be accumulated separately for proper waste management in a suitable container or system.

- *Spent antifreeze solution and waste motor oils recycling.* Both spent antifreeze solution and waste motor oils are generated in substantial quantity at both medium-size and large automotive repair operations. Proper compliance may require collection of these wastes by a registered hazardous waste hauler. Several companies offer off-site recycling services.

- *Spent lead-acid battery recycling.* On a weight basis, spent lead-acid batteries are one of the largest categories of hazardous wastes generated. Recyclers pay auto repair businesses between $1.00 and $1.50 per battery recycled. Spent batteries are either rebuilt for resale or sent to a processor for material salvage. One in six batteries received is rebuildable.

- *Recover and recycle CFC-12.* Rather than venting refrigerant to the atmosphere during air conditioning servicing, recovery and recycling systems are available. The shops assessed in one study reported that the cost of an on-site recovery and recycling system was $3,500.[100] Assuming that it is possible to recover 20 ounces per air conditioner and that the cost of refrigerant is $30 per gallon, a recycling system will pay for itself after servicing approximately 750 air conditioners.

- *Consumer education.* Another way to minimize the generation of automotive maintenance wastes is through promotion of good consumer practices by public agencies and the automotive industry. Consumers should be encouraged to follow specified maintenance schedules and not have service performed needlessly. Simple test methods should be developed to determine if automotive fluids are being replaced prematurely. This could avoid unnecessary fluid changes. Use of synthetic lube oils, which reportedly last 10,000 to 15,000 miles before requiring replacement, could also be a viable waste reduction measure.

---

[100] City of Santa Monica, California. 1989. Hazardous waste minimization audits of automotive repair and refinishing facilities. Prepared by Jacobs Engineering Group Inc., Pasadena, Calif. for the City of Santa Monica Department of General Services. September 1989.

FIBERGLASS AND PLASTICS INDUSTRY

This section discusses waste minimization methods found useful for fiberglass-reinforced plastic and composite plastic (FRP/C) fabrication operations. The primary waste streams associated with FRP/C fabrication are listed in Table 6-3, along with recommended control methods.

The waste streams are: equipment cleaning wastes; scrap solvated and partially cured resin; gelcoat, resin and solvent oversprays; resin- and solvent-contaminated floor-sweepings; empty bags and drums; rejected and/or excess raw material; cleanup rags; laboratory and research wastes; and monomer (resin) emissions due to the polymer-cross linking reaction. The waste minimization methods listed in Table 6-3 can be classified generally as source reduction, which can be achieved through material substitution, process or equipment modification or better operating practices; or as recycling.

Many of the source reduction options available to composite plastic product manufacturers only require better operating practices or minor in-plant process modification to effect significant waste reduction and savings by virtue of less wasted raw materials and off-spec products. Better operating practices are procedural or institutional policies that result in reducing waste. They include:

- Waste stream segregation.

- Personnel practices.
  -Management initiatives.
  -Employee training.
  -Employee incentives.

- Procedural measures.

  -Documentation.
  -Material handling and storage.
  -Material tracking and inventory control.
  -Scheduling.

Many of these measures are used in industry to promote operational efficiency. In addition, they can often be implemented at little or no cost to the facility. When one considers the waste reduction potential, ease of implementation, and little or no implementation cost, better operating practices usually provide a very promising early focus area for any waste minimization effort. They should be addressed before proceeding with more difficult, technology-based measures.

In addition to the specific recommendations discussed below, rapidly advancing technology makes it important that companies continually educate themselves about improvements that are waste-reducing and pollution-preventing. Information sources to help inform companies about such technology include trade associations and journals, chemical and equipment suppliers, equipment expositions, conferences, and industry newsletters. By keeping abreast of changes and implementing applicable technology improvements, companies can often take advantage of the dual benefits of reduced waste generation and a more cost-efficient operation.

The following sections discuss the waste minimization methods for specific waste streams listed in Table 6-3.

Table 6.3. Waste Minimization Options for the Fiberglass-Reinforced and Composite Plastics Fabrication Industry

Waste Stream	Waste Minimization Options
Equipment cleaning wastes	-Restrict solvent issue. -Maximize production runs. -Store and reuse cleaning wastes. -Use less toxic and volatile solvent substitutes. -On-site recovery. -Off-site recovery. -Reduce rinse solvent usage. -Waste segregation.
Scrap solvated and partially cured resins	-Modify resin pan geometry. -Reduce transfer pipe size. -Waste exchange.
Gelcoat resin and solvent oversprays	-Change spray design.

Table 6.3—Continued

Waste Stream	Waste Minimization Options
Rejected and/or excess raw material	-Improve control. -Purchase materials in smaller containers. -Return unused materials to suppliers.
Resin and solvent-contaminated floor sweepings	-Use recyclable floor sweeping compound. -Reduce solvent and resin spillage and oversprays by employing alternate material application and fabrication techniques.
Empty bags and drums	-Cardboard recovery. -Container recycling. -Returnable containers. -Use plastic liners in drums.
Air emissions	-Improve/modify material application. -Cover solvent containers. -Use emulsions or less volatile solvents.
Miscellaneous wastes streams	-Product/process substitution.
Clean-up rags	-Efficient utilization of clean programs. -Auto cleaning process equipment.
Laboratory and research wastes	-Reduce quantities of raw material and products for testing and analysis.

Equipment Cleaning Wastes

Solvents are used to remove uncured resins from spray equipment, rollers, brushes, tools, and finished surfaces. Typical solvents used include acetone, methanol, methyl ethyl ketone (MEK), toluene and xylene.

Acetone and other similar solvents are used for general cleaning, as standard practice for most open-mold fabricators of fiberglass products. To clean the spray equipment, acetone is usually circulated through the lines after the spray operation is shut down for the day. A simple but effective

188 / Pollution Prevention Strategies and Technologies

method practiced by some fabricators to minimize wastes is placing the containers of solvent near the resin spray area to prevent spills and drippage for tool cleaning. Generally, the solvent is reused until the high concentration of resin contamination prevents effective cleaning. However, if the containers are left uncovered, solvent will evaporate, increasing air emissions, as well as resin concentration.

Methylene chloride is an effective solvent for cured resins, and has been used by plastics fabricators. Although many other solvents have been tried, including multi-component mixtures, these have had mixed results. The best way to minimize the need for this chemical is to clean equipment before the resin dries.

Disposal of contaminated solvents represents a major hazardous waste management expense. In addition, fugitive air emissions during the curing and cleaning processes are also of concern. Some of the potential waste reduction methods are described in the following paragraphs.

- *Restrict Solvent Issue.* Many shops have limited the quantity of solvent issued each shift and indicate this has reduced waste, although the savings are difficult to quantify.

- *Maximize Production Runs.* Production runs should be scheduled together to reduce the need for equipment cleaning between batches. Consideration should also be given to the potential for scheduling families of products in sequence, so that cleanup between batches can be minimized.

- *Store and Reuse Cleaning Solvents.* Assessments performed at FRP/C fabricators indicate that some plants collect spent solvents for reuse in cleaning operations.[101] However, the solvents cannot be reused if contaminants build up to levels that do not permit effective cleaning.

- *Use Less Toxic and Less Volatile Solvents.* Relatively less toxic and less volatile solvents that are biodegradable, water-soluble, resin bed compatible and recoverable are commercially available as substitutes for the conventional solvents used in the FRP/C industry. These substitutes can be used in the curing process and/or for cleaning, depending on the type of

---

[101] California DHS. 1989. Waste audit study: fiberglass reinforced and composite plastics products. Report prepared by Woodward-Clyde Consultants, Oakland, California, for the California Department of Health Services, Alternative Technology Section, Toxic Substances Control Division.

solvent. For example, dibasic ester (DBE) based organic solvents do not evaporate as rapidly as acetone. When it spills during an operation, it will remain until it is cleaned up, collected and recovered by distillation, thus reducing VOC emissions and increasing the potential for reuse. One publication claimed a 60% savings by using DBE instead of acetone.[102] DBE also does not have the fire hazard of acetone. Emulsifiers, which can be used instead of solvents in some services, are discussed at the end of this subsection.

- *Reduce Solvent Rinse Usage.* Substantial quantities of solvent are used for cleanout of epoxy pre-treaters, mix tanks and treater pans. Using small labtype wash bottles for treater pan cleanouts can reduce solvent usage. Squeegee tools can also be used for the treater and vessel cleanouts, so that a smaller amount of solvent can be applied to the vessel to dissolve the remaining solvated resin. The squeegee may also be pressed against the vessel walls to force the remaining resin to the bottom of the pan or vessel for collection. One study estimated that using squeegees could reduce solvent requirement by 25%.[103] Additionally, a two-stage cleaning process may be used, where dirty equipment or a tool is first cleaned in dirty solvent (stored in a separate container), followed by a clean rinse with a smaller volume of fresh solvent, which is collected separately. When the dirty solvent approaches the maximum level of contamination, it should be removed for recycle and replaced with the accumulated "clean" rinse solvent.

- *Improving Recyclability of Solvent Waste.* Solvent waste can be more easily recycled if the procedure below is followed:[104]

    -Segregate solvent wastes by separating:

       --chlorinated from non chlorinated solvent wastes;

---

[102] Lucas, D.F. 1990. The effective solvent alternative. In: CMI, Marble Conference '90, Charlotte, N.C., February 22, 1990.

[103] California DHS. 1989. Waste audit study: fiberglass reinforced and composite plastics products. Report prepared by Woodward-Clyde Consultants, Oakland, California, for the California Department of Health Services, Alternative Technology Section, Toxic Substances Control Division.

[104] California DHS. 1986. Guide to solvent waste reduction alternatives. Final report. Prepared by ICF Consulting Associates, Inc., for the California Department of Health Services, Alternative Technology and Policy Development Section.

--aliphatic from aromatic solvent wastes;
--chlorofluorocarbons from methylene chloride; and
--wastewater from flammables.

-Keep water out of the waste solvents.

-Drums should be covered to prevent contamination with water.

-Minimize solids.

--Solids concentrations should be kept at a minimum to allow for efficient solvent reclamation.

-Control solvent concentration.

--Maintain solvent concentration above 40%.

-Label waste.

--Keep a chemical identification label on each waste container. Record the exact composition and method by which the solvent waste was generated.

■ *On-site Solvent Recovery.* Batch-type distillation units have proven to be successful in meeting the needs of firms producing small-to-moderate quantities of contaminated solvents such as acetone. Commercially available sizes range from 5- to 55-gallon units. A basic batch-type system consists of four major components: a contaminated solvent collection tank, a heated boiling chamber, a condenser, and a clean solvent collection container. These units are usually contained within a single compact cabinet, so that the space required is generally less than that required for storage of virgin solvents and contaminated waste. Initial investment ranges from approximately $3,000 for a basic 5 gallon unit to more than $30,000 for a relatively sophisticated 55-gallon unit with labor-saving automatic control systems and pumps.

Large-volume generators of contaminated solvents may find continuous-feed distillation equipment better suited to their requirements than batch recovery units. Capacities for these systems can range from 250 gallons per shift to as much as 200 gallons per hour. Continuous units are not likely to be economical for firms with recovery needs of less than 100 gallons per day, because installation costs for large units are likely to exceed $50,000. The continuous-feed system consists of the same components included in a batch-type distillation unit, with more elaborate controls and materials-handling equipment. An automatic pumping system continuously transfers contaminated solvents from storage to the boiling chamber. Condensers may be either water- or air-cooled. The clean solvent collection system must be equipped with a monitoring system to avoid overflow.

Often, solvated epoxy is the only resin suited to the batch distillation process. Non-epoxy resins (phenolic, polyamide, and polyester) have lower flash points and are more susceptible to runaway reactions. However, some fabricators have reportedly used batch distillation successfully with polyester resins. Reducing the solids content in solvated non-epoxy resin streams may be possible with filtration, yielding the same result without exceeding temperature constraints.

- *Off-site Solvent Recycling.* Commercial solvent recycling facilities offer a variety of services, ranging from operating a waste treatment/recycling unit on the generator's property to accepting and recycling solvent waste at a central facility. Some recyclers accept both halogenated and non-halogenated solvents, while others specialize in one or the other. Off-site commercial recycling services are often well-suited to small quantity generators (SQGs), who may not generate sufficient volume of waste solvent to justify on-site recycling. The off-site services are also attractive to generators who prefer to avoid the technical, safety, and managerial demands of on-site recycling. However, off-site recycling has the disadvantage of potentially high transportation costs and liability.

- *Replace Solvents With Emulsifiers.* Some fabricators now use emulsifiers instead of organic solvents. The emulsifier is an alkaline mixture of surfactants, wetting agents and various proprietary ingredients which can often be disposed of in the sewer. Advantages include: virtually no air emissions, biodegradability, and non-flammability. Some suppliers claim emulsifiers last twice as long as solvents. However, some emulsifier concentrates may contain solvents, dissolved metals, silicates and phosphates that make them unacceptable in some sewage systems. Different cleaning techniques must be employed when using emulsifiers, so adequate instruction of both management and workers is essential. Changing over from solvents to emulsifiers is easiest for hand and tool cleaning, which usually represents the largest consumption of acetone.[105] One study indicated that emulsions are inadequate for cleanup of gelcoat or cured resins.[106]

Scrap Solvated and Partially Cured Resins

- *Modify Resin Pan Geometry.* Pan widths should be no more than 10 inches wider than the fabric. If a narrow width fabric is run in an unnecessarily wide pan, additional solvated resin is wasted, since the wide pan holds a larger quantity at the end of the treater run. To alleviate this problem, simple adjusting devices may be installed that are made to fit into the treater pan to reduce its volume. This could consist of a plastic, wooden, or metal part molded to fit into the end of the treater pan, which would occupy the treater pan volume usually filled with resin but not required when coating the narrow fabric.

- *Reduce Transfer Pipe Size.* Typically, a long pipe connects the mix tank to the treater tank. Each time a run ends the solvated resin in the treater pan is discarded, along with the resin in the interconnecting pipe. Significant resin savings can be realized by installing smaller diameter pipe. However, this requires detailed hydraulic analysis and possibly pump modifications to ensure that an acceptable flow rate is maintained.

- *Waste Exchange.* Participation by a generator in a waste exchange program to reduce the volume of hazardous wastes satisfies the waste minimization certification requirement on the federal Uniform Hazardous Waste Manifest. In addition to helping meet regulatory requirements, participation in a waste exchange program provides the waste generator with an opportunity to explore alternative waste management options that may lead to a more cost-effective waste management program. Waste exchanges an effective vehicle for increasing recycling and resource reuse opportunities, and can be an important pan of a company's overall strategy to manage waste in an environmentally sound and cost-effective manner.

---

[105] Halle, Reidar and J.A. Brennan. 1990. Replace acetone successfully—a practical guide. Fabrication News. Fiberglass Fabrication Association. April 1990.

[106] U.S. EPA. 1990. Assessment of VOC emissions from fiberglass boat manufacturing. U.S. EPA, Air and Energy Engineering Research Laboratory, EPA/600/S2-90/019.

Two wastes of the FRP spray mold and composites industries appear to be particularly well-suited for waste exchange listings: partially-solidified resin and scrap fiber.

- *Improve Material Application Procedures.* Significant waste reduction can be achieved by optimizing material application processes. These processes include spray delivery systems and non-spray resin application methods. The latter include pre-spray fiber reinforcing, in-house resin impregnation, resin roller dispensers, vacuum bag molding processes and closed mold systems. Non-spray resin application methods reduce material waste and other expenses, in particular energy purchase cost. Lower operating pressures for spray delivery systems reduce the cost and maintenance of pressure lines, pumps, controls, and fittings. Routine cleanup of work areas is also reduced in terms of frequency and difficulty.

Gelcoat Resin and Solvent Overspray

Oversprays can be eliminated or reduced to a great extent through simple techniques, such as spray reorientation, and through advanced measures, such as equipment modification. The advantages and disadvantages of both spray and non-spray delivery systems are discussed below.

- *Spray Orientation.* Waste often accumulates around the bottom of sprayed objects because the tip of the spray gun is directed down toward the bottom of the object, rather than horizontally. Likewise, it may be difficult for the operator to shoot the top of high objects. If spraying is directed vertically instead of horizontally to the top of the object, the spray dissipates as a fine mist up to several feet away from the object. Hence, depending upon the shape of the objects, appropriate spray orientations may be developed.

- *Spray Delivery Systems.* Most open-mold fabricators of fiberglass products use spray applicators for transferring and applying coatings, resins, and fibers to the mold. Delivery systems used by FRP fabricators include high-pressure air, medium-pressure airless, and low-pressure air-assisted airless spray guns. In the order listed, the atomization and spray patterns become more efficient, reducing excessive fogging, overspray, and bounce back. Other key issues associated with these delivery systems are as follows:

-The high-pressure air system is practically obsolete due to the large amounts of expensive high pressure compressed air required. Low styrene emissions limits generally cannot be met using a high-pressure air system.

-In the airless method, a pressurized resin stream is electrostatically atomized through a nozzle. The nozzle orifices and spray angle can be varied by using different tips. Orifice size affects delivery efficiency, with larger orifices resulting in greater raw material loss. Airless spray guns are considered to be very efficient in delivering resins to the work surface, although excessive fogging, overspray and bounceback may occur.

-The air-assisted technology modifies the airless gun by introducing pressurized air on the outer edge of the resin stream as it exits the pressure nozzle. The air stream forms an envelope that forces the resin to follow a controllable, less dispersed spray pattern. Lower resin delivery pressure can be used since the air assist helps distribute the resin. Low delivery pressure also reduces fogging, overspray, and bounce back, which in turn reduces raw material waste. Since more resin ends up on the product, the amount of spraying is reduced, leading to a reduction in styrene air emissions. Some vendors claim 5 to 20% savings in the resin spray waste for an air-assisted airless gun, compared to a standard airless gun.

- *Non-Spray Resin Application Methods.* While use of spray delivery of resins has become standard practice for most open-mold fabricators of fiberglass products, alternative applications processes do exist. Conventional gun-type resin application systems are efficient in delivering large quantities of resins to the work surface. Spray delivery systems are also advantageous when the product mold has many recesses or is convoluted. Non-spray application techniques would be messy or even impossible in some cases. However, other delivery techniques merit consideration in other circumstances. The various non-spray resin application methods are as follows:

-Use of fiber reinforcements that are presaturated with resins ("prepregs") offer a number of advantages over conventional spray techniques. In particular, resin-to-fiber ratios can be strictly controlled, atomization of pollutants is practically eliminated, and cleanup and disposal problems are greatly reduced. The disadvantages of this process are high a raw material cost, energy requirements for curing, and the refrigerated storage needs of prepregs. Therefore it is best suited for applications where extremely high strength-to-weight-ratios are required and cost factors are secondary.

-Impregnators appear to have considerable potential for the reduction of pollution associated with open molding operations. They provide the fabricator with some of the advantages offered by prepregs while using lower-cost polyester resins and fiberglass materials. Impregnators can be placed within the lamination area of a plant and can be mounted in such a manner as to feed resin-saturated reinforcing materials directly to the molding operations. Conventional resin pumps and catalyst-metering devices supply resins to a roller-reservoir system. Woven fiberglass is impregnated as it passes through this reservoir system.

-Resin roller dispensers can reduce material losses due to excessive fogging, overspray, turbulence, and bounce back. Low delivery pressures help maintain a cleaner work area External emissions and the need for high levels of make-up air are also reduced with this type of unit operation. Precisely-measured quantities of resin and catalyst are pumped to a mixing head, then to the roller at a relatively low pressure (less than 100 psig). Very often, existing spray gun equipment can be adapted to resin rollers.[107]

-Vacuum bag molding is another technique that offers several benefits. With the exception of the gel coat, resin delivery can be accomplished without atomization. Since final distribution of the resin to all areas of the layup is largely controlled by the vacuum, gel coating is the only step in vacuum bag molding that requires atomization of resin. Pumping or pouring premixed catalyst and resin into a closed mold eliminates fogging, bounce back, and overspray. Vapor emissions and odor are further reduced by confining the resins in the covered mold until curing is complete. Excess resin can be trapped by bleeder material placed under the vacuum bag. Dust-generating secondary grinding operations

---

[107] Davis. D. 1987. Pollution reduction strategies in the fiberglass boatbuilding and open mold plastics industries. Prepared by Department of Manufacturing, East Carolina University, Greenville, N.C.

are minimized because closed molding eliminates most flash removal and edge smoothing requirements.[108]

-Closed mold systems practically eliminate requirements for atomization of resins and may offer a number of production advantages over conventional approaches to molding. In closed mold processes, catalyzed resins are pumped instead of sprayed, which eliminates fogging, bounce back, and overspray. Vapor emissions and odor are further reduced by confining the resins in the mold until curing is complete. There is little, if any, waste of resin. Even dust-producing secondary grinding operations are reduced, because the closed molding system eliminates most trash removal and edge smoothing requirements. The closed molding technologies most frequently applied to production of fiberglass components are compression molding and resin transfer molding.

Rejected and/or Excess Raw Materials

Rejected and excess raw material wastes are generated through improper operating procedures and inventory control. Improper inventory control could result in two waste sources. One is material that has been in stock so long that it has exceeded its shelf life and must be disposed of. The other is material that is in stock but is no longer needed in carrying out the function of the plant. Some of the specific options to minimize wastes generated by way of rejected and excess raw materials are detailed here.

- *Tighter Inventory Control.* The following actions should reduce or prevent the generation of surplus inventory:

  -Purchase materials used in large quantities in returnable or reusable containers.

  -Purchase only the quantity of special-purpose materials needed for a specific production run, so that no material is left over.

---

[108] U.S. EPA. 1990. Assessment of VOC emissions from fiberglass boat manufacturing. U.S. EPA, Air and Energy Engineering Research Laboratory, EPA/600/S2-90/019.

-Use first-in/first-out (FIFO) inventory control.

-Check inventory before approval of new orders.

-Inquire whether suppliers can take back unused or expired materials. It is best done while placing large orders or changing suppliers.

- *Computerized Inventory Control.* Computerized raw material purchases and waste generation data can improve inventory control and identify areas for waste minimization. A basic system can be set up using widely available spreadsheet or database programs. Alternately, more task-specific and user-friendly programs are available from various software companies.

Empty Bags and Drums

Raw material containers, such as 30- and 55-gallon drums, can be cleaned for reuse or nonhazardous waste disposal. Many plants use the uncleaned empty drums to store and dispose of other hazardous wastes, such as contaminated solvents, clean rags and empty packages. Options for minimizing other container waste include container recycling, cardboard recovery, returning containers for reuse, and solid waste segregation.

- *Container Recycling.* Acceptable practices for on-site management of drums include cleaning of reusable containers and selling them to scrap dealers or drum recycling firms. Some drums can be returned to the chemical supplier for refilling. Used containers may also be suitable for the storage of other wastes. The most important aspect in reuse or recycling of drums is that they be completely empty. One way to reduce the volume of waste is to use drums lined with a disposable liner that can be removed when the drum is empty. Disposal of the plastic liner is much easier than disposing of the drum, and eliminates the need for drum cleaning. The number of containers and the associated waste residuals can be greatly reduced by increasing container size or converting to bulk handling altogether.

- *Cardboard Recovery.* Cardboard cartons used to deliver glass roving can be saved and sold to a paper recycling firms instead of being thrown into the dumpster. Other paper waste suitable for recycling includes empty Cab-O-Sil and aluminum trihydrate bags and balsa wood cut-outs discarded from reinforcing operations.

- *Solid Waste Segregation.* An effective way of reducing hazardous waste associated with packaging is to segregate the hazardous materials from the nonhazardous materials. Nonhazardous packaging material may be sold to a recycler. Empty packages that contained hazardous material should be placed in plastic bags (to reduce personnel exposure and eliminate dusting) and stored in a special container to await collection and disposal as a hazardous waste.

Air Emissions

Organic vapor emissions from polyester resin/fiberglass fabrication processes occur when the monomer contained in the liquid resin evaporates during resin application and curing. In addition, cleaning solvent emissions can account for over 36% of the total plant VOC emissions. There also may be some release of particulate emissions from automatic fiber-chopping equipment. Potentially effective air emissions reduction methods include improved material application procedures and changing resin formulation.

- *Improved Material Application Procedures.* Emissions vary according to the way in which the resin is mixed, applied, handled and cured. These factors vary among the different fabrication processes. For example, the spray layup process has the highest potential for VOC emissions because atomizing resin into a spray creates an extremely large surface area, from which volatile monomer can evaporate. By contrast, the emission potential in synthetic marble casting and closed-molding operations is considerably lower, because of the lower monomer content in the casting resins (30 to 38%, versus about 43%) and because of the enclosed nature of these molding operations. It has been found that styrene evaporation increases with increasing gel time, wind speed and ambient temperature, and that increasing the hand rolling time on a hand layup or sprayup results in significantly higher styrene emissions. Thus, production changes that lessen the exposure of mesh resin surfaces to the air should be effective in reducing these evaporation losses.

- *Changing Resin Formulation.* In addition to production changes, resin formulation can be modified to reduce the VOC emissions. In general, a resin with lower monomer content should produce lower emissions. Evaluation tests with low-styrene-emissions laminating resins having a 36% styrene content found a 60 to 70% decrease in emission levels, compared to conventional resin (42% styrene), with no sacrifice in the physical properties of the laminate. Vapor suppressing agents (e.g. Paraffin waxes) also are sometimes added to resins to reduce VOC emissions. Limited laboratory and field data indicate that vapor suppressing agents reduce styrene losses by 30 to 70%.[109]

Other techniques for reducing air emissions have been described above. These include switching to less volatile solvents or emulsifiers and covering solvent containers.

Miscellaneous Waste Streams

Waste streams discussed in this section include floor cleanup waste, equipment cleanup rags and laboratory wastes. Control measures include the use of auto-cleaning equipment, proper purchase of chemicals and reagents, and use of microscale glassware.

■ *Floor Cleanup Waste.* Overspray is material that lands on the floor instead of in the mold. Techniques to reduce the quantity of this waste have been described previously (see Gelcoat Resin and Solvent Overspray in this section). Fabricators employ some type of floor covering to facilitate periodic cleanup of the work area, and this represents an additional source of waste. Most fabricators use heavy paper which has been treated with flameretardant, although some use sand. Since the dried residue is nonhazardous, the coverings may be discarded as a nonhazardous waste. A few fabricators use sawdust to catch overspray, but this practice is very risky. Organic peroxide catalysts react vigorously with sawdust and are likely to cause a fire.

■ *Equipment Cleanup Rags.* Mechanized automatic resin-mixing and dispensing units equipped with air valves to blow out excess materials are commercially available. Contaminated exhaust air can be captured and directed to existing air scrubbers for treatment. Advantages of such units include reduced labor costs and elimination of cleaning rags.

---

[109] U.S. EPA. 1988. Polyester resin plastics product fabrication: compilation of air pollutant emission factors (AP-42). U.S. EPA, Office of Air Quality Planning and Standards, p. 4.12-1, September 1988.

- *Laboratory Wastes.* Purchasing quantities of seldom-used specialty chemicals in the smallest available amount helps to reduce waste by ensuring that the material will more likely be consumed before its shelf life expires. The purchasing agent should consider the cost of disposal of over-age material before deciding to purchase in large quantities. Many tests can be redesigned to use micro-scale glassware to reduce waste generation. Micro-scale testing volumes range from 1 to 10 ml, compared to conventional testing, which may require 50 to 100 ml.[110]

MARINE MAINTENANCE AND REPAIR YARDS

This section discusses potential waste minimization methods for marine maintenance and repair operations. The primary waste streams associated with maintenance and repair operations are listed in Table 6-4 along with potential waste minimization methods. The major waste streams are chemical paint stripping wastes; abrasive blast and surface preparation wastes; painting and solvent wastes; equipment cleaning wastes; engine overhauling and repair wastes; machine shop wastes; specialty shop wastes; and vessel cleaning wastes. The waste from a particular operation often enters wastewaters and air, as well as having a solid component. For example, abrasive blast and surface preparation wastes may become air particulate matter, trace metal pollutants of potable and navigable water, and contaminants in soil and groundwater.

The waste minimization methods in Table 6-4 can be classified generally as source reduction, which can be achieved through material substitution, process or equipment modification, or better operating practices; or as recycling.

Source reduction is achieved through better operating practices by employing procedural or administrative policies that result in a reduction of waste. They include:

- Waste stream segregation.

- Personnel practices.

    -Management initiatives.
    -Employee training.
    -Employee incentives.

---

[110] See also discussion on p. 268.

- Procedural measures.

    -Documentation.
    -Material handling and storage.
    -Material tracking and inventory control.
    -Scheduling.

In addition to the specific recommendations provided below, rapidly advancing technology makes it important that companies continually educate themselves about improvements that are waste reducing and pollution preventing. Information sources to help inform companies about such technology include trade associations and journals, chemical and equipment suppliers, equipment expositions, conferences, and industry newsletters. By keeping abreast of changes, and implementing applicable technology improvements, companies can often take advantage of the dual benefits of reduced waste generation and a more cost-efficient operation.

Table 6.4. Waste Minimization Options for Marine Maintenance and Repair Yards

Waste Stream	Waste Minimization Options
Chemical paint stripping wastes	-Waste segregation. -Nontoxic stripping agents. -Maximize stripper usage by reuse. -Recycling stripper using appropriate recycling techniques. -Better operating practices.
Abrasive blast waste (wet solids and wastewater)	-Use alternate blasting media and techniques. -Segregation and recycle of blast media. -Use blast dust collection systems.
Paint and solvent wastes	-Tighter inventory control and good housekeeping. -Use water-based and less toxic coatings. -Solvent reuse and recycling. -On-site waste exchange. -Off-site waste exchange. -Waste segregation. -Minimization of fugitive overspray.

Table 6.4—Continued	
Waste Stream	Waste Minimization Options
Equipment cleaning wastes	-Segregation of cleaning agents. -Recycling of cleaning solvents. -Replace solvent cleaners with detergents wherever possible.
Engine repair wastes	-Use aqueous degreasers. -Use dedicated solvent sinks for parts washing. -Segregation of spent engine and lube oils. -Recycling of oils and solvents. -Waste exchange.
Machine shop wastes	-Use of water soluble cutting fluids. -Recycling of cutting and lubrication fluids -Segregation and waste exchange of metal and plastic turnings and scrap.
Specialty shop wastes	-Good operating practices. -Process and equipment modifications. -Use of detergent cleaners instead of solvent cleaners.
Vessel cleaning wastes	-Use of detergent cleaning agents.
Spills and floor washdowns	-Good housekeeping. -Proper storage. -Spill control measures.

Chemical Stripping Wastes

Methylene chloride is the most commonly used paint stripping agent, although it is increasingly being replaced by solvents, such as dibasic esters, which are less volatile and hazardous. Chemical stripping wastes consist primarily of the stripping agent and paint sludges. The following waste reduction methods are suggested to minimize the generation of chemical stripping wastes.

- *Waste segregation.* Segregating the stripping agents from other waste streams will help facilitate cost-efficient reuse and recycling of contaminated strippers. Extreme care must be taken to avoid cross-contaminating the stripping agents.

- *Use of less toxic stripping agents.* Inorganic strippers, usually consisting of aqueous solutions of caustic soda, can substitute for methylene chloride-based strippers in many applications. Although the waste stripper is still hazardous, it is relatively less toxic and easy to treat on-site, generating a nonhazardous waste that can possibly be disposed of to the public sewer. Several new less hazardous and toxic degreasing and stripping agents are currently available in the marketplace. Substitutes include dibasic esters (DBE), semi-aqueous terpene-based products, detergent-based products and C9- to C12-based hydrocarbon strippers. Manufacturers claim that these products are non-chlorinated, biodegradable, exhibit low volatility and are not listed as hazardous substances.

- *Store and reuse stripping agents.* Solvent strippers particularly stripping baths, can generally be reused several times before their effectiveness is compromised. Appropriate collection and storage systems need to be installed if the stripper is to be reused.

- *Recycle spent strippers.* Both spent organic and caustic stripper solutions can be treated to remove contamination. One method of treatment uses centrifuge or filtration systems to separate the paint sludge from the stripper; then makeup chemicals are added to the stripper, which can be reused. Stripping baths equipped with such treatment systems can often be used almost indefinitely. While the paint sludge that is separated may have to be treated as a hazardous waste, the quantity is much smaller than the quantity of combined sludge and stripper usually discarded. In addition, recycling and reuse considerably reduces the need for fresh or make-up stripping solution and thus results in substantial savings.

- *Better operating practices.* Chemical stripping of small parts employs dip tanks and generates wastes consisting of spent stripper that clings to the part after rinsing (drag-out). Some reduction in drag-out and contaminated rinse water can be achieved by allowing the dipped parts to drain longer above the dip tank, or by improving the way in which dipped parts are stacked before draining to prevent "pooling" of stripper on the parts.

Abrasive Blast Wastes

Abrasive blasting is often used in preference to chemical stripping for removing paints. While this procedure avoids disposal of chemical strippers, it does not eliminate wastes altogether. The most commonly used blasting media is sand or grit with a large volume of water.

The presence of paint chips containing hazardous metallic and organometallic biocides makes abrasive blasting wastes potentially hazardous. Blast wastewater generally constitutes the largest single waste stream from many repair yards. For instance, wet abrasive blasting of an average-sized naval vessel (DDG class) can generate up to 180 tons of wet abrasive and 500,000 gallons of contaminated water.[111]

- *Use alternate blasting media and techniques.* Research and testing is underway on a number of innovative alternatives to both grit blasting and chemical stripping. The alternate techniques include: plastic media blasting, water jet stripping, thermal stripping, dry ice pellets, laser paint stripping, and cryogenic stripping. Descriptions of each of these alternatives are provided below.

    -Plastic media blasting. When chemical stripper is applied to a large area and then washed off, large volumes of hazardous wastewater can be created. The military has experimented extensively with plastic media blasting (PMB) as a substitute for chemical stripping, with mixed results. Three disadvantages are that PMB will not work well on epoxy or urethane paints, and the blasting equipment is more expensive than conventional grit blasting equipment and requires more highly trained operators. On the positive side, the same types and quantities of solid wastes are generated as with grit blasting, but the plastic media tend to be more easily recyclable through the use of pneumatic media classifiers that are part of the stripping equipment. Thus, the main waste to be disposed of is the paint waste itself. Abrasion eventually turns the plastic media into fine dust, that must be disposed of. Based on research conducted by the military, chemical stripping a fighter aircraft will

---

[111] Adema, C.M., and G.D. Smith. 1987. Development of cavitating water jet paint removal system. In: Waste minimization and environmental programs within DOD. Proceedings of the 15th Environmental Symposium. April 1987. pp. 270-275.

generate about 250,000 gallons of stripper waste; by contrast, PMB will generate two 55-gallon drums of paint chips, and 200 pounds of nonrecoverable dry spent plastic medium.[112] In addition, PMB can be used on fiberglass boats, which cannot be stripped chemically.[113]

-Water jet stripping. A cavitating water jet stripping system has been developed to remove most paints, separate the paint chips from the water, and treat the water to eliminate dissolved toxic materials. Relatively little hazardous waste is generated by this process. However, it is not as efficient as conventional grit blasting and the equipment has higher capital and operating costs.

-Thermal stripping. Thermal stripping utilizes a flame or stream of superheated air to heat and soften the paint layer, thus allowing it to be peeled relatively easily. This method is applicable only to some situations; for instance it is not applicable to surfaces that might be heat-sensitive. In addition, this process is more labor-intensive than other stripping methods. The advantage of this method is that it generates only one waste stream, namely, a waste paint.

-Dry ice pellets. Carbon dioxide (CO_2) dry ice pellets can be used as a blast medium that generates no media waste. After use, the dry ice evaporates, leaving only paint chips that can be swept up and placed in containers for disposal. The cost of the dry ice, storage, and handling equipment could be substantial, limiting applicability of this method.[114]

-Laser paint stripping. Laser paint stripping has been developed that generates "zero" residue. A pulsed CO_2 laser controlled by an industrial robot, is used as the stripping agent. This method is complex, capital intensive and requires highly-skilled operators.[115]

---

[112] California DHS. 1989. Waste audit study: marine maintenance and repair. Report prepared by SCS Engineers, Inc., Long Beach, CA, for the California Department of Health Services Alternative Technology Section, Toxic Substances Control Division.

[113] Ballard, R. 1991. Safe and gentle stripping. Products Finishing. January 1991.

[114] Yaraschak, P.J. 1989. Hazardous waste minimization—making it happen. AIChE National Conference, Washington, D.C., December 4-5, 1989.

[115] Id.

-*Cryogenic stripping.* In cryogenic stripping, parts are immersed in liquid nitrogen, followed by gentle abrasion or plastic shot blasting to remove the brittle paint coating. This process requires special equipment for handling the liquid nitrogen and is applicable only for small objects.

- *Segregate and recycle blast media.* Most abrasive blast media are recyclable or reusable. In many cases, the contaminated grit can be reused several times to blast additional vessels before becoming too contaminated or worn for further use. Because of the difference in density between the grit materials and the waste paint chips, it is possible to separate the grit from the paint waste. Cyclone separators, gravity shakers, air separators, water separators, and other systems can separate the paint residues from the grit, which can then be reused.

Recycling is routinely performed when using steel shot lead shot, or plastic media because of the cost of the blast media. In the case of "sand" blasting, the sand does not have enough value to justify recycling. However, if the sand blast waste is determined to be hazardous, requiring off-site disposal, then this medium may become costly enough to justify recycling. Natural organic abrasives, like walnut shells or rice hulls, do not recycle well and are susceptible to biological growth and deterioration.

- *Use dust collection systems.* Dedicated bag-house filters may be installed on enclosed blasting stations to collect the blast dust emissions. One approach to prevent fugitive dust emissions from open space blasting operations would be to enclose the area with plastic sheeting or screening, thus confining the waste to the immediate vicinity of the blasting. After blasting has been completed, the waste should be collected, transferred to dumpsters or other containers, and transported off-site.

Paint and Solvent Wastes

Methods for minimizing paint and solvent wastes include tighter inventory control and good housekeeping, input material substitution, solvent recycling and minimizing fugitive oversprays.

- *Tighter inventory control and good housekeeping.* Rigid inventory control provides a very effective means of source reduction at virtually no cost to the operator and can be implemented in several ways. In smaller yards, an owner may monitor employee operations and make verbal or written comments on product usage and suggested limits. In larger yards where monitoring of employees can be more difficult, the owner or manager can limit access to storage areas containing raw materials, forcing employees to stretch the use of raw materials and providing the owner/manager a means of monitoring raw material use. In an effort to minimize paint waste, many small yards either purchase paint specifically for each job or require the vessel owner to supply the paint.

Good housekeeping can provide very effective source reduction. Examples of good housekeeping for paints and solvents include storage area leak control and containment, which can be easily implemented at no cost, and improvements in drum location, product transfer, leak collection, and drum transport, which can limit product loss.

- *Raw material substitution.* The bottoms of all vessels are coated with special antifouling paints, which are highly toxic and, thus, hazardous materials. The purpose of such paints is to prevent, or at least retard, growth of marine organisms (e.g. barnacles). Hence, there are no non-toxic alternative materials for this service. Cuprous oxide and copper flake rank among the least toxic, but effective, antifoulants, while arsenic and mercurials rank among the most toxic. Tributyltin compounds are also extensively used and are suspected to be highly toxic.

Paints for parts of the vessel that are not immersed in water can be non-toxic. Part of the paint waste problem can be alleviated by using water-based instead of solvent-based paints. This has become standard practice in many industries and helps to reduce not only hazardous paint wastes, but also solvent cleanup wastes. Waterbased coatings may not provide the same protection that solvent-based coatings do. Solvent-based coatings are generally more durable, tend to be less corrosive to the metals being coated, and dry quickly. The longer drying time of water-based coatings is exacerbated by the high relative humidity of marine air. Water-based coatings are suitable for areas where decoration is more critical than protection, such as vessel interiors or decorative topside work.

Powder coatings, based on finely pulverized plastics, have been substituted for paints in some industrial applications. This technique uses no solvent and eliminates VOC emissions, but requires that the coated surface be oven-cured at about 400^0F. Hence, it is not suitable for large or heat sensitive components and is not used to any significant degree for marine maintenance or repair. In addition, the application equipment is more expensive than conventional paint applicators, and highly-trained operators are necessary.

- *Solvent recycling.* Aside from raw material substitution, the best way to minimize solvent waste generation is to directly reuse the solvents as much as possible, and then recover and recycle them. Processes for recycling thinners and solvents are well established and widely used in many industrial sectors, especially the automotive paint industry. A waste minimization study of the automotive paint industry indicated that all companies contacted used some form of on-site or off-site thinner reclamation. Those generators who did not find it economical to recycle contaminated thinners on-site sent their solvents to commercial recyclers for recovery.[116] Reclaimed thinners were often sold back to the generators after recovery. Thinner recyclers can reprocess a high percentage of the incoming spent thinners into reusable products.

  There are several alternatives for resource recovery and recycling on-site. Gravity separation, for example, is inexpensive and easy to implement. This method of reclamation separates a thinner or solvent from the contaminant solids under quiescent conditions. The clear supernatant thinner can be decanted using a drum pump and a float valve and can be used as a cleaning solvent, or where thinner purity is not critical, as in parts washer systems. Gravity separation can provide valuable savings to any marine or boat yard by reducing both waste quantities and new solvent purchases.

  For larger shipyards, on-site distillation may be cost effective. Distillation of solvent and thinner wastes can significantly reduce the quantities of waste solvents disposed of, and the purchase of new materials.[117]

- *On-site waste exchange.* Larger shipyards should also consider on-site waste exchange. Solvents contaminated in one process or shop may be usable as cleaning solvents in a less demanding operation elsewhere in the shipyard. Yard and shop managers need to discuss their individual solvent needs and waste characteristics to determine the potential for this type of on-site reuse.

---

[116] California DHS. 1987. Hazardous waste minimization audit studies on the paint manufacturing industry. Prepared by Jacobs Engineering Group, Inc., for California Department of Health Services Alternative Technology Section, Toxic Substances Control Division. April 1987.

[117] See also discussion of leading solvent recovery systems used for the automotive repair industry beginning on p. 179.

- *Off-site services.* In general, three types of off-site services are offered. The simplest involves collecting all recyclable wastes and hauling them to a commercial recycler, who either recovers them or processes them into fuel, depending on the quality of the waste. This is similar to the practice currently being followed at most yards for bilge wastes.

In a second type of service, the marine maintenance and repair yard purchases thinners from suppliers who also collect and recycle the wastes, a common practice among automobile painting companies. Some suppliers may include the cost of waste collection and recycling in the price of their thinner. This increases the thinner cost but eliminates separate hauling and disposal or recycle costs, and also reduces the administrative burden on the owner or manager of the yard.

Other companies provide a third type of service, in which a parts washer system is leased to the client. The parts washer system can be as simple as a sink atop a drum of solvent. Solvent is pumped out of the drum into the sink for washing parts and equipment; used solvent drains back into the drum, and solids settle to the bottom of the drum. Either on a fixed schedule or whenever the solvent becomes too contaminated for further use, the service company removes the drum and leaves a drum of fresh solvent in its place. This service is widespread among automotive repair shops and other shops where large quantities of solvent are used to clean small parts.

- *Waste segregation.* Regardless of whether on-site recycling, on-site reuse, or off-site recycling is adopted, an essential waste management practice is waste segregation—placing different wastes into different containers for recovery or disposal. This practice is critical to the success of any program designed to reduce or recycle waste solvents, because solvents are much easier to directly reuse or recycle if they are segregated, minimizing solvent contamination. Mixing different solvents or putting waste waters, oils, excess paints, or paint strippers into common liquid waste drums can make solvent reuse or recycling, difficult and impractical. For instance, when an incompatible solvent or water is added to a chlorinated solvent, hydrochloric acid can form slowly, due to hydrolysis. This renders the solvent unsuitable for direct reuse until the acid is neutralized or otherwise removed.

One related practice that can facilitate segregation and reduce the potential for solvent contamination is to standardize the use of solvents at a yard. This would be useful primarily at larger shipyards, where different shops might purchase their materials separately. Centralizing and controlling solvent purchases would reduce the potential for cross contamination, minimize the number of different solvents purchased and increase the potential for on-site reuse.

Waste segregation can also reduce the overall quantities of hazardous waste generated. When only a single container is provided for collecting all waste materials, it is common for nonhazardous wastes to be placed in the same container with hazardous wastes, increasing the amount of hazardous waste being generated. By providing separate, clearly-labeled containers for each waste type, nonhazardous waste will not be added to hazardous wastes. Many companies have noticed a decrease in the total amount of hazardous waste being sent off-site after implementing waste segregation programs.

- *Minimization of fugitive oversprays.* Paint overspray is not usually collected or managed. However, at those marine yards that conduct painting operations at the water's edge or in uncontained areas, overspray can be a major component of run off into adjoining surface waters. Overspray in non-marine industries is controlled by improved painting techniques, including air-assisted airless; high volume, low pressure turbine; air-atomized electrostatic; and airless electrostatic application techniques. However, the compatibility of such techniques in marine applications needs to be evaluated. Operators should be trained in ways to minimize paint usage, such as maintaining a fixed distance from the surface while triggering the paint gun and releasing the trigger when the gun is not aimed at the target. Overspray can be confined by the use of plastic sheeting under and around the vessel being painted.

Equipment Cleaning Wastes

Painting sprayers, brushes, and equipment must be cleaned after use. Whenever possible, water-based coatings should be used. If solvents are needed, the best way to minimize solvent waste generation is by reusing the spent solvents as much as possible, and then recycling them. These options are discussed above.

Machine Shop Wastes

The major hazardous wastes from metal machining are waste cutting oils and degreasing solvents. The currently preferred method to reduce quantities of both is to substitute a water-soluble cutting fluid. This practice has been adopted in many machine shops without adversely impacting work efficiency or quality, and without increasing the annual operating cost appreciably.

Many machine shops successfully recycle spent coolant from machining operations, and a number of proprietary systems are available. This option may require monitoring coolant strength (using a refractometer), removing tramp oil and adding stabilizer and inhibitor chemicals. Recycling is most easily implemented when a standardized type of coolant is used throughout the shop.

Most shops already collect scrap metals from machining operations and sell these to metal recyclers. Metal chips which have been removed from the coolant by filtration should be drained and included in this collection.

The solvents used in machine shops are those used in automotive repair and other surface cleaning operations, and they should be segregated for reuse as discussed under the subsection "Waste Segregation." The waste cutting oils are amenable to the same types of off-site oil recycling as engine lube oils. A more detailed discussion on minimizing wastes from machine shops can be found in a waste audit report for finished metal products.[118]

Engine Repair Shop Wastes

Most marine engine repair work is similar to automobile repair work. Typical wastes include solvents, waste turbine oils, fuels, and batteries. Of these, the solvents are generally the only wastes suitable for recovery and recycle on-site.

In qualifying new or rebuilt turbines, thousands of gallons of lightly-used turbine oil may be generated as waste. The oil may be used for only a few minutes to a few hours and holds excellent potential for recycling. There are a number of recycling operations equipped to re-refine contaminated oil. Some states, such as North Carolina, operate portions of their motor fleet on this waste oil. In addition, there are several waste exchanges that use the oil as a feed stock for other processes. However, to derive maximum economic benefit, some care must be exercised in handling the waste product. Waste oil containers should be clearly labeled and kept secure to avoid cross-contamination with other chemicals and to keep water

---

[118] California DHS. 1989. Hazardous waste minimization audit studies on the finished metal products industry. Prepared by Jacobs Engineering Group, Inc., for California Department of Health Services Alternative Technology Section, Toxic Substances Control Division.

and general trash out. Incompatible products should be kept separate, as directed by the firm that will be accepting the waste. If the waste is badly contaminated, then its value for burning, recycling, or other use is greatly diminished, and the waste generator will be forced to pay a much higher price for disposal.

Vessel Cleaning Wastes

Vessel cleaning wastes are generated on board. Upon return to port, however, these wastes are removed and managed by the repair yard, making the yard the generator of record. Since the yard is not the source of vessel waste, there is little that the yard can do to reduce that generation. Wastes may also be generated from the cleaning of boiler tanks, sanitary systems, and other tank systems on larger ships. If chemical cleaners are used, these wastes will probably be hazardous. It may be possible, in some cases, to substitute a detergent cleaner or, in the case of a sanitary system, a bacterial enzyme cleaner, either of which may be disposed in sanitary sewers. The generator must contact the local sewer authority to verify whether the wastes may be disposed in sanitary sewers.

In order to reduce the risks associated with on-site storage of liquid wastes, some yards subcontract the entire cleaning operation to an outside firm specializing in the collection and appropriate disposal of such wastes.

Spill Control

Spill control is especially important at marine maintenance and repair yards, because most yards abut the ocean. It is common for yards to be designed so that they slope toward the ocean. Consequently, any spilled materials will eventually be washed into the ocean. Unless spills can be prevented or contained in the yard, it may become necessary to implement run-on/run off controls for the yard, consisting of curbs or berms around the yard perimeter to confine all contamination to the yard, and having surfaces sloping toward a collection dump to allow all contaminated materials to be collected, stored and disposed of properly. A detailed discussion on waste containment and confinement strategies, including design illustrations, can

Industry-Specific Pollution Prevention Strategies / 213

be found in Marine Maintenance and Repair: Waste Reduction and Safety Manual.[119]

The potential for spills and leaks of thinners and solvents is highest when a product is transferred from bulk drum storage to the point of use. Spigots or pumps should always be used when dispensing new materials, and funnels should be used to transfer waste materials into storage containers. Materials should never be poured directly from drums into smaller containers.

If drum handling is necessary, the drums should be moved correctly using powered equipment or hand trucks to prevent damage or punctures to the drums. Under no circumstances should drums be tipped or rolled, even when empty. Negligent handling may damage the seams, which could leak or rupture in future use.

The risk of spills increases in both frequency and magnitude when fueling services are included with yard services. Storage tanks should have secondary containment. Other ways to reduce the risk and to minimize spills include:[120]

- Watch the fuel tank vent to avoid overfilling.

- Be sure fuel flow has stopped before removing the fuel nozzle from the fill pipe.

- Provide a drip pan for the fuel nozzle.

- Be sure the proper type of fuel is selected, to avoid cross-contamination.

- Practice preventive maintenance on the entire fueling system.

Derelict vessels can be one of the most intractable waste problems facing a marine yard. Usable parts are generally salvaged and reused, leaving a useless hull. Metal hulls often can be sold as scrap, but wood and fiberglass hulls have almost no commercial value. Burning of wooden hulls is illegal in many places, and strongly discouraged in most others, so usually the

---

[119] Davis, D., and L. Piantadosi. 1988. Marine maintenance and repair: waste reduction and safety manual. Prepared for Pollution Prevention Pays Program, North Carolina Department of Natural Resources and Community Development. Raleigh, N.C., December 1988.

[120] Id.

wreck is left for eventual natural decomposition. Fiberglass vessels can sometimes be used in constructing artificial reefs, although this may be expensive. The best advice to a yard regarding minimization of this type of waste is to prevent derelicts from accumulating, since they tend to attract even more derelicts.[121]

COMMERCIAL PRINTING INDUSTRY

The list of individual primary lithographic waste streams and their sources along with a list of source reduction methods is presented in Table 6-5. Recommended waste reduction methods and identified procedures are discussed in the following sections.

In addition to the waste reduction measures that are classified as process changes or material/product substitutions, a variety of waste reducing measures labeled as "good operating practices" has also been included. Good operating practices are defined as procedures or institutional policies that result in a reduction of waste. The following describes the scope of good operating practices:

- Waste stream segregation.

- Personnel practices.

    -Management initiatives.
    -Employee training.

- Procedural measures.

    -Documentation.
    -Material handling and storage.
    -Material tracking and inventory control.
    -Scheduling.

- Loss prevention practices.

    -Spill prevention.
    -Preventive maintenance.
    -Emergency preparedness.

    Good operating practices apply to all waste streams.

---

[121] Id.

Table 6.5. Waste Minimization Options for the Commercial Printing Industry.

Operation	Waste Minimization Options
Material handling and storage	-Material pre-inspection. -Proper storage of materials. -Restrict traffic through area. -Inventory control. -Purchase quantities according to needs. -Recycle empty containers. -Recycle photographic film and paper.
Image processing	-Electronic and laser plate making. -Material substitution. -Extend bath life. -Use of squeegees. -Employ countercurrent washing. -Recover silver and recycle chemicals.
Plate processing	-Reduce solution loss. -Replace metal etching/plating operations. -Use nonhazardous developers and finishers.
Makeready	-Implement accurate counting methods. -Automated plate benders, scanners. -Automatic ink key setting system. -Computerized registration. -Ink/water ratio sensor.
Printing	-Install web break detectors. -Automatic web splicers. -Use automatic ink levelers. -Store ink properly. -Use less hazardous inks. -Standard ink sequence. -Recycle waste ink. -Use alternative fountain solutions. -Use alternative cleaning solvents. -Automatic blanket cleaners. -Reduce the need to clean.

Table 6.5—*Continued*

Operation	Waste Minimization Options
Printing *(cont'd)*	-Improve cleaning efficiency. -Collect and reuse solvent. -Recycle lube oils. -Alternative printing techniques.
Finishing	-Reduce paper use and recycle waste paper.

Material Handling and Storage

Improper storage and handling can result in spoilage and obsolescence of the raw materials. Good operating practices can reduce or eliminate waste resulting from obsolescence and improper storage.

- *Material Preinspection.* Materials should be inspected before being accepted, and unacceptable or damaged materials should be returned to the manufacturer or supplier. This avoids both disposal of a nearly full container of unusable material and printing an unacceptable product.

- *Proper Storage of Materials.* Many photo processing and plate developing chemicals are sensitive to temperature and light. Photosensitive film and paper storage areas should be designed for economical and efficient use. Some shops waste up to one-fourth of these materials due to improper storage.[122] The chemical containers list the recommended storage conditions. Meeting the recommended conditions will increase their shelf life. Even more important to the efficient use of paper is proper handling to avoid damage.

Paper waste can also be reduced through proper handling and storage of rolls or packages of paper. Webs need to be handled so that the outer paper wrapper is not damaged. Paper should be stored in a space having proper temperature and humidity, since it can easily absorb moisture; sheet fed paper should be conditioned to the temperature and humidity of the press

---

[122] Campbell, M.E., and Glenn, W.M. 1982. Profit From Pollution Prevention; A Guide to Industrial Waste Reduction and Recycling. Toronto, Canada: Pollution Probe Foundation.

room for one day before printing; this requires additional space in the press room.

- *Restrict Traffic through Storage Area.* To prevent raw material contamination, the storage area should be kept clean. Also, the storage area should not be open to through traffic. Through traffic will increase dust and dirt in the storage area, increasing possible contamination. In addition, spills in the storage area will be easier to contain if traffic is restricted.

- *Inventory Control.* Inventories should be kept using the "first-in, first out" practice. This will reduce the possibility of expired shelf life. This practice may not work for specialty materials that are seldom used. Computerized inventory systems can track the amounts and ages of the raw materials.

- *Purchase Quantities According to Needs.* Raw material order quantities should be matched to usage. Small printers should order ink in smaller containers according to use. This avoids having a large, partly used container of ink going bad in storage because it wasn't properly sealed. Residual ink in sheet fed ink cans should be smoothed and covered in order to prevent loss by skinning. Large printers should order materials in large containers, which may be returnable, thereby eliminating or reducing the need to clean them. It takes less time to scrape out the large single container than several small ones. Ordering materials in returnable tote bins may maximize these advantages. However, the size of container chosen by the printer is determined by economics of volume, as well as consideration of waste minimization.

- *Recycle Empty Containers.* Most ink containers are scraped free of ink and discarded in the trash. Since the degree of cleanliness is a function of operator effort, the amount of ink discarded can vary widely. By purchasing ink in recyclable bulk containers, the container can be returned to the ink supplier for refilling instead of being thrown away. In addition, the use of bulk containers also cuts down on the amount of cleaning required since the surface area of the container per unit volume of ink stored is reduced.

- *Recycle Spoiled Photographic Film and Paper.* It is a current practice in the industry to send used and/or spoiled film to professional recyclers for recovery of silver. However, this option might not be practical to small scale producers or available to facilities located far away from recyclers.

- *Test Expired Material for Usefulness.* Materials having expired shelf-life should not automatically be thrown out. Instead, this material should be tested for effectiveness. The material may be usable, rather than becoming a waste. A recycling outlet should be found for left over raw material that is no longer wanted.

Image Processing

The major waste stream associated with image processing is wastewater that contains photographic chemicals and silver removed from film. Much work has been done by the printing and photographic industries to abate this waste.

- *Electronic Imaging and Laser Platemaking.* A recent advance in image processing is the use of computerized "electronic pre-press systems" for typesetting and copy preparation. Text, photos, graphics, and layout are fed into the system through an electronic scanner. The copy is edited on a display monitor rather than on paper. This reduces the quantity of film, developing chemicals, and paper used. Only the final edited version is printed out.

Electronic pre-press systems were initially restricted to large printers, such as major newspapers, because of the high initial equipment expenses. As the prices of computer hardware and software drop, smaller printing operations are beginning to use them. Electronic pre-press systems should reduce waste and improve productivity.

- *Material Substitution.* Nonhazardous chemicals and films can be substituted for hazardous ones. This can reduce hazardous waste generation at the image processing step. In particular, the wastes from photo processing using silver films are occasionally hazardous due to silver compounds in the wastewater. Some sanitation districts will accept photo processing wastewater with silver, if the silver concentration is low enough.

Photographic materials are available that do not contain silver, but these are slower in speed than silver halide films. Diazo and vesicular films have been used for many years. Vesicular films have a honeycomb-like cross-section and are constructed of a polyester base coated with a thermoplastic resin and a light-sensitive diazonium salt. Photopolymer films contain carbon black as a substitute for silver. These films are processed in a weak alkaline solution that is neutralized prior to disposal. As such, they produce a nonhazardous waste.

More recently, photopolymer and electrostatic films are being used. Electrostatic films are non-silver films having speeds comparable to silver films and having high resolution as well. An electrostatic charge makes them light sensitive, after which a liquid toner brings out the image.

Some photographic intensifiers and reducers contain hazardous compounds, such as mercury or cyanide salts. Nonhazardous substitutes are available.

- *Extended Bath Life.* Wastes from photographic processing can be reduced by extending the life of fixing baths. Techniques include (1) adding ammonium thiosulfate, which doubles the allowable concentration of silver buildup in the bath; (2) using an acid stop bath prior to the fixing bath; and (3) adding acetic acid to the fixing bath as needed to keep the pH low.

Accurately adding and monitoring chemical replenishment of process baths will cut down chemical waste. Stored process bath chemicals should be protected from oxidation by reducing exposure to air. Some smaller photo developers store chemicals in closed plastic containers. Glass marbles are added to bring the liquid level to the brim each time liquid is used. In this way, the amount of chemical subject to degradation by exposure to air is reduced, thereby extending the chemical's useful life and the life of the bath.

- *Squeegees.* Squeegees can be used in non-automated processing systems to wipe excess liquid from the film and paper. This can reduce chemical carryover from one process bath to the next by 50%.[123] Minimizing chemical contamination of process baths increases recyclability, enhances the lifetime of the process baths, and reduces the amount of replenisher chemicals required. Most firms, however, use automated processors. Also, using squeegees may damage the film image if it has not fully hardened, so a squeegee should be used after the film image has hardened.

- *Countercurrent Washing.* In photographic processors, countercurrent washing can replace the commonly used parallel tank system. This can reduce the amount of wastewater generated. In a parallel system, fresh water enters each wash tank and effluent leaves each wash tank. In countercurrent rinsing, the water from previous rinsings is used in the initial filmwashing stage. Fresh water enters the process only at the final rinse stage, at which

---

[123] Id.

point much of the contamination has already been rinsed of the film. However, a countercurrent system requires more space and equipment.

- *Recovery of Silver and Recycling of Spent Chemicals.* Basically, photo processing chemicals consist of developer, fixer, and rinse water. Keeping the individual process baths as uncontaminated as possible is a prerequisite to the successful recycling of these chemicals. Silver is a component in most photographic films and paper and is present in the wastewaters produced. Various economical methods of recovering silver are available (e.g., metallic replacement, chemical precipitation, electrolytic recovery), and a number of companies market equipment that will suit the needs of even the smallest printing shop.

The most popular method of silver recovery is electrolytic deposition. In an electrolytic recovery unit, a low voltage direct current is created between a carbon anode and stainless steel cathode. Metallic silver plates onto the cathode. Once the silver is removed, the fixing bath may be able to be reused in the photographic development process by mixing the desilvered solution with fresh solution. Recovered silver is worth about 80% of its commodity price.

Another method of silver recovery is metallic replacement. The spent fixing bath is pumped into a cartridge containing steel wool. An oxidation-reduction reaction occurs and the iron in the wool replaces the silver in solution. The silver settles to the bottom of the cartridge as a sludge.

Some of the companies that buy used film or cartridges containing recovered silver can be located under "Gold and Silver Refiners and Dealers" in a business telephone directory. These firms may pick up directly or may purchase through dealers. To recycle used film, it may be worthwhile to sort the film into "largely black" versus "largely clear" segments, since the rate of payment for mostly black film may be twice that for mostly clear.

Technologies for reuse of developer and fixer are available and include ozone oxidation, electrolysis, and ion exchange.

Plate Processing

Recent advances in plate processing techniques—some made specifically to reduce the quantity and/or toxicity of hazardous wastes and to improve worker safety—have reduced waste generation.

- *Reducing Solution Loss.* In gravure printing, metal etching and metal plating operations typically involve chemical compounds that are deemed hazardous. Waste solutions from metal etching or metal plating operations usually require treatment before discharge to the municipal sewer. The same is true for all wastewater used in rinsing operations.

The amount of wastewater generated by rinsing plates can be reduced by using multiple countercurrent rinse tanks. The toxicity of wastewater from plating operations can be reduced by reducing drag out from plating tanks. Examples of drag out reduction techniques include (1) positioning the part on the rack to drain more easily, (2) using drain boards to collect drag out and return it to the plating tank, and (3) raising the temperature to reduce the surface tension of the solution.

- *Replacement of Metal Etching/Plating Operations.* Because of increasing environmental regulations and higher costs of hazardous waste treatment and disposal, the printer should replace metal etching or plating processes whenever possible. Alternative processes include pre-sensitized lithographic, plastic or photopolymer, and hot metal, which do not present the hazardous material problems associated with metal etching and plating operations.

The wastes generated by using pre-sensitized lithographic plates are (1) wastewater from developing and finishing baths, (2) used plates, and (3) trash, such as plate wrappers and empty plate processing chemical containers. Chemicals consumption can be reduced by frequently monitoring the bath for pH, temperature, and solution strength. This can extend the bath life, and solution change-outs can be reduced to several times a year. Automatic plate processors can maintain bath conditions. Pre-sensitized plates should be stored at the recommended conditions to maintain effectiveness. The used plates are not a hazardous waste and they should be collected and sold to an aluminum recycler.

- *Nonhazardous Developers and Finishers.* Nonhazardous developers and finishers are available. Presensitized plates are available that are processed with water only. A company makes plates that are water-resistant until exposed; upon exposure, the coating of the non-image becomes water soluble. The same company markets a platemaking system that produces offset lithography plates directly from copy or artwork, eliminating the need for photo processing. The system is economical for large printing operations.[124]

---

[124] Id.

Makeready

Paper represents the largest supply item that printers buy and is probably the most expensive component of their work. The printed paper produced in makeready is frequently the largest waste a printer generates and is nonhazardous. Paper waste at this step is determined by the efficiency of the quality control press adjustments needed to achieve the desired print quality, specifically through proper ink density and accurate registration.

Both in makeready and printing operations, printers need to know how much waste paper is generated relative to the quantity of acceptable pieces. One method that can be used by both sheet fed and offset printers is to weigh discarded paper and discarded product signatures and express the weight of waste as a percent of total paper used. Press counters are available, but under some circumstances, such as when a jam occurs on the press and the counter is not turned off, the count may be inaccurate.

A number of specific devices have been developed to automate press adjustments. With proper use, most of these are promoted by manufacturers as speeding up the makeready step and, thus, saving paper and ink. However, their direct benefit is to increase quality control. Described briefly below are the use of automated plate benders; automated plate scanners; automatic ink key setting systems; computerized registration; and ink/water ratio sensors. More detailed information, including costs, can be obtained by consulting product manufacturers and manufacturers' representatives. Also, the Graphic Arts Technical Foundation in Pittsburgh can provide additional information about these products. Each printer must weigh the cost of these items against the potential increases in quality and reduced paper and ink waste that may result.

- *Automated Plate Benders.* Automated plate benders are designed to prevent all or some of the problems that occur in fitting a plate to a cylinder: plate cracking; non-straight plate bending along the length of the bend; curvature of the plate differing from that of the cylinder; and other plate fitting parameters that affect proper registration.

- *Automated Plate Scanners.* Automated plate scanners have been developed for both web and sheet fed offset presses that take advantage of microprocessor technology and high quality optics. Finished plates are scanned to determine the relative density of the printing image across the plate's surface. This information is then used to set the ink fountain keys—in some systems by a press operator in others using remote controlled ink keys that are automatically pre-set to compensate for variations in image density.

- *Automatic Ink Key Setting System.* Automatic ink key setting is usually accomplished as part of a system that includes scanning densitometry to determine ink density. Information about the ink density is then transmitted to a computer controlled inking system, so that automatic adjustments are made to the ink profile for each ink slide position.

- *Automated Registration.* Optical scanners and microprocessors also form the basis for automated registration systems. One manufacturer's system employs motorized scanners that move laterally in search of registration marks and lock in to the marks for the entire press run. These systems have been developed for gravure, web offset, and other printing processes.

- *Ink/Water Sensors.* Press operators need to know whether ink flow or water flow needs to be adjusted to result in proper ink/water ratio. This ratio must be optimal to produce a sharp dot and strong contrast without the risk of tinting. One manufacturer has developed an ink/water sensor that is part of an automated press control system. An optical system detects light reflected from the ink form roller and measures the surface water and the amount of water emulsified in the ink. Both water feeds and ink keys are linked to the system, so both can be modified and any deviation in the ink/water ratio can be corrected at once.

Printing and Finishing

The major wastes associated with printing and finishing are scrap paper, waste ink, and cleaning solvents. The clean-up solvent waste stream consists of waste ink, ink solvents, lubricating oil, and solvent. In many printing establishments, excess ink and solvent is collected in a drip pan underneath the press. This waste is typically drummed and hauled away to a landfill area.

- *Standard Ink Sequence.* Adopting a standard ink sequence can reduce the amounts of waste ink and waste cleaning solution. If a standard ink sequence is employed, the ink rotation is not changed with the job and you do not have to clean out the fountains in order to change the ink rotation.

- *Installation of Web Break Detectors.* This device detects tears in the web as it passes through a high speed web press. It automatically shuts down the press to prevent damage to the press. Otherwise the broken web begins to wrap around the rollers and forces them out of the bearings. Web break detectors are primarily used to avoid severe damage to the presses. However, they also reduce the waste that would otherwise be generated if a web break damaged a machine. Electronic systems are available for detecting web breaks in a non-contact fashion that will neither smear ink nor crease the web, thereby reducing waste from these sources.

- *Automatic Web Splicers.* Automatic web splicers have become almost standard on web offset presses. The splice can be made while the paper is running at operating speed (flying web splicer) or while the paper is stationary (zero-speed splicer). Either option can result in significant savings in time and paper waste reduction.

- *Use of an Automatic Ink Leveller.* Ink waste and spoilage is reduced by maintaining the desired ink level in the fountain for optimum inking conditions. Automatic ink levellers are commercially available.

- *Proper Ink Storage.* Ink containers should be resealed after using. Open containers are subject to contamination with paper dust and dirt, as well as forming a "skin" on the surface, loss of solvent, or eventual hardening. It was noted previously that sheet fed offset lithographic inks should be levelled in the can before placing a liner over the ink. Other offset lithographic inks may have low enough viscosity to level themselves. Ink should be scraped from emptied containers with a spatula or knife to get as much of the ink out as practical. This prevents the empty containers from becoming a hazardous waste.

- *New Ink Systems.* Currently-used heat set offset lithographic inks contain: pigments used as colorants; resins; film formers; and 25% to 45% ink oils. Most of the ink oil is removed during heating while passing through a dryer, and is either incinerated, recovered and recycled, or emitted to the air. New types of inks are described below. These were formulated with pollution reduction and energy conservation in mind.

Water-base or water-borne inks are usually composed of pigmented suspensions in water and film formers. These inks find their greatest application in flexographic printing on paper substrates and their use has been recommended for gravure.

One factor stifling the development of water-base inks is that they require more energy to dry than do solvent borne inks. Another difficulty results from the necessity to shut presses down for short periods of time. When water-base ink dries, water is not a solvent for the dried ink; therefore more frequent equipment cleaning is required. Other problems besetting water-borne inks are low gloss and paper curl.

UV inks consist of one or more monomers and a photo synthesizer that selectively absorbs energy. Benefits of using UV inks are that the inks contain no solvent. The paper is not heated above 50^0C, and a minimum of moisture is lost in the process. Since the inks do not "cure" until exposed to UV light, and may therefore be allowed to remain in the ink fountains (and plates) for long periods of time, the need for clean-up is reduced. UV inks are particularly recommended for letter press and lithography.

Electron-beam-dried (EB) inks are similar in concept to UV inks, offer the same operational advantages as UV inks, and have no solvents. However, the EB systems require operator protection from X-rays created by the process. Also, the system often degrades the paper.

- *Automatic Blanket Cleaners.* Automatic blanket cleaners, which consist of a control box, a solvent metering box for each press unit, and a cloth handling unit, have increased press efficiency for a number of printers that have installed them. They are available for many types of presses, including both web and sheet fed offset. The increased speed of automated washing compared with manual cleaning results in fewer wasted impressions during the shorter time period needed for wash up.

There is an added benefit of increased safety. Using the automatic cleaning system eliminates the risk of bodily injury to employees holding a rag against a running blanket; also the amount of solvent used in cleaning is controlled, so the possibility of ignition in the dryer of paper soaked with solvent is less likely.

- *Recycling of Waste Inks.* Most waste inks can be recycled. One recycling technique relies on blending waste inks of different colors together to make black ink. Small amounts of certain colors or black toner may be needed to obtain an acceptable black color. Recycling to get black ink is generally more practical than recycling to get the original color. This reformulated black ink is comparable to some lower quality new black inks, such as newspaper ink. For this reason, much of the black ink for newspaper printing contains recycled ink. Waste ink recycling equipment is available and advertised in the printing trade magazines.

226 / Pollution Prevention Strategies and Technologies

Off-site ink recycling, either by ink manufacturers or by large printers, may be more economical for smaller printers. The waste ink is reformulated into black ink and sold back to the printer. The small printer can avoid landfill disposal of the ink and the large printer can reduce purchases of new black inks.

Labor time necessary to fill, operate, and empty the ink recycler is about the same as labor required to pack waste ink into drums with vermiculite adsorbent, and to manifest it. Therefore, the labor savings is not significant. The major operating cost savings are the reductions in raw materials costs and waste disposal costs.

- *Alternative Fountain Solutions.* The fountain solution contains water, isopropyl alcohol (IPA), gum arabic, and phosphoric acid, all of which end up on the paper or evaporate; they do not become hazardous waste. However, the evaporation of the IPA may create an air emissions problem. In states or regions with stringent air quality limits on VOCs, this may result in a requirement for air pollution control equipment. Alternatively, fountain solution concentrates are available that contain no VOC or very small amounts of VOC.

- *Alternative Cleaning Solvents.* Dangerous chemicals such as benzene, carbon tetrachloride, trichloroethylene, and methanol have been used as components of cleaning solutions. Specially-made blanket washes are now available that are less toxic and less flammable. These blanket washes typically contain mixtures of glycol ethers and other heavier hydrocarbons that have a high flash point and low toxicity. Many printers feel that these solvents do not work as well as the solvents mentioned above. However, because of the less hazardous nature of the material, they are gaining in popularity. These blanket washes are typically used for all cleaning operations in the printing step. General cleanup should be done with detergents or soap solutions whenever possible. Solvent cleanup should be used only for cleaning up inks and oils.

- *Reducing the Need to Clean.* Most presses are cleaned by hand with a rag wetted with cleaning solvent. The dirty solvent remains on the rag. A separate waste solvent is not produced. To reduce the amount of solvent and the number of rags used, ink fountains should be cleaned only when a different color ink is used or when the ink might dry out between runs. Aerosol spray materials are available to spray onto ink fountains to prevent overnight drying, so that the ink can be left in the fountain without cleaning at the end of the day. This reduces the amount of waste ink produced and the amount of cleaning solvent and rags used.

- *Increasing Cleaning Efficiency.* Where cleaning of rollers is accomplished with solvent and roller wash-up blade, several factors affect cleaning efficiency: condition of the rollers; condition of the blade; the blade's angle of attack against the roller; and press speed during wash up. Both rollers and blade should be in good condition. The blade's angle of attack should be adjusted so that sufficient pressure is exerted on the roller, but the angle should not be so coarse that the blade can be "grabbed" and "pulled under" the roller. Too slow a press speed means long wash up times, and generally increased solvent use.

- *Collect and Reuse Solvent.* In this practice, solvent is poured over the equipment and then wiped clean with a rag. The solvent is collected in drip pans under the equipment and becomes waste solvent which can be reused. If one container of solvent is used for each color printing unit, the solvents can be reused without cross-contaminating the inks. Used solvent can be reused in cleaning most of the ink from the rollers and blankets, with only a small amount of fresh solvent needed for the final cleanup. Small solvents stills can be used to reclaim the solvent. In some cases, used solvents having one particular ink color can be used to make up the solvent content of new inks of the same color.

- *Recycle Lube Oils.* If the printing presses are lubricated with oil, the used oil should be collected and turned over to a recycler. The recycler can either re-refine the oil into new lubricating oil, create fuel grade oil, or use it for blending into asphalt.

- *Reduce Paper Use and Recycle Waste Paper.* Because paper is the largest supply item a printer company buys and it may be the most expensive component of its work, paper use and the disposition of waste paper is a critical concern. Many printers segregate and recycle paper according to grade: inked paper is one grade and is recycled separately; unprinted white paper is sent separately to recycling; and wrappers for paper, which are a lower grade, are disposed of in trash.

PRINTED CIRCUIT BOARD INDUSTRY

This section discusses recommended waste minimization methods for printed circuit board manufacturers. The primary waste streams associated with manufacturing are listed in Table 6-6 along with recommended control methods. Many control measures associated with photo processing and cleaning wastes are not discussed here. The reader should consult other sources for information regarding these waste streams.

The waste minimization methods listed in Table 6-6 can be classified generally as source reduction, or recycling. Source reduction can be achieved through material or product substitution, process or equipment modification, or better operating practices. Recycling can include recovery of part of the waste stream or reuse of all of it, and can be performed on-site or off-site.

Better operating practices are procedural or institutional policies that result in a reduction of waste. They include:

- Waste stream segregation.

- Personnel practices.

    -Management initiatives.
    -Employee training.
    -Employee incentives.

- Procedural measures.
    -Documentation.
    -Material handling and storage.
    -Material tracking and inventory control.
    -Scheduling.

- Loss prevention practices.

    -Spill prevention.
    -Preventive maintenance.
    -Emergency preparedness.

- Accounting practices.

    -Apportion waste management costs to departments that generate the waste.

Better operating practices apply to all waste streams. In addition, specific better operating practices that apply to certain waste streams are identified in the appropriate sections that follow.

Table 6.6. Waste Minimization Options for the Printed Circuit Board Industry

Operation	Waste Minimization Options
PC board manufacture	-Product substitution: --Surface mount technology. --Injection molded substrate and additive plating.
Cleaning and surface preparation	-Materials substitution: --Use abrasives. --Use non-chelated cleaners. -Increase efficiency of process: --Extend bath life. --Improve rinse efficiency. --Countercurrent cleaning. -Recycle/reuse: --Recycle/reuse cleaners and rinses.
Pattern printing and masking	-Reduce hazardous nature of process: --Aqueous processable resist. --Screen printing versus photo lithography. --Dry photoresist removal. -Recycle/reuse: --Recycle/reuse photoresist stripper.
Electroplating and electroless plating	-Eliminate process: --Mechanical board production. -Materials substitution: --Non-cynide baths. --Non-cynide stress relievers.

Table 6.6—*Continued*

Operation	Waste Minimization Options
Electroplating and electroless plating *(cont'd)*	-Extend bath life to reduce drag-in: --Proper rack design/maintenance. --Better precleaning/rinsing. --Use of demineralized water as makeup. --Proper storage methods. -Extend bath life to reduce drag-out. --Minimize bath chemical concentration. --Increase bath temperature. --Use wetting agents. --Proper positioning on rack. --Slow withdrawal and ample drainage. --Computerized/automated systems. --Drain boards. -Extend bath life to maintain bath solution quality. --Monitor solution activity. --Control temperature. --Mechanical agitation. --Continuous filtration/carbon treatment. --Impurity removal. -Improve rinse efficiency: --Closed-circuit rinses. --Spray rinses. --Fog nozzles. --Increased agitation. --Countercurrent rinsing. --Proper equipment design/operation. --Deionized water use. -Recovery/reuse: --Segregate streams. --Recover metal values.
Etching	-Eliminate process: --Differential plating. -Materials substitution: --Non-chelated etchants. --Non-chrome etchants.

Table 6.6—*Continued*	
Operation	*Waste Minimization Options*
Etching *(cont'd)*	-Increase efficiency: --Use thinner copper cladding. --Pattern vs. panel plating. --Additive vs. subtractive method. -Reuse/recycle: --Reuse/recycle etchants.
Wastewater treatment	-Reduce hazardous nature: --Alternative treatment chemicals that generate less sludge. --Use of ion exchange and activated carbon for recycling wastewater. -Reuse/recycle: --Waste stream segregation.

Product Substitution

While not under the control of most printed circuit board manufacturers, improvements in the techniques used in the packaging of microchips can result in a decrease of waste associated with printed circuit board manufacturing. Two new techniques include:

- *Increased use of surface mount technology.* Dual-in-line package (DIP) accounts for about 80% of all packaging of integrated circuits. More efficient packages, however, are being developed which utilize a relatively new method of attaching packages to printed circuit boards. One important method is called surface mount technology (SMT). The use of SMT instead of the conventional through-hole insertion mounting allows for closer contact areas of chip leads, and therefore reduces the size of printed circuit boards required for a given number of packages or DIPS. For a fixed number of packages, the printed circuit board needs to be only 35% to 60% as large as a printed circuit board designed for the old style package.[125] As the metal

---

[125] Bowlby, R. 1985. The DIP may take its final bows! *IEEE Spectrum.* June 1985, pp. 37-42.

area on which cleaning, plating and photoresist operations are performed is decreased, the wastes associated with these operations can also be reduced. At present, however, SMT uses considerably higher quantities of chlorofluorocarbons for degreasing than through-hole mounting. CFC-113 is one of the major degreasing agents in current use. Because of the danger that some chlorofluorocarbons present to the atmospheric ozone layer, the overall environmental risks of SMT must be carefully examined, and alternative degreasing solvents identified, before replacing through-hole technology with SMT.

- *Use of injection molded substrate and additive plating.* The development of high-temperature, high-performance thermoplastics has introduced the use of injection molding into the manufacturing of printed circuit boards. In this process, heated liquid polymer is injected under high pressure into precision molds. Since the molded substrates are unclad, semi-additive or fully additive plating is used to produce metalized conductor patterns. Injection molding, coupled with a fast-rate electrode position (FRED) technique, can be used to manufacture complex three-dimensional printed circuit boards with possible reduction in hazardous waste generation due to the elimination of spent toxic etchants.

Cleaning and Surface Preparation

Information is provided below on: abrasive cleaning; use of non-chelated cleaning chemicals; extending bath life and improving rinse efficiency; use of countercurrent cleaning arrangements; and reuse/recycle of cleaning agents and rinse water.

- *Use Abrasive Instead of Aqueous Cleaning.* Mechanical cleaning methods offer an alternative to aqueous techniques and generate less hazardous waste; however, these methods can only be employed before electronic components have been added to the boards. Abrasive blast cleaning uses plastic, ceramic, or harder media, such as aluminum oxide, to remove oxidation layers, old plating, paint and burrs from workpieces, and to create a smooth surface. The aim is to select a blast medium that is harder than the layer to be stripped, but softer than the substrate, in order to prevent damage to the part. Abrasives can also be used in vibratory cleaning (in which parts are immersed in a vibrating tank containing abrasive material and water), in tumbling barrels, or applied via a buffing wheel.

- *Use Non-Chelated Cleaning Chemicals.* The use of non-chelate process chemicals instead of chelated chemical baths can reduce hazardous waste generation. Chelators are employed in chemical process baths to allow metal ions to remain in solution beyond their normal solubility limit. This enhances cleaning, metal etching, and selective electroless plating. Once the chelating compounds enter the waste stream, they inhibit the precipitation of metals, and additional treatment chemicals must be used. These treatment chemicals end up in the sludge and contribute to the volume of hazardous waste sludge.

Ferrous sulfate is a common reducing agent used to treat wastewaters that contain chelators. The ferrous sulfate breaks down the complex ion structures to allow metals to precipitate. However, the iron added to the treatment process also precipitates as a metal hydroxide. Since enough ferrous sulfate is usually added to the wastewater to achieve an iron to metal ratio of 8:1, a significant additional volume of sludge is generated.

Common chelators used in printed circuit board manufacturing chemicals include ferrocyanide, ethylenediaminetetraacetic acid (EDTA), phosphates, and ammonia. Chelating agents are commonly found in cleaning chemicals and etchants. Non-chelate alkaline cleaners are available; however, laboratory tests have shown that some of these products still have the ability to chelate metals.[126]

In addition to using non-chelated chemistries, the use of mild chelators can also reduce the need for additional treatment of wastewaters. Mild chelators are less difficult to break down. Therefore, metals can be precipitated out of solution during treatment without using the volume of treatment chemicals that is often necessary with strong chelators.

- *Extend Bath Life and Improve Rinse Efficiency.* This method applies to nearly any tank of processing solution used in the facility. See the discussion of electroplating waste reduction methods for detailed information.

- *Use Countercurrent Cleaning Arrangement.* A common hazardous waste stream generated by printed circuit board manufacturers is waste nitric acid from the cleaning of electroplating workpiece racks. Typically, racks are placed in a nitric acid bath to clean off the plated copper. When the copper content in the bath gets too high to effectively clean the racks, the nitric acid

---

[126] Couture, S.D. 1984. *Source Reduction in the Printed Circuit Industry. Proceedings - the Second Annual Hazardous Materials Management Conference*, Philadelphia, Pennsylvania, June 5-7, 1984.

is containerized for disposal. Use of a cascade cleaning system can significantly reduce nitric acid waste generation.

- *Reuse/Recycle of Cleaning Agents.* Peroxide/sulfuric acid solution is used as a mild etchant for cleaning copper and removing oxides prior to plating. When the solution is brought off-line and cooled, the copper crystallizes as copper sulfate. The supernatant can then be returned to the tank, replenished with oxidizers, and reused. The copper sulfate crystals can be used as copper electroplating bath makeup. The practice is only advisable, however, if the crystals are first dissolved into solution and treated with activated carbon to remove the organics. Otherwise, the organics present in the crystals could ruin the plating bath.

In addition to recovering metals from the spent bath, spent acid can be regenerated by means of ion exchange. Sodium phosphate salts, formed in nickel/copper electroless plating, can be converted into useful hypophosphite salts by ion exchange resins activated with hypophosphorous acid. The use of ion exchange resins for regeneration, however, suffers from the disadvantage of generating additional wastes, such as spent resins and resin regeneration solutions.

- *Reuse/Recycle of Rinse Water.* After rinse solutions become too contaminated for their original rinse process, they may be useful for other rinse processes. For example, rinses containing high levels of process chemicals can be concentrated through evaporation and returned to the process baths as makeup. Closed-circuit rinsing of this type can dramatically reduce the hazardous chemicals content of the waste stream.

Effluent from a rinse system that follows an acid cleaning bath can be reused as influent water to a rinse system following an alkaline cleaning bath. If both rinse systems require the same flow rate, 50% less rinse water would be used to operate them. In addition, using the effluent from the rinse solution that follows an acid cleaning process as the feed to the rinse system that follows an alkaline cleaning process rinse system can actually improve rinse efficiency for two reasons. First, the chemical diffusion process is accelerated because the concentration of alkaline material at the interface between the drag-out film and the surrounding water is reduced by the neutralization reaction. Second, the neutralization reaction reduces the viscosity of the alkaline drag-out film.

Adding acid rinses to alkaline rinses can result in problems, however. Unwanted precipitation of metal hydroxides onto the cleaned workpieces can occur in some instances. Before being implemented, a combined acid and alkaline rinse system must be thoroughly investigated in the particular environment of the process line.

Other rinse water recycling opportunities are also available. Acid cleaning rinse water effluent can be used as rinse water for workpieces that have gone through a mild acid etch process. Effluent from a critical or final rinse operation, which is usually less contaminated than other rinse waters, can be used as influent for rinse operations that do not require high rinse efficiencies. Spent cooling water or steam condensate can also be employed for rinsing if technically permissible and economically justified. Printed circuit board manufacturers should evaluate the various rinse water requirements for their process lines and configure rinse system arrangements that take advantage of rinse water reuse opportunities.

Pattern Printing and Masking

Many of the source reduction techniques discussed for the photo processing industry apply to this phase of printed circuit board manufacturing. Listed below are several techniques that deal with circuit board fabrication.

- *Use aqueous processable resist instead of solvent processable resist.* Aqueous processable resists can be used in place of solvent processable resists whenever possible to eliminate the generation of toxic spent solvents. Hundred of facilities are now employing these aqueous processable films for the manufacturing of printed circuit boards.

- *Use screen-printing instead of photolithography to eliminate the need for developers.* Screen-printing has conventionally been used only to produce printed circuit boards which require very low resolution in the width and spacing of the circuit lines. Some companies have recently developed screen-printing techniques which can provide higher degrees of resolution. For example, General Electric has developed a method for screen-printing down to 0.01 inch resolution which can be used to manufacture printed circuit boards for appliances.[127] The majority of printed circuit board manufacturers, however, are still using the photo lithographic technique for printed circuit boards having circuitry finer than 12 mil lines and spaces.

---

[127] Greene, R., ed. 1985. CE Alert, New Technology. Chem. Eng. March 4. p. 85.

- *Use Asher dry photoresist removal method to eliminate the use of organic resist stripping solutions.* Although this method is increasingly popular in the semiconductor industry, its use has not been reported by printed circuit board manufacturers, probably because the printed circuit board resists are usually much thicker than the corresponding semiconductor resist layers.

- *Recycle/reuse photoresist stripper.* Photoresist stripper is used to remove photoresist material from the board. This photoresist is a polymer material that remains in the stripper tank in small flakes that slowly settle to the bottom. When the sludge formed at the bottom of the stripper tank builds up, the flakes begin to adhere to circuit boards and the stripper solution is considered spent. Increased use of the solution can be achieved by decanting and filtering the stripper solution out of the tank into a clean tank. This is feasible because the stripper usually becomes spent as a result of the residue buildup long before it becomes spent as a result of a decrease in chemical strength.

Electroplating and Electroless Plating

Source reduction methods associated with electroplating and electroless plating center around eliminating the need for the operation, reducing the hazardous nature of the materials used, extending process bath life, improving rinse efficiency, and recovering/reusing spent materials.

- *Eliminate Need for Operation.* Use mechanical board production methods/systems. For facilities that produce low-volume prototype circuit boards, mechanical board production systems are available which bypass all operations involving chemicals. Circuit boards are designed on a computer and the pattern is then etched by means of a mechanical stylus on a copper-clad board. While this system is not viable for producing boards in large quantities, it is highly suited for use in development/ research settings.

- *Reduce Hazardous Materials Used.* Use non-cyanide plating baths and non-cyanide stress relievers. In the case of electroless copper plating, water soluble cyanide compounds of many metals are typically added to eliminate or minimize the internal stress of the deposit. It has been found that polysiloxanes are also effective stress relievers.[128] By substituting polysiloxanes for cyanides, the hazardous nature of the spent bath solution can be reduced.

- *Extend Process Bath Life.* Process baths may contain high concentrations of heavy metals, cyanides, solvents and other toxic constituents. They are not discarded frequently but rather are used for long periods of time. Nevertheless, they do require periodic replacement due to impurity build-up resulting from drag-in or decomposition and the loss of solution constituents by drag-out. When a solution is contaminated or exhausted, the resulting waste solution may contain high concentrations of toxic compounds and require extensive treatment. The source control methods available for extending process bath life include reducing or removing impurities formed in the bath, reducing the loss of solution (drag-out) from the bath, and maintaining bath solution quality.

- *Reduce Impurities.* Impurities come from five sources: racks, anodes, drag-in, water or chemical make-up, and air. The buildup of impurities can be limited by the following techniques:

   -*Proper rack design and maintenance.* Corrosion and salt buildup deposits on the rack elements contaminate solutions if they chip away or fall into the solution. Proper design and regular cleaning will minimize this form of contamination. Fluorocarbon coatings applied to the racks have also been found to be effective.[129] Such a coating lowers drag-out as well since less bath solution remains in the corroded crevices on the racks or barrels.

   -*Use purer anodes and anode bags.* During the plating process, metal from the anode dissolves in the plating solution and deposits on the cathode (workpiece). Some of the impurities contained in the original anode matrix stay behind in the plating solution, eventually accumulating to prohibitive levels. Thus, the use of purer metal for the

---

[128] Durney, L.J., ed. 1984. *Electroplating Engineering Handbook.* 4th ed. New York, N.Y.: Van Nostrand Reinhold Co.

[129] Lane, C. 1985. Fluorocarbon coating eliminates corrosion of acid bath racks. Chem. Process. 48(10): 72.

anode extends the plating solution life. Anode bags can also be used to prevent pieces of decomposed anodes from falling into the tank.

-Drag-in reduction by better rinsing. Efficient rinsing of the workpiece between different process baths reduces the drag-in of plating solution into the next process bath.

-Use of deionized or distilled make-up water. To compensate for evaporation, water is required for makeup of plating solutions. Using deionized or distilled water is preferred over tap water, since tap water may have a high mineral or solids content, which can lead to impurity buildup.

-Proper storage of chemicals. Proper storage of the process solutions can also reduce waste generation. Usually, the process solutions are stored as a two-part solution and are mixed when a batch is needed. Prolonged storage of mixed solutions may allow some chemical reactions to occur that could generate contaminants that reduce bath life. In electroless copper plating, if formaldehyde (a reducing agent) is stored with a hydroxide, the hydroxide can cause the formaldehyde to break down into formic acid and methyl alcohol. Thus, it is better to only store nonreactive mixtures of materials or to store each item separately.

Once you have reduced impurity buildup in the bath, you need to concentrate on reducing solution losses through drag out.

- **Reduce Drag-Out.** Several factors contribute to drag-out. These include workpiece size and shape, viscosity and chemical concentration, surface tension, and temperature. By reducing the volume of drag-out that enters the rinse water system, valuable process chemicals can be saved and sludge generation can be reduced. More discussion of the impact on sludge generation due to dragout is presented under "alternative treatment methods."

Process chemical suppliers assess drag-out using a standard rate of 10 to 15 ml/ft2 of circuit board. However, this standard rate does not take into account the various process bath operating parameters that can be used or the effects of various workpiece rack withdrawal methods. Nevertheless, this standard drag-out rate is a good starting point for determining the impact of drag-out on waste generation.

Factors affecting drag-out are:

-High surface tension.
-Highly viscous plating solution.
-Larger workpiece size.
-Faster workpiece withdrawal.
-Shorter drainage time.
-Orientation of workpiece during removal so that drainage is reduced.

Generally, drag-out minimization techniques include:

-Minimize bath chemical concentration. Controlling the chemical concentration of the process bath can reduce drag-out losses in two ways. Reducing toxic chemical concentrations in a process solution reduces the quantity of chemicals and the toxicity in any drag out that occurs. Also, greater concentrations of some of the chemicals in a solution increase the viscosity. As a result, the film that adheres to the workpiece as it is removed from the process bath is thicker and will not drain back into the process bath as quickly. Therefore, the volume of drag-out loss is increased and a higher chemical concentration in the drag-out is created.

Chemical product manufacturers may recommend an operating concentration that is higher than necessary to perform the job. A printed circuit board manufacturer should determine the lowest process bath concentration that will provide adequate product quality. This can be done by mixing a new process bath at a slightly lower concentration than normal. As fresh process baths are mixed the chemical concentration can continue to be reduced until product quality begins to be affected. At this point, the manufacturer can identify the process bath that provides adequate product quality at the lowest possible chemical concentration.

Fresh process baths can often be operated at lower concentrations than used baths. Makeup chemicals can be added to the used bath to gradually increase the concentration. This procedure allows newer baths to be operated at lower concentrations and older baths to be maintained for longer periods of time before requiring disposal.

-Increase bath operating temperature in order to lower viscosity. Increased temperature lowers both the viscosity and surface tension of the solution, thus reducing drag-out. The resulting higher evaporation

rate may also inhibit the carbon dioxide absorption rate, slowing down the carbonate formation in cyanide solutions. Unfortunately, this benefit may be lost due to the formation of carbonate by the breakdown of cyanide at elevated temperatures. Additional disadvantages of this option would include higher energy costs, higher chance for contamination due to increased makeup requirement, and increased need for air pollution control due to the higher evaporation rate.

-Use wetting agents. Wetting agents can be added to a process bath to reduce the surface tension of a solution and, as a result, reduce the volume of drag-out loss. However, most printed circuit board manufacturers prefer using process chemicals that are free of wetting agents because they can create foaming problems in the process baths. Although the process bath chemistries of a printed circuit board manufacturing line may not always allow the addition of wetting agents, their use should be evaluated.

-Position workpiece properly on the plating rack. When a workpiece is lifted out of a plating solution on a rack, some of the excess solution on its surface (drag-out) will drop back into the bath. Proper positioning of the workpiece on a rack will facilitate maximum drainage of drag-out back into the bath.

While positioning of the printed circuit board offers little variability—the boards are generally placed upright in a rack—a board that is tilted at an angle, allowing it to drip down onto an adjacent board instead of directly into the bath, may lead to increased drag-out loss. The operator must ensure that the workpiece is positioned properly to prevent unnecessary drag-out loss.

-Withdraw boards slowly and allow ample drainage. The faster an item is removed from the process bath, the thicker the film on the workpiece surface and the greater the drag-out volume will be. However, since workpieces are usually removed from a process bath manually, it is difficult to control the speed at which they are withdrawn. Nevertheless, supervisors and management should emphasize to process line operators that workpieces should be withdrawn slowly.

Workpiece drainage once the part is removed from the bath also depends on the operator. The time allowed for drainage can be inadequate if the operator is rushed to remove the workpiece rack from the process bath area and place it in the rinse tank. However, installation of a bar or rail above the process tank, and the requirement that all workpieces be hung from it for at least 10 seconds, may help ensure that adequate drainage time is provided prior to rinsing. Printed circuit board manufacturers express concern that increasing workpiece rack removal and drainage time will allow for chemical oxidation on the board. Although some process steps may not be amenable to these drag-out reduction techniques, increased workpiece rack removal and drainage time can still be effective for many process steps.

-Use computerized/automated control systems. Computerized process-control systems can be used for board handling and process bath monitoring to prevent unexpected decomposition of the plating bath. Since the use of a computerized control system not only requires a large capital outlay for initial installation but also increases the demand for skilled operations and maintenance personnel, only very large companies which manufacture both printed circuit boards and other electronic components are incorporating this change in their manufacturing process.

-Recover drag-out from baths. In addition to reducing the volume of drag-out that is lost from the process bath, printed circuit board manufacturers can recover drag-out losses by using drain boards and close-circuit rinsing. Drain boards are used to capture process chemicals that drip from the workpiece rack as it is moved from the process bath to the rinse system. The board is mounted at an angle that allows the chemical solution to drain back into the process bath. Drainage boards should be installed if there is space between the process bath tank and the rinse tank where chemical solutions would otherwise drip onto the floor and enter the wastewater system when the floor is washed down.

- *Maintain Bath Solution Quality.* Once the amount of drag-in and drag-out from the process bath has been reduced, attention should focus on ways to maintain the bath at optimum operating conditions. Many facilities rely on drag-out from the bath as the way of purging impurities that would otherwise build up and interfere with operation. From an environmental viewpoint, this is a poor technique since it does not directly address the issue of impurity formation, results in high losses of valuable process solutions, and moves the problem downstream to the treatment unit.

The following methods are noted as ways of increasing bath life and minimizing the impact on existing treatment systems:

-Monitor solution activity. By frequent monitoring of the bath activity and regular replenishment of reagents or stabilizers, bath life can be prolonged. These reagents or stabilizers differ from process to process. Stabilizers such as 2-mercaptobenzothiozole and methanol are found effective in electroless copper plating used for manufacturing printed circuit boards. The addition of stabilizers can sometimes decrease the deposition rate, but can still be economical in the long run.

-Control bath temperature. Good control of the bath temperature is important from the view point of performance predictability and is another method of prolonging bath life. Many surface treatment operations use tanks with immersed cooling/heating coils. As the salts precipitate and form scales on the coils, the heat transfer is impeded and temperature control becomes increasingly difficult. Heat transfer efficiency can be maintained by periodic cleaning of the coils or by using jacketed tanks instead of coils.

-Use mechanical agitation. Many process baths employ air agitation to increase and maintain the efficiency of the bath. This practice can introduce contaminants into the bath. The two principal contaminants are oil from the compressor or blower and carbon dioxide. The oil will lead to undue organic loading while the carbon dioxide can lead to carbonate buildup in alkaline baths. A viable alternative is to use mechanical agitation.

-Use continuous filtering/carbon treatment. To avoid surface roughness in the plating resulting in high reject rates, baths should be continuously filtered to remove impurities. The flow rate to the filter should be as high as practical to prevent particles from settling on the parts. Since filters can seldom remove solids at the same rate that they are introduced by way of drag-in, filtering should be performed even when the bath is not in use. Install as coarse a filter as practical, since coarse filters allow higher loading before requiring replacement, allow for higher flow rates and hence greater tank turn-overs, and require less servicing. When organic buildup is a problem, use of carbon filter cartridges is appropriate.

-Regenerate solution through impurity removal. There are methods that have been successfully used to increase the longevity of plating solutions through impurity removal. More efficient filtering of a plating solution has kept levels of impurities low and extended solution life.[130] Metallic salts can sometimes be removed by temporarily lowering the bath temperature so as to form solid crystals. In the case of electroless nickel plating, the sodium sulfate that forms can be crystallized by lowering the bath temperature to 41-50°F.[131] The crystals can then be removed by filtration.

- *Improve Rinse Efficiency.* Most hazardous waste from a printed circuit board manufacturing plant comes from the treatment of wastewater generated by the rinsing operations that follow cleaning, plating, stripping, and etching processes. Three basic strategies are used to provide adequate rinsing between various process bath operations. These are (1) turbulence between the workpiece and the rinse water, (2) sufficient contact time between the workpiece and the rinse water, and (3) sufficient volume of water during contact time to reduce the concentration of chemicals rinsed off the workpiece surface. The third strategy is most commonly employed by printed circuit board manufacturers. Reliance on this strategy causes printed circuit board manufacturers to use significantly more rinse water than is actually required.[132]

Many techniques are available that can improve the efficiency of a rinsing system and reduce the volume of rinse water used. These techniques include:

-Use of closed-circuit rinses. Installing one or more closed-circuit still or counter-flow rinsing tanks immediately after a plating bath allows for metal recovery and lowered rinse water requirements. The contents of the rinses are used to replenish the upstream plating bath. A major problem with the use of still rinses is that while they are commonly

---

[130] McRae, G.F. 1985. In-process waste reduction: part 1. Plat. Surf. Finish. 72(6): 14.

[131] Durney, L.J., ed. 1984. Electroplating Engineering Handbook. 4th ed. New York, N.Y.: Van Nostrand Reinhold Co.

[132] Couture, S.D. 1984. *Source Reduction in the Printed Circuit Industry. Proceedings - the Second Annual Hazardous Materials Management Conference*, Philadelphia, Pennsylvania, June 5-7, 1984.

installed at many plants, operators typically do not return the solution to the bath due to concern over solution contamination.

-Use spray rinsing. Although spray rinsing uses between one-eighth and one-fourth the volume of water that a dip rinse uses, it is not always applicable to printed circuit board manufacturing because the spray rinse may not reach many parts of the circuit board. However, spray rinsing can be performed along with immersion rinsing. This technique uses a spray rinse as the first rinse step after the workpieces are removed from the process tank. The spray rinsing typically takes place while the parts are draining above the process tank. This permits lower water flows in the rinse tank because spray rinsing removes much of the drag-out before the workpiece is submerged into the dip rinse tank.

-Use fog nozzles. A variation on the spray nozzle is the fog nozzle. A fog nozzle employs water and air pressure to produce a fine mist. Much less water is needed than with a conventional spray nozzle. It is more often possible to use a fog nozzle rather than a spray nozzle directly over a heated plating bath to rinse the workpiece, because less water is added to the process bath using the fog nozzle.

-Increase degree of agitation. Agitation between the workpiece and the rinse water can be performed either by moving the workpiece rack in the water or by creating turbulence in the rinse water. Since most printed circuit board manufacturing plants operate hand rack lines, operators could easily move workpieces manually by agitating the hand rack. However, the effectiveness of this system depends on cooperation from the operator.

Agitating the rinse tank by using forced air or water is the most efficient method for creating effective turbulence during rinse operations. This is achieved by pumping either air or water into the immersion rinse tank rinsing operations. Air agitation provides the best rinsing because the air bubbles create the best turbulence for removing the chemical process solution from the workpiece surface. This type of agitation can be performed by pumping filtered air into the bottom of the tank through a pipe distributor (air sparger). Great care should be exercised, however, to ensure that the air is free of dust or oil so as not to contaminate the boards being cleaned.

-Use countercurrent rinse stages. Multiple stage rinse tanks increase contact time between the workpiece and the rinse solution and thereby improve rinsing efficiency compared to a single-stage rinse. If these multiple tanks are set up in series as a countercurrent rinse system, water usage can also be reduced. Manufacturers do not need to rely on large volumes of rinse water to prevent chemical concentrations in the rinse solution from becoming excessive. Multiple rinse tanks can be used to provide sufficient rinsing while significantly reducing the volume of rinse water used. A multistage countercurrent rinsing system can use up to 90% less rinse water than a conventional single-stage rinse system.[133]

-Proper equipment design/operation. Printed circuit board manufacturers can use excessive amounts of rinse water if their water pipes are oversized or if the water is left on even when the rinse tanks are not being used. Rinse water control devices can be installed to increase the efficiency of a rinse water system. Flow restrictors limit the volume of rinse water flowing through a rinse system. These are used to maintain a constant flow of fresh water into the system once the optimal flow rate has been determined. Also, since most small and medium-sized printed circuit board manufacturers operate batch process lines in which rinse systems are manually turned on and off throughout the day, pressure activated flow control devices, such as foot pedal activated valves, can be helpful for assuring that the water is not left on after the rinse operation is completed. If the water lines are over-sized at a plant, pressure-reducing valves can be installed upgradient of the rinse water in fluent lines. This is also helpful for controlling water use in the rinse tanks.

A conductivity probe or pH meter can also be employed to control fresh water flow through a rinse system. A conductivity/pH cell is used to measure the level of dissolved solids or hydrogen ions in the rinse solution. When this level reaches a pre-set minimum, the conductivity probe activates a valve that shuts off the flow of fresh water into the rinse system. When the concentration builds to the preset maximum level, the probe again activates the valve, which then opens to continue the flow of fresh water. This control equipment is especially valuable to the printed circuit board manufacturing industry.

---

[133] Id.

-Use deionized water for rinsing. Natural contaminants found in water used for production processes can contribute to the volume of waste generated. During treatment of wastewater, these natural contaminants precipitate as carbonates and phosphates and contribute to the volume of sludge. The extent to which these contaminants increase sludge volume depends on the hardness of the rinse water. In addition to the direct effect on sludge volume, the presence of natural contaminants in the water may reduce rinse water efficiency and the ability to reuse/recycle rinse water. Therefore, rinse systems may require more water than would be necessary if the water were pretreated.

- *Recovery/Reuse of Spent Materials.* Recycling and resource recovery includes technologies that use waste as raw material for another process or that recover valuable materials from a waste stream before the waste is disposed of. Opportunities for both the direct use of waste materials and the recovery of materials from a waste stream are available to the printed circuit board manufacturing industry. Many of the spent chemical process baths and much of the rinse water can be reused for other plant processes. Also, process chemicals can be recovered from rinse waters, and valuable metals such as copper can be recovered from waste streams.

  A printed circuit board manufacturer must understand the chemical properties of its waste stream before it can assess the potential for reusing the waste raw material. Although the chemical properties of a process bath or rinse water solution may become unacceptable for their original use, these waste materials can still be employed in other applications. Printed circuit board manufacturers should therefore evaluate waste streams for properties that make them useful, as well as properties that render them waste.

- *Segregate Streams to Promote Recycling.* In a typical facility, the mixing of different rinse streams is not uncommon, and until relatively recently, rinse waters and spent baths were frequently mixed and treated together. By segregating various rinses, their reuse or recycling can be promoted. Metal reclamation by electrolysis from various streams is made easier if they are not mixed.

- *Recover Metal Values from Bath Rinses.* In the past, copper and other metal recovery from printed circuit board manufacturing has not proven to be economical. However, effluent pretreatment regulations have made the cost of treatment an economic factor. Also, the cost of management of sludges containing heavy metals has increased significantly because of the increased regulatory requirements placed on the handling and disposal of hazardous wastes. As a result, board manufacturers may now find it economical to recover copper and other metals and metal salts lost due to drag-out from process chemical baths.

Etching

Most of the source control techniques listed under plating and electroplating apply as well to waste produced by etching. Special source reduction methods associated with etching operations are discussed below.

- *Use differential plating instead of the conventional electroless plating process.* If the concentrations of certain stabilizers in the electroless copper bath are controlled, copper deposits three to five times faster on the through-hole walls than on the copper cladded surface. This reduces the amount of copper that must be subsequently etched away in the subtractive method. The use of differential electroless plating has not been reported by printed circuit board manufacturers, and it may require significant developmental work before commercialization is possible.

- *Use non-chelated etchants.* Non-chelate mild etchants, such as sodium persulfate and hydrogen peroxide/sulfuric acid, can be used to replace ammonium persulfate chelate etchant.

- *Use thinner copper foil to clad the laminated board.* This change reduces the amount of copper which must be etched, and thus reduces the amount of waste generated from the etching process. Printed circuit board manufacturers are switching to boards cladded with thinner copper as their starting materials.

- *Use pattern instead of panel plating.* Since panel plating consists of copper plating the entire board area, while pattern plating requires copper electroplating only the holes and circuitry, the use of the latter technique reduces the amount of non-circuit copper which must be subsequently etched away. This practice can therefore reduce the amount of waste generated from the etching operation. The switch from panel to pattern plating has been made by a large number of printed circuit board manufacturers. Customers demanding applications for a uniform cross section of circuitry in computer and microwave printed circuit boards, however, may dictate the use of panel plating to provide highly uniform copper thickness.

- *Use additive instead of subtractive method.* This change eliminates the copper etching step and, therefore, eliminates the generation of substantial volumes of spent etchant, as well as reducing the amount of metal hydroxide sludges generated. Although the subtractive method is still the most widely used in the manufacturing of printed circuit boards, the additive method is gaining in popularity since it results in less waste and lower manufacturing costs.[134] A noted drawback to the additive method, however, is the requirement for solvent processable instead of aqueous processable photoresists. Furthermore, the spent additive plating bath often contains heavily complexed copper which may result in waste treatment problems.

- *Use non-chrome etchants.* Whenever possible, ferric chloride or ammonium persulfate solution should be used instead of chromic-sulfuric acid etchants. Non-chromium etching solution has reportedly been used by printed circuit board manufacturers in an effort to reduce the toxicity of the waste generated.

- *Recycle spent etchants.* Use of an electrolytic diaphragm cell for regenerating spent chromic acid from etching operations has been reported.[135] The electrolytic cell oxidizes trivalent chromium to hexavalent chromium and removes contaminants.

Wastewater Treatment

Process chemical loss due to drag-out is the most significant source of chemicals entering wastewater. Treatment of this wastewater is a major source of hazardous waste in PC board operation because of the resulting sludge. The volume of sludge generated is proportional to the level of contamination in the spent rinse water.[136] The major ways of reducing waste associated with treatment (in addition to those associated with drag-out

---

[134] Brush, P.N. 1983. Fast track for printed circuit boards. Prod. Finish. November 1983, pp. 84-5.

[135] AESI. 1981. American Electroplater's Society, Inc. Conference on Advanced Pollution Control for the Metal Finishing Industry (3rd) held at Orlando Hyatt House, Kissimmee, Florida on April 14-16, 1980. EPA-600-2/8 1-028. Cincinnati, Ohio: U.S . Environmental Protection Agency.

[136] Couture, S.D. 1984. Source Reduction in the Printed Circuit Industry. Proceedings - the Second Annual Hazardous Materials Management Conference, Philadelphia, Pennsylvania, June 5-7, 1984.

reduction, reduction in the use of rinse water, and use of deionized water) include waste stream segregation, use of alternative treatment chemicals, and alternative treatment technologies.

- *Waste Stream Segregation.* Segregating waste streams can improve the efficiency of a waste treatment system. An example of waste stream segregation is the separation of chelating agent waste streams from nonchelating agent streams. Since most small printed circuit board manufacturing plants use treatment systems that can be operated as a batch process, they can implement waste stream segregation and selective treatment with minimal impact on the production system. The main drawback to this alternative is usually the limited storage capacity for the segregated waste streams.

If waste streams containing chelating agents are treated in a batch process separately from other waste streams, the use of ferrous sulfate to break down the chelators can be minimized. Since the iron in ferrous sulfate will precipitate out in the sludge, reduction in its use will also reduce the volume of sludge generated.

By isolating cyanide-containing waste streams from waste streams containing iron or complexing agents, the formation of cyanide complexes is avoided, and treatment made much easier.[137] Segregation of wastewater streams containing different metals also allows for metals recovery or reuse. For example, by treating nickel-plating wastewater separately from other waste streams, a nickel hydroxide sludge is produced which can be reused to produce fresh nickel plating solutions.

Another waste alternative is to separate noncontact cooling water from industrial wastes. It is likely that this cooling water can bypass the treatment system and be discharged directly to the sewer because it does not come in contact with process chemicals. This practice can reduce wastewater volume and, as a result, reduce the amount of treatment chemicals used. Also, acidic or alkaline waste streams that do not contain metals can simply be neutralized prior to discharge; therefore, if they are segregated from other wastes that require metal removal, the volume of treatment chemicals can be reduced. This, in turn, will reduce the volume of sludge generated.

- *Use of Alternative Waste Treatment Chemicals.* The selection of

---

[137] Dowd, P. 1985. Conserving water and segregating waste streams. Plat. Surf. Finish. 72(5): 104-8.

chemicals used in the waste treatment process can affect the volume of sludge generated. This selection should, therefore, consider a chemical's effect on sludge generation rates. For example, lime and caustic soda are two common chemicals used for neutralization and precipitation.

Alum and ferric chloride are commonly employed as coagulating agents to improve floc formation. When used, they convert to hydroxides and contribute to the volume of sludge. Polyelectrolyte conditioners can also be used as coagulants, but they are more expensive than inorganic coagulants. However, polyelectrolytes do not add to the quantity of sludge and may actually be less expensive overall when considering waste handling costs.

The selection of alternative treatment chemicals depends on specific waste characteristics and removal efficiency needs for a particular treatment facility. The potential use of various treatment chemicals should be discussed with chemical manufacturers' representatives and experimented with to determine their effectiveness.

- *Alternative Wastewater Treatment Ion Exchange.* Ion exchange systems can be employed to treat the entire waste stream prior to discharge to the publicly owned treatment works. When used for this purpose, the ion exchange units do not recover process chemicals for reuse because all sources of wastewater are mixed prior to treatment. The units can be used to recycle rinse water, however, by utilizing an activated carbon treatment system following ion exchange treatment. The costs for operating an ion exchange system depend on the volume and chemical concentrations of the wastewater.

AUTOMOTIVE REFINISHING INDUSTRY

This section discusses recommended waste minimization methods for automotive refinishing shops. The primary waste streams associated with automotive refinishing are listed in Table 6-7 along with recommended control methods. Waste streams include body repair wastes, paint application wastes, and shop cleanup wastes. A discussion of waste minimization methods for shop cleanup wastes, which include handling of automotive fluids leaking from damaged cars, is presented in the section on Automotive Repair Industry.

The waste minimization methods listed in Table 6-7 can be classified generally as source reduction, which can be achieved through material substitution, process or equipment modification, or better operating practices; or as recycling. Better operating practices are procedural or shop policies that result in a reduction of waste. They include:

- Waste stream segregation.

- Personnel practices.

    -Management initiatives.
    -Employee training.
    -Employee incentives.

- Procedural measures.

    -Documentation.
    -Material handling and storage.
    -Material tracking and inventory control.
    -Scheduling.

- Loss prevention practices.

    -Spill prevention.
    -Preventive maintenance.
    -Emergency preparedness.

- Accounting practices.

    -Apportion waste management costs to departments that generate the waste.

Better operating practices apply to all waste streams. In addition, specific better operating practices that apply to certain waste streams are identified in the appropriate sections that follow.

The following waste minimization measures are aimed at reducing the generation of wastes associated With body repair and paint applications. For ways to reduce waste associated with the handling of various automotive fluids, read the section on Automotive Repair Industry.

In addition to the specific recommendations provided below, rapidly advancing technology makes it important that shops continually educate themselves about improvements that are waste reducing and pollution preventing. Information sources to help inform companies about such technology include trade associations and journals, chemical and equipment suppliers, equipment expositions, conferences, and industry newsletters. By keeping abreast of changes and implementing applicable technology improvements, shops can often take advantage of the dual benefits of reduced waste generation and a more cost-efficient operation.

Table 6.7. Waste Minimization Options for the Automotive Refinishing Industry

Waste Stream	Waste Minimization Options
Body Repair	
Filler waste	-Rigid inventory control to minimize Bondo use.
Sand dust	-Sweep or vacuum up. -Don't flush to street or clarifier.
Painting	
Paste waste	-Rigid inventory control to reduce thinner use. -Better housekeeping to reduce leaks and spills. -Mix paint according to need. -Use high transfer efficiency equipment. -Provide operator training. -Practice proper equipment cleaning methods. -Use enclosed cleaners. -Recycle solvent off-site by means of thinner leasing agreements. -Recycle solvent on-site. -Contact waste exchanges. -Make leftover paint available to customer.
VOC emissions	-Use high transfer efficiency equipment. -Use enclosed cleaning devices. -Use low VOC coatings.
Booth filters	-Use high transfer efficiency equipment. -Use styrofoam filters.
Shop Cleanup	
Various	-Manage waste automotive fluids properly.

Body Repair

Polyester/fiberglass filler is used to fill in dents that cannot be removed by mechanical methods. After filling and hardening, the filler is sanded to create a smooth surface. Filler dust collects on the shop floor and is either swept up and disposed of in the trash or washed down the storm drain. Ways to reduce this waste include rigid inventory control and use of dry cleanup methods.

- *Rigid inventory control.* Rigid inventory control is often an effective way of reducing the indiscriminate use of raw materials. In one shop, records are kept on the amount of "Bondo" each worker checks out from the storeroom. These records can be checked against the number of cars the worker repairs, and wasteful use of materials can be quickly spotted. This type of information is very useful in determining trouble spots or problem areas that need careful attention. Comparison of usage rates among workers and facilities allows a manager to determine if the problem is worker-related (correct procedures improperly performed) or facility-related (improper procedures specified and implemented).[138]

- *Use of dry cleanup methods.* Shops that operate a clarifier unit to remove oil, grease, and solids from sewer discharges should use dry collection methods, such as sweeping or vacuuming for filler dust. Clarifier sludges may be classified as a hazardous waste and the introduction of nonhazardous solids into the clarifier needlessly increases sludge volumes and disposal costs. Combination sanding and dust collection systems are commercially available, but they are reportedly very expensive. For shops performing wet sanding, use of a "wetvac" to collect and pick up the filler particles might be a viable option.

Paint Application

Paint application wastes include leftover paints, dirty thinner due to cleaning of spray guns and paint cups, air emissions of volatile organic compounds (VOCs) and pigments, and dirty spray booth filters. Ways to

---

[138] Hazardous waste minimization audits of automotive repair and refinishing facilities. Prepared by Jacobs Engineering Group, Inc. for the City of Santa Monica Department of General Services. September 1989.

reduce these wastes include rigid inventory control; better housekeeping practices; mixing paint according to need; better operator training; proper cleaning methods; recycling solvents on and off-site; and waste exchanges. Also, options for minimizing waste in paint application include using alternative coatings and using styrofoam filters.

- *Rigid inventory control.* Rigid inventory control provides a very effective means of source reduction at virtually no cost to the operator. This alternative can be implemented in several ways. The owner may monitor employee operations and make verbal or written comments on product usage and suggested limits. In larger shops where monitoring of employees is not a viable alternative, the owner or manager can limit access to storage areas containing raw materials. This inaccessibility forces the employee to stretch the use of raw materials farther. Moreover, through this practice, the owner/manager can monitor the use of raw materials.

Not surprisingly, there is a high positive correlation between the amount of paint thinner used and the amount of waste generated. There is a hypothetical minimum amount of thinner that is essential to paint an average car; thinner use above that amount may be presumed to be waste. While it is difficult to generalize because each firm's thinner usage varies, more stringent inventory control and restrictions on thinner use can result in potential savings.

- *Better housekeeping practices.* Basic housekeeping techniques can be very effective as a means of source reduction. There are a wide variety of methods available to control and minimize leaks which can be implemented easily at no cost to the operator. Specific approaches to drum location, material transfer methods, leak collection, and drum transport can effectively limit product loss.

There are two predominant patterns of drum location. If inventory control is necessary to minimize product usage, drums should be stored together in an area of limited accessibility, such as indoor/outdoor sheds, lockers labeled "flammable", or locking storage rooms. If employees take individual responsibility for regulating product use and if inventory control is not a problem, it may be more effective to separate drums and place them at points of highest use in the facility. This alternative reduces the chance of product leaks and spills during transport from storage to work areas.

The potential for accidental spills and leaks is highest at the point of transfer of thinners from bulk drum storage to process equipment. Spigots or pumps should always be used to transfer waste materials to storage containers. Material should never be poured directly from drums to smaller containers.

Evaporation is a material loss that can be controlled through the use of tight-fitting lids, spigots and other appurtenances. The reduction of evaporation will increase the amount of available material and result in lower solvent purchase cost.

If drum transport or movement is necessary, it is essential that drums be moved correctly to preserve the integrity of the containers and to prevent damage or punctures. Drums should be lifted by means of powered equipment or hand trucks. Under no circumstances should drums be tipped or rolled even when empty. Negligent transport procedures will cause drum damage, particularly to seams, which could lead to leaks or ruptures during future use.

- *Mix paint according to need.* In the practice of automotive refinishing, many operators prepare a fixed amount of paint for each job (e.g., one pint or quart). Any paint not used for the job is considered to be a hazardous waste and must be disposed of as such. In particular, for small jobs, which are most common, the amount of paint prepared will often exceed the amount of paint actually applied.

Most small cars can be painted entirely with one quart of paint; touch-ups and damage repair would use substantially less than one quart. Availability of variously graded sizes of paint mixing and sprayer cups would enable operators to use the equipment best suited to the size of a particular job. Varying paint cup sizes could be an effective means of source reduction in two important ways. It would limit overmixing of paint to be used on a specific project, and decrease the amount of solvent needed for equipment cleanup when doing spot painting and small jobs.

A disadvantage of mixing smaller quantities is that color matching becomes more difficult. As the amount of paint mixed decreases, weighing accuracy becomes more critical Special attention would need to be given to the purchase, installation, maintenance, and use of higher precision weighing equipment.

- *Use high transfer efficiency equipment.* Another way to reduce VOC emissions is to reduce the amount of paint sprayed for a given job. The standard method of applying paint is the air spray gun. Typical transfer efficiency is on the order of 20 to 40%. Many of the newer spray application systems have transfer efficiencies greater than 65%. Since with lower efficiency, more paint is wasted, higher efficiency systems are being promoted for use.

In a recent study,[139] nine different spray painting techniques in current use were studied. The nine techniques included: air-atomized conventional (AAC); airless conventional (ALC); air-assisted airless (AAL); high volume low pressure (HVLP) turbine; high volume, stepped down low pressure (HVSDLP); low pressure, low volume (LPLV); thin film atomization (TFA); air-atomized electrostatic (AAE); and airless electrostatic (ALE). Automotive refinishing is generally performed by manual spray painting using conventional air-atomization/spray equipment.

A comparison of the methods showed that HVLP turbine, HVSDLP, LPLV, and TFA appeared to be potential candidates for replacement of AAC in the automotive refinishing industry. One automotive refinisher in the San Diego area reported very favorable results with an HVLP turbine system. Paint usage had been reduced by one-third and the finish quality was very good to excellent. The firm has experienced no operational problems with the equipment and it responded that overall operating costs were very similar to those incurred with conventional air-atomized systems. In another study,[140] the shop reponed that High Volume Low Pressure (HVLP) spray guns were tested but that the resulting paint job suffered from "halo" effects when spraying metal flake paints. Operator training has been reported to be a key element in the successful conversion to a high efficiency system. Given the regulatory drive to reduce VOC emissions from automotive refinishing operations, high transfer efficiency spray guns will see increasing use in the near future.

- *Better operator training.* Often overlooked, transfer efficiency is also a function of operator skill and training. Operators may be very skilled at producing high quality finishes but be poorly trained in the ways of reducing paint usage. Operators should be trained not to arc the spray gun and blow paint into the air. The practice of maintaining a fixed distance from the painted surface while triggering the gun should be encouraged. Air pressure (often set too high) should be well regulated. When the pressure is set too

---

[139] Alternative automotive refinishing technique study Phase I. Final report. Prepared by Jacobs Engineering Group, Inc. for the County of San Diego Air Pollution Control District. June 1989.

[140] Hazardous waste minimization audits of automotive repair and refinishing facilities. Prepared by Jacobs Engineering Group, Inc. for the City of Santa Monica Department of General Services. September 1989.

high, most of the paint bounces off the car and forms a fog. The proper adjustment of air pressure can increase transfer efficiency by 30 to 60%.

- *Proper cleaning methods.* In reducing solvent use, greater attention should be paid to the methods employed in equipment cleaning. Paint cups should first be scraped free of residual paint using a plastic spatula and then rinsed with solvent. The practice of filling the cup with solvent, stirring until the paint dissolves, and then repeating the procedure as needed should be discouraged. New Teflon-lined metal paint cups are available, which should provide for easier cleaning.

The typical way of cleaning a spray gun is to fill the spray cup with solvent and then spray the solvent into the booth filters or into the air. This results in a large waste of thinner and considerable air emissions. To recover the thinner for reuse and prevent undue air emissions, an enclosed gun cleaning station should be used. Thinner is sprayed through the gun and into the cleaning station where it is condensed for recovery and reuse. To simplify operation, the cleaning station uses compressed air instead of electricity to produce refrigeration/condensation. Several air quality agencies are requiring the use of enclosed spray gun cleaners at automotive refinishing shops.

- *Recycle solvent off-site.* Sludges from drum cleanup and thinner recovery from solvent-based paints contain as much as 50% organic thinners, such as volatile hydrocarbons, ketones, esters, and alcohols, and about five percent inorganic pigments. Processes for recycling thinners are well established and widely used. Small quantity generators and those generators that do not possess the technical expertise, or find it uneconomical to recycle contaminated thinners on-site, usually send thinner wastes to commercial recyclers for recovery. Commercial recyclers have versatile distillation processes and can handle large volumes and varieties of thinners. Reclaimed thinners are often sold back to the generators after the thinners are reconstituted.

In general, suppliers who offer recycling services include the cost of waste collection and recycling in the price of their thinner. This increases the thinner cost, but effectively eliminates separate hauling and disposal or recycle costs. It also reduces the administrative burden on the owner or manager of the auto painting firm.

- *Recycle solvent on-site.* Several alternatives are available to operators who wish to conduct recycling processes on-site. Gravity separation is inexpensive and relatively easy to implement by allowing the thinner/sludge mixture to separate under quiescent conditions. The clear thinner can then be decanted using a drum pump and used for equipment cleaning, reducing requirements for purchased wash thinner. Use of reclaimed thinner for formulating primers and base coats is possible, but might create problems if the thinner is not sufficiently pure.

For the larger-quantity generators in the auto body repair and painting industry, on-site distillation may provide a more cost-effective alternative. The batch distillation of all high-grade thinner wastes can virtually eliminate the need for purchasing lower quality thinners for use in preliminary painting operations and cleanup. From 5 gallons of paint and thinner wastes, the operator can reclaim 4 ½ gallons of thinner, with ½ gallon left as sludge. This ratio varies depending on operations. Addition of a commercial additive to the ½ gallon of paint sludge yields a viscous material which can be used as underseal.

- *Use waste exchange.* Waste exchanges provide another waste removal alternative for auto body and painting companies. Waste exchanges are organizations that manage or arrange the transfer of wastes between industries, such that one producer's waste material might be another industry's feed stock. Most exchanges exist as information clearinghouses, which provide information on waste availability. Opportunities exist for the direct transfer (without processing) of waste solvents from industries requiring ultra-high-purity solvents (e.g., the electronics industry) to industries that do not have such stringent purity requirements (e.g., the machinery and painting industries). Waste solvents are available through the waste exchanges, which could potentially be used as a substitute for new wash thinner. In addition, several generators have recently found new opportunities to ship residual still bottoms to cement industries for use as supplemental fuels.

- *Use alternative coatings.* There are four basic types of paint used in automotive refinishing: acrylic lacquer, synthetic enamel, acrylic enamel, and catalyzed acrylic enamel. A fifth type, color and clear polyurethane, is used primarily in truck-fleet finishes. Painting usually proceeds with application of a primer/surfacer followed by one or more coats of paint. To achieve a high gloss finish, many cars are painted with a color base coat followed by a clear top coat. Since several coatings of various compatible materials may be required to achieve a desired finish, coating materials are often referred to as "systems."

The quantity or amount of VOC emissions is related to the type of paint used since each material varies in solvent content and the number of coats necessary for a high quality finish. Acrylic lacquers are typically thinned with solvent by 125 to 150%. To achieve enough buildup for sanding and buffing, at least four or five double coats are applied.

With synthetic enamels, solvent thinning amounts to 15 to 33%. Since enamel dries to a gloss and is not sanded, only two or three medium coats are required. Base coat/top coat systems usually require two or three coats of each.

Three ways to minimize or eliminate VOC emissions from automotive refinishing operations are to substitute solvent-based paint with water-borne paint, use high-solids paints, or switch from high solvent to medium solvent paints.

- *Use styrofoam filters.* Replacement and disposal of dirty spray booth filters is currently performed by the thinner supplier/recycler. Filters must be disposed of as hazardous waste if they contain wet paint (i.e., solvents), due to their potential flammability. Filters may also be hazardous due to their potential toxicity if the overspray contains lead or chromium pigments. To reduce filter waste, a cleanable styrofoam filter element has been developed. When dirty, the filter can be blown clean with compressed air and reused (removed paint would require collection and still be classified as hazardous if it contained lead or chromium compounds). When the filter is no longer reusable, it can be disposed of with dirty thinner waste by dissolving it in the drum of waste thinner. Before using this filter, shops should check with their thinner recycler to determine if dissolved styrofoam will interfere with their dirty thinner recycling operation.

Shop Cleanup Wastes

The human aspects of industrial activity can be very important in waste reduction. Often termed "good operating practices" or "good housekeeping," these methods can be very effective in reducing the amount of shop cleanup wastes generated. Typical wastes include dirty rags, sawdust, clarifier sludges, area wash downs, and disposal of out-dated supplies. Good housekeeping methods include improved employee training, management initiatives to increase employee awareness of the need for and benefits of waste minimization, and preventive maintenance to reduce the number of leaks and spills that occur. Additional ways to reduce or minimize waste are discussed in the Automotive Repair Industry section.

PESTICIDE FORMULATION INDUSTRY

This section discusses recommended waste minimization methods for pesticide formulating operations. The primary waste streams associated with pesticide formulation are listed in Table 6-8 along with recommended control methods. In order of occurrence at a facility, the waste streams are equipment cleaning wastes; spills and area washdowns; off-spec product; empty bags and drums; air emissions and wastes from air emission control equipment; wastewater associated with laundering protective clothing; water used for aerosol leak testing, and storm water run off.

The waste minimization methods listed in Table 6-8 can be classified generally as source reduction, which can be achieved through material substitution, process or equipment modification, or better operating practices; or as recycling. Source reduction through product substitution is not an easily implemented procedure. This is because of the high level of effort and cost associated with registering a new pesticide with EPA as required by the Federal Insecticide, Fungicide, and Rodenticide Act (FIFRA). Registration of a new pesticide is required before it may be bought, solid, distributed, or otherwise handled. Pesticide registration involves the development and submission of a health and ecological risk database to EPA for review. Consequently, the waste minimization assessments of pesticide formulating firms have focused on source reduction that relied on good operating practices and process modifications.

Better operating practices are procedural or institutional policies that result in a reduction of waste. They include:

- Waste stream segregation.

- Personnel practices.

    -Management initiatives.
    -Employee training.
    -Employee incentives.

- Procedural measures.

    -Documentation.
    -Material handling and storage.
    -Material tracking and inventory control.
    -Scheduling.

Many of these measures are used in industry to promote operational efficiency. In addition, they can often be implemented at little or no cost to the facility. When one considers the effects of reduced waste, increased efficiency, and little or no implementation cost, good operating practices usually provide a very high return on investment.

Table 6.8. Waste Minimization Options for the Pesticide Formulating Industry

Waste Stream	Waste Minimization Options
Equipment cleaning wastes	-Maximize production runs. -Store and reuse cleaning wastes. -Use of wiper blades and squeegees. -Use of low volume high efficiency cleaning.
Spills and area washdowns	- Use of plastic or foam "pigs". -Use of dedicated vacuum system. -Use of dry cleaning methods. -Use of recycled water for initial cleanup. -Actively involved supervision.
Off-specification products	-Strict quality control and automation. -Reformulation of off-spec batches.
Containers	-Return containers to supplier. -Triple rinse containers. -Drums with liners vs. plastic drums or bags. -Solid waste segregation.
Air emissions	-Control bulk storage air emissions. -Dedicated dust collection system. -Automatic enclosed cut-in hoppers.
Miscellaneous wastewater streams	-Pave high spillage areas. -Wastewater treatment*

*This method can only be viewed as waste minimization if it allows the continued use of spent cleaning solutions.

Equipment Cleaning Wastes

One of the most concentrated wastes produced at pesticide formulation plants results from the cleaning of process equipment. A typical formulation plant produces a variety of different pesticides, all on a batch basis. Between batches, the mixing tanks and all other equipment exposed to the pesticide must be cleaned to avoid contamination between different products.

If powders or other "dry" pesticides are formulated, then cleaning is accomplished using a dry, inert material, such as clay or sand. These inert materials are passed through the system where they pick up traces of pesticide dusts. Many facilities save this flush material and use it in the next production run of the same product.

In the formulating of liquid solvent-based pesticides, cleaning is normally performed by rinsing or flushing the equipment with the same type of solvent used in the formulation followed by a "bail-out" and rinse using water. Waste solvents are usually saved and reused in the next batch of similar product while the wastewater is disposed of as hazardous waste. For water-based pesticide formulations, only water is employed for cleaning.

- *Maximize production runs.* Production runs of a given formulation should be scheduled together so as to reduce the need for equipment cleaning between batches. Consideration should also be given to the potential for scheduling families of products in sequence; while some cleanup is still needed between batches, it can be minimized.

- *Store and reuse cleaning wastes.* Many facilities collect cleaning dusts and solvents for reuse as make-up in the next compatible formulation. Rinse water has also been saved and reused when the facility produces water-based formulations.

If it is not practical to use the waste rinse water as make-up during a later formulation, it can be reused as rinse water. In those instances where more than one rinse is needed to clean the equipment, the first rinse can be performed using old rinse water from a previous formulation. This rinse will remove the bulk of the pesticide residue from the equipment, then a second rinse with fresh water can be used to complete the cleaning.

- *Use of wiper blades and squeegees.* After a mix tank has been drained, some residual formulation remains clinging to the walls. To remove this clingage and to reduce the subsequent level of cleaning solvent/water contamination, mechanical wipers can be employed. For facilities where tanks are wiped with rags, use of wipers would reduce or eliminate waste rags. Use of wipers and squeegees usually requires manual labor; hence, the extent of waste reduction depends on the operator. Since the benefits will be offset by increased labor, mechanization/automation should be considered. Mixers designed with automatic wall scrapers are available. These mixers can be used with any cylindrical mix tank (flat or conical bottom).

- *Use of low-volume high efficiency cleaning systems.* High pressure spray nozzles can be used in place of the standard rinsing hoses. Other types of low-volume high efficiency cleaning systems include water knifes and portable steam cleaners. Steam cleaners are a viable alternative to the "boil-out" procedure whereby a tank is filled with water and then heated to effect cleaning.

- *Use of a plastic or foam "pig" to clean lines.* Many industries use "pigs" (fluid propelled pipe inserts) to clean piping. The "pig" is forced through the pipe from the mixing tank to the filling machine hopper. The "pig" pushes ahead paint left clinging to the walls of the pipe. This, in turn, increases yield and reduces the subsequent degree of pipe cleaning required. Inert gas is used to propel the "pig" and minimize drying of paint inside the pipe. The equipment (launcher and catcher) must be carefully designed so as to prevent spills, sprays, and potential injuries, and the piping runs must be free of obstructions so that the "pig" does not become stuck or lost in the system.

- *Self-draining piping design.* Proper piping design in any liquid processing operation should be self draining and free of pockets.

Spills and Area Washdowns

The cleanup of spills and area washdowns often contributes significantly to the total waste volume produced at formulation plants. Spills are caused by the accidental discharge of pesticides during transfer operations or from equipment failures, such as leaks. Area wash downs with water hoses are performed routinely at some formulation plants, and are necessary in the event of contamination of the working area. Waste reduction methods available for these wastes include:

- *Use of recycled water for initial cleanup.*

- *Use of impervious coatings on floors.*

- *Use of dedicated vacuum system.* For facilities producing dry formulations, spilled powders are usually cleaned up by vacuuming. The vacuum system employed ties in to the facility's main dust collection system where all collected dusts and powders are removed from the air stream by a bag house. Recovered dust cannot be reused due to cross contamination and the material must be disposed of as hazardous waste. If a dedicated vacuum system were available for collecting spilled material, most (if not all) of the recovered material could be returned directly to the process for use.

- *Use of dry cleanup methods for liquid spills.* Rather than cleaning spills with water and producing a hazardous waste, many formulating plants use dry absorbents for spill cleanups. This greatly decreases the waste volume associated with the cleanup. In addition, floor sweeping, mopping and use of squeegees can collect spills for product reformulation.

Off-Specification Products

Off-specification batches of pesticide formulations are produced as a result of poor process control and operation. Ideally, this waste source could be eliminated totally by making use of the following source control techniques:

- *Strict quality control and process automation.* The formulation of pesticides is a relatively simple process. Nevertheless, process automation and control during formulation ensures repeatable high quality products and avoids generation of off-spec batches due to operator error.

- *Reformulation of off-specification batches.* If a batch of off-specification pesticide is produced, it should be reformulated to an acceptable quality rather than discarded as a waste.

Containers

Pesticide ingredient containers can be cleaned for reuse or nonhazardous waste disposal. These include 30 and 55-gallon drums. Many pesticide formulating plants use the uncleaned empty drums to store and dispose of other hazardous wastes, such as pesticide contaminated dusts and empty paper bags and cartons. Generally, the following options exist for minimizing container waste:

- *Return containers to the pesticide supplier for refilling with the same pesticide.* While this option can be very effective and economical, few suppliers will accept used drums from formulators. Instead of drums, formulators should investigate the feasibility of receiving raw materials in returnable bulk containers. In the paint industry, where users of the formulation often return the empty container to the formulator for cleaning and disposal, use of recyclable "Tote bins" is becoming more common. The formulator is able to clean the container, use the cleaning waste in the formulation, and fill the container with a new batch that goes back to the user.

- *Triple rinse and (1) dispose of drums as nonhazardous; (2) sell drums to a scrap dealer or recycling firm (approved by the Department of Transportation) for reconditioning; or (3) recondition drums on-site.* Triple rinsing of the drums is usually performed using a water spray or, more effectively, with steam. The volume of wastewater generated by the triple rinsing typically varies from about 10 to 20 gallons per drum.

Following triple rinsing, the drums are subject to firing to about 1200^0C, sandblasting, and repainting. (Reconditioned drums may be reused for the same chemical class of pesticides previously contained and if reconditioned by a DOT-approved facility. However, there are no regulations covering reconditioning of smaller (1- to 2-gallon) containers, so this is not permitted at present.) By regulation (40 CFR 162.10), reconditioned containers may not be used to contain food, animal feed, beverages, drugs, or cosmetics.

The feasibility of recycling used pesticide drums varies from plant to plant based on-site specific conditions. The following factors will influence drum recycling feasibility:

—cost of transporting and disposing of drums at a Class I landfill;

—cost of triple rinsing, transporting, and disposing of drums at a Class III landfill, including treatment/ disposal cost of rinse water;

—income to be realized by sale of triple-rinsed drums to scrap dealer or drum recycling firm;

—potential savings that may be realized by on-site reconditioning, and reuse or sale of reconditioned drums; and

—use of drums with plastic liners in place of plastic drums or paper bags.

The use of plastic drums presents many difficult disposal problems. It has been reported that triple rinsing is often ineffective at removing traces of pesticide from the container.[141] The plastic may absorb the pesticide and will always be hazardous. Another problem is that plastic drums have a "memory;" they retain their shape when crushed. This creates a large volume low bulk waste stream that is very expensive to dispose of. One way to reduce the volume of waste is to use drums lined with a disposable liner that can be removed when the drum is empty. Disposal of the plastic liner would be much easier than disposing of the drum and it eliminates the need for drum cleaning.

- *Solid waste segregation.* The most effective way of reducing hazardous waste associated with bags and packages (or any other waste stream) is to segregate the hazardous materials from the nonhazardous materials. Economic benefits include reduced disposal costs and the potential sale of nonhazardous scrap paper to a recycler. Empty packages that contain hazardous materials should be placed into plastic bags (so as to reduce personnel exposure and eliminate dusting) and should be stored in a special container to await collection.

Air Emissions

The two major types of air emissions that occur in the pesticide formulating process are volatile organic compounds (VOCs) and pesticide dusts. VOCs may be emitted from the bulk storage of solvents and from their use in open tanks during formulation. Dusts generated during handling, grinding, and other formulation operations are a potential waste source. It is common practice to install dust collection equipment, such as hoods served by a bag house filter, on all dust-generating operations. Some potentially effective waste reduction methods include:

---

[141] Lewis, D.A. 1988. Waste minimization in the pesticide formulation industry. Draft report by FMC Corporation, Fresno, California. August 1988. 20 pp.

- *Control bulk storage air emissions.* Many methods are available for reducing the amount of emissions resulting from fixed roof storage tanks. Some of these methods include use of conservation vents, conversion to floating roof, use of nitrogen blanketing to suppress emissions and reduce material oxidation, use of refrigerated condensers, use of lean-oil or carbon absorbers, or use of vapor equilibration lines. When dealing with volatile materials, employment of one or more of these methods can result in cost savings to the facility by reducing raw material losses and improve compliance with local air quality requirements.

- *Dedicated dust collection systems.* At Daly-Herring Co., in Kinslon, N.C., dust streams from several different production areas were handled by a single bag house. Since all of the streams were mixed, none of the waste could be recycled to the process that generated them. By installing separate dedicated bag houses for each production line, all of the collected pesticide dust could be recycled.[142]

At FMC Corp. in Fresno, CA., common dust collectors were used by multiple production systems. Due to the cross contamination of materials, recycling was impossible. To promote recycling, the company compartmentalized the dust collectors with each compartment serving a single source. All collected materials are analyzed for cross contamination and if none exists, they are reused in the succeeding product batch. Other work involved the installation of self-contained dust collectors at each inlet hopper dump station so that captured dust can be returned to the system.[143]

- *Use of automatic enclosed cut-in hoppers.* The manual opening and emptying of pesticide dust containers leads to the generation of dust which must be collected. One way to reduce the amount of dust generated is by use of an enclosed cut-in hopper which allows the bags to be opened and emptied while avoiding the release of dust.

---

[142] Huisingh, D., L. Martin and H. Hilge. 1985. Proven profit from pollution prevention. Washington, D.C.: The Institute for Local Self-Reliance.

[143] Lewis, D.A. 1988. Waste minimization in the pesticide formulation industry. Draft report by FMC Corporation, Fresno, California. August 1988. 20 pp.

Miscellaneous Waste Streams

Waste streams included in this section include wastewater associated with laundering of protective clothing, water used for aerosol leak testing, storm water run off, and laboratory wastes. Control measures include:

- *Wastewater reuse after treatment.*

- *Paving of high spillage areas.* For facilities located outdoors, paving can be an effective means of reducing rainwater contamination by allowing the recovery of spilled materials.

- *Proper purchasing of chemicals and reagents for lab use.* Purchase quantities of specialty items that are seldom used in the smallest available amount. This helps to reduce waste by ensuring that the material will more likely be consumed before its shelf life expires. Purchasing agents should factor in the cost of disposal before deciding to purchase items in large quantities just because the per unit purchase price is less.

Chapter 7

Pollution Prevention Case Studies: Industrial Facilities

INTRODUCTION

This chapter outlines various waste minimization methods that are being evaluated and implemented at different types of facilities.

In the beginning sections, from Nuclear-Powered Electrical Generating Station to Automotive Air Conditioning Condenser Manufacturer, case studies are provided of waste reduction evaluations undertaken by the New Jersey Department of Environmental Protection (NJDEP) pursuant to a waste minimization assessments program referred to as "Assessment of Recycling and Recovery Opportunities for Hazardous Waste (ARROW)." This project is designed to evaluate the use of waste minimization assessments in thirty hazardous waste generating facilities (across ten industries) in New Jersey. The assessments are being initiated by the New Jersey Institute of Technology (NJIT) personnel and follow the EPA recommended procedure outlined in the Waste Minimization Opportunity Assessment Manual.[144]

In the sections from Railcar Refurbisher through Commercial Ice Machine and Ice Storage Bin Manufacturer, case studies are provided of waste minimization assessments conducted pursuant to the University-Based Assessment Program. This program is a pilot program between EPA and the University City Science Center (UCSC) to assist small and medium-sized manufacturers who wish to minimize formation of hazardous wastes, but lack the in-house expertise to do so. USCS's Industrial Technology and Energy Management division has established three waste minimization assessment centers (WMACs) at Colorado State University in Fort Collins, the University of Louisville (Kentucky), and the University of Tennessee in Knoxville. The WMACs conduct waste minimization assessments for small

---

[144] EPA/625/7-88/003 (1988).

and medium-sized manufacturing companies at no out-of-pocket cost to the company. To be eligible for WMAC assistance, the manufacturer must meet the following criteria:

- Standard Industrial Classification Code 20-39;
- Gross annual sales not exceeding $50 million;
- No more than 500 employees; and
- Lack of in-house expertise on waste minimization.

WMAC staff carry out waste minimization assessments for eligible companies by following the procedures outlined in the EPA Waste Minimization Opportunity Assessment Manual. WMAC staff visits the manufacturing site, locates the sources of hazardous waste at each plant, and identifies current disposal and treatment methods and their associated costs. They then identify and analyze a variety of ways to reduce or eliminate the company's hazardous wastes. Specific measures are recommended to achieve waste minimization goals. Essential technological and economic information is developed. Finally, a confidential report is prepared for the manufacturer, which details WMAC findings and recommendations, including cost savings, implementation costs, and payback times.

NUCLEAR-POWERED ELECTRICAL GENERATING STATION

The facility is an electrical power generating plant. The energy is produced by a nuclear generator. The product of the facility is energy. Hazardous wastes are generated predominantly during the times when power generation is not in operation. (Radioactive wastes are not included in this study.) Moreover, it is apparent from the results of the assessment that the bulk of the hazardous waste from the facility is produced from construction and maintenance activities largely when the energy generation activity is shutdown.

Existing Waste Management Activities

The facility has implemented several effective steps to reduce waste generation at the facility. One successful idea involved making surplus materials available to employees for their personal use. Also, the facility increased its investigation into opportunities for selling surplus materials to commercial users. Ordering and warehouse procedures were improved to reduce overstocking and surplus materials. Innovative material handling procedures were developed, such as purchasing materials in large containers and dispensing them in "just the right amount" containers.

Waste Minimization Opportunities

This facility produces electrical energy by a process which depends upon heating water by a nuclear source. The operation of the facility results in the formation of radioactive waste which is managed according to the appropriate federal regulations. The high costs of waste management for this type of waste has encouraged significant waste reduction efforts in this area throughout the industry. The focus of this assessment—non-radioactive waste—has similarly benefited from waste reduction efforts, although the assessment has identified additional options which could be implemented. Three departmental operations have been found to be associated with the generation of waste: Maintenance, Site Services, and Operations. In addition, a significant source of waste for disposal is off-specification and partially used materials which are not easily associated with any specific operation or job process. Major waste streams identified were:

- Oil and Oil/Water Mixtures
- Coatings (Paints, Epoxy, Enamels)
- Solvents
- Grease
- Laboratory Reagents

Much of the waste oil stream results from a remediation project at the site and not directly from the operation of the facility. The other materials result frequently from regular equipment and facility repair and upgrade activities. Significant quantities of off-specification and partially used containers of materials are presented for waste management or disposal. Options identified for waste reduction included strengthening inventory controls, encouraging "just-in-time" delivery of supplies, directing charge back of waste treatment expenses to the unit or project responsible for the waste, encouraging the use of materials with reduced hazard level, and changing frequency or material used for coating of surfaces.

Additional Options Identified

In addition to the options previously discussed, other options were suggested. It was observed that occasionally containers of hazardous waste are found on the site away from the active secured sections which cannot be identified according to source. It is presumed that these materials are discarded by contractors or other non-employees. It is suggested, therefore, that vehicles entering the facility be examined to ensure that they do not leave such containers at the site.

A clear correlation was observed between the amount of full containers and usable materials presented for waste disposal and the scheduled inspections of the facility. It is postulated that such materials are discarded in order to demonstrate a neater appearance to the inspection team. Alternate storage arrangements for such situations should be developed.

MANUFACTURER OF FINISHED LEATHER

The plant produces finished leathers which are sold to manufacturers of leather goods, such as handbags, belts, shoes, and other items. The operation of the plant varies according to customer demand. Many different colors, textures, and designs must be incorporated into the product to meet varying customer requirements, forcing the operation of several special production steps on an irregular basis. The facility formerly tanned raw hides, but that process has been phased out as a result of changing supply and market conditions.

Existing Waste Management Activities

This facility receives tanned leather from various sources and transforms it into a product of higher commercial value by applying various coatings and other surface modifications to make it more usable and appropriate for finished consumer products. The raw materials include, in addition to the leather itself, various water- and solvent-based coatings, as well as some specialized colorants and other surface modification products. The solvents typically are aromatic and aliphatic hydrocarbons, esters, and alcohols.

A typical hide in the manufacturing process might receive one of several finishing steps. Newly received hides are prepared for finishing by washing, retanning if necessary, and drying. The aqueous wastes from these steps are sent to the POTW with regular monitoring to ensure compliance. Some hides undergo surface modification by mechanical buffing. The resulting dust (<100 lbs/yr) is classified as a hazardous waste and is disposed of off-site. The back coating step applies essentially the final finish to the back of the leather while the base coating of the smooth side serves as the primer for additional finishes to be applied. The coatings are applied using an automated spray system. The facility has shifted largely to water-based coatings for these steps resulting in a significant decrease in solvent use. Any overspray is captured by a water screen or by filters and disposed of off-site. The next coating steps are accomplished using solvent-based materials. No satisfactory non-solvent based coatings have yet been identified for these finishing steps. The applied finishes are thermally dried with venting of solvent vapors to the atmosphere. Approximately 130 tons/yr of evaporated coating solvent are produced from oven drying and as a result of spills and leaks. The final steps in the manufacturing process are ironing, grading, measuring, and shipping operations which are not significant waste-generating activities.

The facility has shifted to the use of water-based coatings, whenever possible; moreover, the technical staff continues to evaluate new commercial reduced-solvent products in order to make further reductions. An optical/computer interfaced system has been used to determine the shape and position of each hide presented for coating which is used to control the automated spray coating system, resulting in significant reduction of overspray.

Waste Minimization Opportunities

Several waste minimization options were enumerated. A 100% reduction in the disposal of buffing dust could be achieved by selling the material as a filler in a resin-based composite product. Solvent losses could be reduced by implementing several options. As satisfactory water-based materials appear on the market, solvent-based coatings can be replaced by water-based coatings. The automated spray coating equipment could be reprogrammed to compensate for required angle spraying. This could reduce waste by 65% when angle spraying is performed. A solvent capture system could be installed to allow for the capture and reuse of solvent. Depending on the type of solvent used, a 90% waste reduction could be achieved. Other improved operating procedures and minor equipment modifications were discussed.

In addition to the options previously discussed, another option was suggested. It was observed that the wooden pallets and cardboard used for shipping hides to the facility might have increased value if recycled.

LOCAL SCHOOL DISTRICT

The facility is a school district with a range of activities with potential for generation of waste, which include vehicle maintenance and repair, building cleaning and maintenance, grounds keeping, instructional programs, and specialized programs, such as science laboratories and art classes. The operations in the district are not centrally located. There is a common administration building. In addition, there is a high school for about 1,000 students, a middle school for about 500 students, and six elementary schools.

The assessment focused on the administration building and the high school. Located at the administration building is a central warehouse for building and maintenance supplies, including cleaners, floor care products, paints, and similar materials. Also at the administration building is the vehicle maintenance and repair facility. Further, there is a wood shop, which has responsibility for building and repairing furniture and related items for use within the district. At the high school, paper-, computer-, and video-based instructional activities occur. In addition, hands-on instruction in areas with potential for waste generation also occurs in science laboratories, art classes, and vocational educational areas.

Existing Waste Management Activities

The waste minimization opportunities assessment carried out at a local school district identified empty paint cans, broken or spilled containers of hazardous materials, solvent wastes from motor parts degreasing, used oil, motor engine antifreeze solution, white paper, cardboard, aluminum cans, glass containers, waste chemicals from teaching laboratories, and vapors from art projects as primary sources of waste.

The district has already instituted several practices which have a positive impact on pollution prevention. As a result of the Community and Worker "Right-to-Know" initiatives, the following procedures were emphasized:

- Ordering only the quantity of materials that can be used in a single year
- Stocking the materials near the point of use
- Conversion to the use of dry copiers, replacing the former solvent-based systems.

In addition, there has been a concerted effort to change to water-based paints and cleaners from solvent-based products whenever possible and to identify and use other products with reduced potential toxicity factors in all areas. Moreover, in keeping with municipal initiatives that encourage recycling, cardboard, white paper, aluminum cans, glass containers, and used motor oil are collected and recycled. In the industrial arts metal shop at the high school, cutting oil is recovered by allowing the metal fragments to settle and then filtering the decanted oil. No new oil has been purchased for this purpose since 1966. Wastes, such as laboratory wastes, are treated as hazardous wastes and collected by a contractor for off-site treatment.

Waste Minimization Opportunities

Several waste minimization options were described. The elimination of hundreds of empty paint cans could be realized by the purchase of paint in returnable containers. A 100% reduction in degreasing solvent wastes could be achieved by enlisting the services of a solvent supply and recycling contractor or acquiring a distillation apparatus. Utilization of antifreeze recycling technology would eliminate 300 gallons of waste antifreeze solution annually. Laboratory wastes could be minimized by using smaller amounts of hazardous chemicals and improved inventory control. Hazardous art project wastes could be minimized by increased substitution of nonhazardous materials for various projects.

Additional Options Identified

Other options were identified which could be considered by the district but may be more pertinent when commercial technology improves. The district uses chlorofluorocarbons in refrigeration equipment and to a limited extent in motor vehicle air conditioning. There is already a commitment to change to substitutes with reduced impact upon the upper atmosphere. In addition, as mobile air conditioning becomes more common in district vehicles, a refrigerant recovery and reuse capability should be considered. In some areas, such equipment may become a legal requirement. Consideration could be given to joint acquisition of recycling equipment with the municipal government, such as antifreeze recycling or degreasing solvent distillation equipment. Ideally, the equipment should be easily movable to allow it to be taken to the facility where the need exits.

STATE DEPARTMENT OF TRANSPORTATION MAINTENANCE FACILITY

The major activity at the state Department of Transportation (DOT) facility is the maintenance of vehicles used by the Department, including automobiles and trucks, and to a more limited extent, large machinery used by the Department, such as mowers. Other activities that are carried out at the facility include wood shop, metal shop, and collection and reuse or disposal of no longer useful materials. A waste reduction opportunity assessment was carried out in order to identify specific operations that generate waste at the facility and to propose a list of options for operational changes which have potential to reduce the waste that is generated and requires treatment or disposal. Because of the diversity of the activities at the facility, each individual operating area was examined for the purpose of identifying waste reduction opportunities.

Oil Waste Reduction Opportunities

From the twelve maintenance facilities in the DOT system, approximately 14,000 gallons of used oil are produced each year. This facility generated 2,700 gal of used oil during fiscal year 1989. The facility practices recycling as the preferred management technique for waste oil. The oil handling procedure at present is to collect used oil in small drums near work stations and to periodically transfer the contents to a larger storage tank. Once the storage tank is full, a contracted recycler removes the material from the site where it is prepared for other beneficial uses.

There are three concerns that generate waste in the oil recycling operation. One is the sporadic appearance of oil/water mixtures in the storage tank. The second concern is the relationship between the generator and the contractor who collects the waste oil. The third area of concern is oil spills. The problems of occasional oil/water contamination and oil spills can be addressed by implementing improved housekeeping and materials handling procedures. The problem of timely removal of the waste oil by the contractor is being addressed by reexamining the purchasing process in terms of bidding and contracting to make the process more responsive to the time needs of the capacity of the oil storage tank. A future option may be to investigate methods for reducing oil usage.

Antifreeze Waste Reduction Opportunities

Approximately 2,000 gallon/yr of commercial antifreeze is used at this maintenance facility. The volume of the waste stream is larger because the coolant in the engine is a water solution, often about a 1:1 mixture. Current practice is to drain the cooling system of the vehicles periodically and replace the antifreeze solution with fresh liquid, discarding the old.

One waste minimization option for antifreeze use is to acquire the use of a commercial system which will prepare the antifreeze solution for reuse by filtration, pH adjustment, and additive addition, if necessary. The recycling of antifreeze would cause a decrease in new antifreeze purchases and a decrease in disposal costs.

Freon/CFC Waste Reduction Opportunities

Freon and other chlorofluorocarbons are present at the facility because of their use in vehicle air conditioning systems. Based upon purchase data, the use of CFCs at the facility is about 140 lbs/yr. There are two substantial pathways for the loss of the material to the atmosphere. The first is loss through leaks which develop in the air conditioning systems in the vehicles. The second pathway results from the industry-wide repair procedure of recharging the system with fresh CFC, locating the leak, discharging the CFC to the atmosphere, and so on.

Two waste minimization options would address the CFC problem. Development of a regularly scheduled preventative maintenance inspection of all vehicle air conditioning systems would create loss prevention through leak prevention, by avoiding major leaks. Second, during the repair stage, use of a commercial CFC capture and reuse device would be advantageous. Such devices are capable of connection to the vehicle system for recovery of the CFC and have the ability to purify the material to quality standards, thus qualifying it for reuse.

Paint Waste Reduction Opportunities

The largest quantity of wastes come from the painting operations themselves. Constructive steps have been taken towards pollution prevention by shifting from solvent-based paints to water-based paints whenever possible. Painting operations are also investigating alternative paint application systems.

278 / Pollution Prevention Strategies and Technologies

Recycling of Used Tires

There is an active program for recycling used tires. A contractor periodically picks up the collected tires (from throughout the DOT system) at this facility and takes them off-site for recycling. A possible waste reduction option may be to increase the useful life of the tire in service. This could be done by investigating a new modified tire rotation procedure and by educating employees on driving and parking techniques to reduce tire wear.

The DOT maintenance facility has also taken steps to minimize its drum/container wastes and serves as a collection point for other forms of scrap material. It is clear that this facility has a commitment to the concept of pollution prevention and is putting it to work in their operations.

LEGAL SUPPLY PRINTING COMPANY

The printing company produces, on a quick turnaround basis, legal forms, business cards, and office supplies for the legal profession. The manufacturing operations of the facility involve two major procedures. Impressions are made using either an engraving process or a printing process. These activities and related procedures, including photo processes and etching, present potential opportunities for waste reduction.

An objective of the company for beginning a waste minimization opportunities assessment is to identify additional areas within the operation that may be candidates for waste reduction initiatives, as well as to identify various technical options to address these opportunities. One objective of this study was to make the most efficient use of limited technical/time resources by developing a concise listing of opportunity areas and technology options.

Waste Reduction Opportunities

Within the manufacturing process there are two major options—engraving or printing—which are used for different purposes and products. The first step, a photographic operation, is common to both.

After the creative design, artistic and layout work are completed by the design group, a photographic negative is produced using a normal photographic process with typical development techniques. Subsequently, a photo transfer step is used to reproduce the image on a metal plate. Copper plates are used in the engraving process and aluminum plates in the printing process.

Currently, the developer and related solutions are managed as hazardous waste. Because of the silver content of the photographic process, it is possible that the liquid waste streams, particularly the spent developer solution, contain enough silver to support a silver recovery operation.

Engraving Process

The primary step within the engraving process where waste reduction opportunities occur is in the etching operation. Fundamentally, the etching step accomplishes the chemical removal of unprotected copper from the copper plate, creating depth differences on the plate that can be used to transfer the image to the paper. The chemical system uses a solution consisting of 55% ferric chloride and 45% hydrochloric acid. The spent acidic iron and copper chloride solution is currently disposed of at an annual cost exceeding $10,000.

Three waste reduction options can be proposed for this operation. The first option is to identify and use an off-site vendor that would regenerate the bath solution by copper removal, and return the renewed solution to the company for reuse. The second option would encourage the acquisition of electrolytic equipment to carry out the bath regeneration on site. The third option is to shift to a new chemical system using a cupric chloride solution as the etchant, rather than the ferric chloride solution now used.

The final step in engraving plate preparation is plate cleaning. Removal of the polymeric photoresist protective coating is accomplished by immersing the plate in a bath of N-methylpyrolidone. Currently, the spent solvent from this cleaning process is handled as a hazardous waste. Two options provide opportunities for waste reduction. One is recovery and reuse of the organic solvent via distillation. The other option is to switch from a chemical cleaning process to a mechanical cleaning technique, such as polishing, brushing, or sandblasting.

The final operation in the engraving process is the impression itself. Ink sludge is generated by the cleaning of equipment at the rate of approximately 110 gallon/yr. Two waste reduction options exist for this waste: dewatering via filtration, centrifugation, or drying; or use of the ink solids as raw material in the manufacturing of the ink.

Printing Process

Two fundamental differences between the engraving process and the printing process are the type of plate used and the composition of the ink used. The two areas within the printing process that present the most promising pollution prevention opportunities are in the impression step and in the equipment cleaning step. The impression step would involve a change from solvent-based inks to a water-based ink system. The equipment cleaning waste reduction option would involve a switch to water-based cleaner.

AUTOMOTIVE AIR CONDITIONING CONDENSER MANUFACTURER

A plant annually producing 400,000 condensers and evaporators for automotive air conditioners was evaluated as part of EPA's University-Based Assessments Program.

Waste Generation and Management Activities

On the condenser production line, aluminum coils, tube stock, header assemblies, and extrusions; steel coils; and miscellaneous hardware, are cut, bent, pierced, welded, brazed, and painted. These parts and assemblies go through many degreasing, rinsing, blowing off, and oven-curing sequences. In the course of these assembly line procedures, the following wastes are generated: cutting oil, 1,1,1-trichloroethane, methyl ethyl ketone (MEK), brazing slurry, and paint solids, liquids, and ash. No costs are associated with the parts of these wastes that evaporate: cutting oil, 1,1,1-trichloroethane, and MEK. The management costs to dispose of the gallons and pounds of waste that are generated amount to $53,000; contaminated paint solids and liquids are the most costly ($42,270).

The evaporator process line is similar to that of the condenser line, except that paint is not applied to the evaporators. The parts do, however, undergo a chromate treatment process. Although, with these exceptions, the waste disposal costs are similar ($11,300 for the evaporator line and $10,730 for the condenser line), the disposal costs for the chromate surface treatment exceed $98,000. On both process lines, much wastewater is carried to a nonchromate wastewater facility where treatment produces 576 cubic yards of wastewater sludge with an annual management cost exceeding $158,500.

The plant employs several other waste management measures. It operates an extensive wastewater treatment plant. The plant also captures some of the slurry run off from the spray process in small run off troughs on the spray brazing booths. Fume scrubbers on the brazing oven stacks capture slurry particulates from the stack gases. Toxic hexavalent chromic acid waste is converted to the trivalent form before removing it from the plant.

Waste Minimization Opportunities

The greatest waste management cost on the condenser line concerns disposal of contaminated paints; the greatest savings proposal also concerns the liquid and solid paint waste. By replacing the dip paint system with an electrostatic epoxy powder paint coating system, not only would there be more even coating of the complex surfaces, the waste could be reduced 100% and the savings, including costs of raw materials, would be $133,820. The payback period for the $130,320 implementation cost would be one year.

An alternative to this waste paint disposal opportunity could be to modify the dip paint system. By increasing the time the parts drip into the paint tank and by tilting them back and forth, 40% of the pounds of paint waste could be reduced at a savings of almost $20,000. The payback period for the $25,440 implementation cost would be 1.3 yr.

By covering the 1,1,1-trichloroethane troughs on the condenser and evaporator process lines, the evaporative loss could be reduced 50% and the annual savings would be $14,000+, including raw materials. The payback period for the $1,880 implementation cost would be 0.1 yr.

Wastewater sludge could be reduced 75% by installing a dry-off oven to reduce the volume of hauled-off sludge, and a 1% reduction could be achieved by modifying the brazing slurry run off to reuse 40% of the slurry. The first change would annually save almost $24,000 (over and above oven operating cost); cost $28,440 to implement; and be paid back in 1.2 yr. The second change could save almost $4,000, cost almost $5,000, and be paid back in 1.3 yr.

Additional recommendations, not completely analyzed but presented to the plant management, include:

- Pumping the 1,1,1-trichloroethane to cleaning troughs rather than transferring it manually in buckets;
- Using an alternative fluxing system with less hazardous material; and

- Analyzing treated water from the nonchromate system to determine if more could be reused.

Railcar Refurbisher

A company rebuilding approximately 2,000 rail cars a year was evaluated as part of EPA's University-Based Assessments Program.

Waste Generation and Management Activities

Refurbishing a rail car requires several steps which generate waste. The rail cars are mechanically shaken to remove dirt and other residue from the cars. Next the cars are cleaned with a high pressure water spray. After the removal of any damaged parts, the paint removal process takes place. The paint removal process employs a steel grit blast system which slings steel grit against the car's metal surface. The paint chips and grit are conveyed to an outdoor cyclone. The cyclone separates the reusable steel grit from the paint dust and spent steel grit. 225 tons of steel grit and 214 tons of paint dust are shipped off-site for disposal as hazardous waste at a cost of $95,560 annually.

After the old paint is stripped away, primer is applied using hand-held spray guns. Once the cars are reassembled, the paint application takes place. The paint is also applied using hand-held spray guns. 56,042 lbs of overspray from paint and primer application processes are disposed of as hazardous waste at a cost of $36,860 per year. Additionally, 6000 gallon/yr of solvents used as paint thinners are lost due to evaporation.

To recycle axles from the rail cars, the axle is washed in caustic solution to remove dirt and grease. 2400 lb/yr of sludge from this process are disposed of as hazardous waste at a cost of $5,350/yr. The plant also rebuilds air brake components. In the rebuilding process, external debris is removed from the components with a plastic bead blast system. As a result, 900 lb/yr of spent beads and paint residue are disposed of as a hazardous waste, costing $5,200/yr.

The plant had already installed a water treatment facility, discontinued use of methylene chloride to wash the axles, and contracted with a vendor to reclaim solvent used to clean the air brakes.

Waste Minimization Opportunities

Three of the recommended minimization opportunities centered on the painting operation:

- By installing an electrostatic spray system, primer and paint overspray losses would be reduced by 15% and solvent losses by 1.4%, at an annual savings of $11,080. The payback period for the $58,320 implementation cost would be 5.3 years.

- By improving the painters' techniques, use of paint and primer would be reduced 5% and solvent 5%, at an annual savings of $4,820. The payback period for the $3,500 implementation cost would be 0.7 year.

- By covering the painting areas with plastic sheets to collect paint and primer residue, the waste would be reduced 5% and $1,540 would be saved annually. There would be an annual operating cost but no capital implementation cost.

Another recommendation concerned the steel grit operation. Presently, 90% of the coatings are removed. Plant personnel indicated that removing 75% would not affect quality. By modifying this operation, 17% of the annual waste could be reduced at a savings of $24,980, including the cost of raw material. The payback period for the $13,500 implementation cost would be 0.5 year.

Additional recommendations (not included because of insufficient data, or difficult implementation, or lengthy payback period) that may become attractive include:

- Custom design a system to separate spent steel grit (nonhazardous) and paint residue;

- Mechanically preclean the axle to remove some concentrated, relatively dry residue that otherwise becomes part of the wash system sludge;

- Install an ultrasonic axle wash system to eliminate the caustic wash; and

- Eliminate air drafts to reduce paint overspray in the railcar painting sheds.

Permanent-Magnet DC Electric Motor Manufacturer

A plant that annually manufactures approximately 13 million permanent-magnet DC electric motors and replacement parts was evaluated as part of EPA's University-Based Assessments Program.

The raw materials used in the various processes include iron and aluminum castings; steel tubing, shafting, and laminations; copper wire and commutators; poles; epoxy coating; varnish; adhesive; and cleaning chemicals. Fan components are purchased. The armature and the stator are made separately and then assembled.

Waste Generation and Management Activities

The armature assembly generates several sources of waste. During cutting and machining of steel shafts, 17,160 lb/yr of waste coolant is generated. This material is disposed of as a hazardous waste at a cost of $5,504/yr. The plant has already installed a coolant recovery system that recovers almost all coolant for reuse in machining tools. Scrap metal is also generated. 1-1/2 truck loads of this material are reclaimed annually.

Later, the assembled armatures are dipped into an epoxy powder dip tank. This process creates 5,175 lb/yr of waste epoxy dust which is collected in bag filters. This dust is disposed of as a hazardous waste at a cost of $3,948 annually.

Near the end of the armature assembly, varnish is applied. Occasionally, varnish becomes too thick for proper application. This results in the generation of 920 lb/yr of hazardous waste that is disposed of at a cost of $524/yr.

The stator assembly begins with the machining of steel tubing. Afterwards, the parts are cleaned in a 5-stage tank washing line to prepare the surface for adhesive fastening of magnetic poles. This cleaning process annually generates 3,520 lbs of zinc phosphate and caustic sludge which is disposed of as a hazardous waste. Wastewater from the tanks is neutralized and pumped to the sewer.

After final assembly of the motors, approximately 2% of the plant's products are painted black for cosmetic purposes. The spray-painting process generates several wastes. 7,920 lb/yr of spray paint booth wastes, in the form of paint overspray, plastic sheets, and air filters, are disposed of as hazardous at a total cost of $6,151/yr. The water from the water curtain is too contaminated to be reused so it is disposed of as a hazardous waste (8,840 lb/yr) at a cost of $3,463/yr. 2,640 lb/yr of spent solvent from cleaning paint spray guns is disposed of as a hazardous waste at a cost of $719/yr.

Waste Minimization Opportunities

The two operations generating the most waste are the painting operations and the five-stage washer assembly line. Alternatives to control liquid and solid paint wastes include:

- Replacing the water curtain spray booth with an electrostatic powder system to reduce solid paint wastes 93% and liquid wastes 100%, at an annual savings of $10,230. The payback period for the $78,440 implementation cost would be 7.7 yr.

- Replacing compressed-air paint spray guns with air-assisted airless paint spray guns to reduce paint overspray and increase paint adhesion. This would reduce both the liquid and the solid paint wastes 50%, with an annual savings of $5,850. The payback period for the $10,000 implementation cost would be 2.6 yr.

- Replacing the water curtain spray paint booth by using a presently inactive electrostatic spray paint booth to reduce raw material costs and the amount of waste: 47% of the paint solids and 50% of the paint liquids at a savings of $9,970. The payback period for the $7,000 implementation cost would be 0.7 yr.

90% of spent epoxy powder could be recycled by installing an air-tight collection system at an annual savings of $14,470. The payback period for the $6,480 implementation cost would be 0.5 yr.

For a 100% reduction in the waste generated at the five-tank washer line in the stator assembly line, the use of the pole adhesive could be discontinued. Instead, the poles could be mechanically attached to the inside of the stators. The annual savings would amount to $31,760, and the payback period for the $110,880 implementation cost would be 3.5 yr.

Several additional recommendations, not completely analyzed, were brought to the manufacturer's attention for future reference. One recommendation was to install drag-out boards on the five-step cleaning operation tanks to drain solutions back into the tanks. Another recommendation suggested collecting the paint spray-gun cleaning solvent for reuse. There is the possibility of using a detergent or a water-based solvent to clean dirty, metallic raw material. In the future, consideration should be given to converting the varnish spray system to a robotic dip system. Finally, as an alternative to discontinuing the adhesive fastening process, an automatic metering system could be installed to reduce the amount of excess adhesive used to attach poles.

METAL BANDS, CLAMPS, AND TOOLS MANUFACTURER

A plant that annually produces 2 million pounds of metal clamps, bands, retainers, and tools was evaluated as part of EPA's University-Based Assessments Program. The raw materials used for these products include stainless steel, carbon steel, forged iron blanks, zinc electroplating solution, brightener, and all the necessary cutting, cleaning, and rinsing chemicals.

Waste Generation Activities

To form the bands, 24-inch stainless-steel coils are cut and the sharp edges removed and beveled. The stainless-steel scrap is sold to a scrap metal dealer. To form the buckles, stainless and carbon steels are punched and crimped onto one end of the bands. For cylindrical bands, the open end is inserted into the buckle. For open-ended clamps, the buckle is only crimped onto one end. After inspection, the bands are packaged and stored for final shipment.

For customers who purchase the bands and buckles separately, the company fabricates specialized tools to apply and install the clamps and fittings. It is the manufacturing of these tools that creates the greatest portion of the plant's waste.

In the tool manufacturing process, many wastes are generated. In the ensuing discussion, all waste and disposal cost data are presented in terms of annual rates. First, iron blanks (which are forged off-site) are machined into tools on-site. Waste cutting fluid and hydraulic oil are generated from this process step at the rate of 660 gallon/yr. This material is incinerated off-site at a cost of $70. After further processing, the tools are placed in electroplating barrels where they enter a metal cleaning line.

Initially, the tools undergo caustic cleaning. 1,120 gallons of spent caustic cleaner (containing sodium hydroxide and sodium metasilicate) is pH-balanced and discharged to the sewer system as industrial wastewater. Next, the tools receive acidic electrocleaning. 560 gallons of alkaline electrosoap solution is also pH-balanced and discharged as industrial wastewater. 170 gallons of sludge from the caustic cleaning tank and the electrosoap tank is drummed and disposed of as hazardous waste at a cost of $1,200. After acidic electrocleaning, the tools proceed through the tap water rinse stage. This stage generates 650,000 gallons of wastewater which is disposed of in the sewer system at a cost of $750. The next stage of the cleaning process is the acidic cleaning. 560 gallons of spent cleaner containing sodium fluoride is pH-adjusted and disposed of in the sewer system. Following this step, 650,000 gallons of wastewater is generated with a disposal cost of $750 during the cascade rinse phase. The final step in the metal cleaning line is the acid stripping step. During this step, 170 gallons of acid solution is disposed of as hazardous waste at a cost of $900.

Once the metal cleaning phase is completed, the tools pass through an electroplating line. This line involves acid zinc plating, rinsing and brightening. From the acid zinc plating line, 280 gallons of solution containing sulfuric acid and ammonium chloride is disposed of as hazardous waste at a cost of $1,200. Wastewaters from the rinsing steps are recycled and reused; however, 230 gallons of sludge is disposed of as hazardous waste, costing $1,000. During the brightening step, 170 gallons of hazardous waste, in the form of solution containing nitric acid, chromium nitrate and ammonium bifluoride, is disposed of at a cost of $700.

Existing Waste Minimization Practices

The plant has already taken the following waste minimizing steps:

- Segregated excess metal for sale to a scrap dealer.

- Treated plating-line rinse water (to remove metal contaminates) for reuse.

- Employed air-agitation for zinc-plating-line rinses.

- Filtered zinc plating solution to remove solid contaminants.

- Discontinued using leaded steel.

- Employed cascade rinses in the metal cleaning and zinc plating lines.

Waste Minimization Opportunities

Presently the first rinse in the metal cleaning line uses tap water as make up. By redirecting the cascade rinse overflow to replace the tap water make up, 650,000 gallons less water will need to be purchased and disposed of in the sewer system. The net annual savings would be $1,090. The payback period for the $470 implementation cost would be 0.4 yr.

By segregating high-grade from lower grade scrap, no waste reduction would occur, but the cash received from the recycler would increase.

To generate less sludge in the caustic cleaner tank, use deionized rather than tap water in the reagent baths on the metal cleaning and electroplating lines. The waste would be reduced by 150 gallon/yr for a net savings of $1,370 after the rental cost of the ion exchange unit is considered. With no implementation cost, the payback is immediate.

To reduce water use in the tap-water and cascade rinses, flow reducers and flow meters could be installed on the metal cleaning line. This reduced use (124,800 gallon/yr) would save $220/yr. The payback period for the $130 implementation cost would be 0.6 yr.

Increasing the drainage time over the caustic cleaner and electrosoap tanks would reduce waste by 250 gallon/yr and save $340/yr with no required implementation cost.

Several other recommendations made to the company include:

- Having a formal cutting-fluid management program to reduce the volume needing disposal.

- Replacing kerosene with a nonhazardous cleaner.

- Installing an automated pH-adjuster for metal cleaning line effluent to prevent compliance problems.

- Installing splash guards on some machines to reduce loss of cutting fluids and lessen unnecessary cleanup.

ALUMINUM EXTRUSIONS MANUFACTURER

A plant that annually manufactures over 36 million lbs of aluminum extrusions was evaluated as part of EPA's University-Based Assessments

Program. These extrusions are made for the use of other product manufacturers.

Waste Generation Activities

Virgin aluminum ingots, scrap aluminum, and alloying metals (e.g., copper, zinc, nickel) are melted in natural-gas-fired furnaces. The molten metal is then cast into logs in a water-quench hydraulic cast system, heat-treated and extruded into desired shapes; sheared; heat-treated again; and then buffed, anodized, or painted. Each year approximately 1.3 million lbs. are buffed and shipped; 14 million lbs. are anodized, colored, and sealed or anodized and sealed, then shipped; and 21 million lbs. are painted and shipped. These three procedures (buffing, anodizing, and painting) create the wastes of interest.

Existing Waste Minimization Practices

The buffing procedure annually generates 26,000 lbs. of buffing compound sludge which is disposed in an on-site landfill along with 7,271 used buffing pads. The anodizing procedure annually is a major contributor to wastewater generation. Wastewater, fume scrubber wastewater, and overflow water are sent to caustic or acid waste lagoons or to a reaction pit for further treatment and disposal. The painting operation also generates wastewater. Along with wastewater, painting generates 50,000 lbs. of contaminated air filters and plastic sheeting which are landfilled off-site as hazardous wastes. The cleaning of paint lines results in the evaporation of 13,130 gallons of toluene annually and 1,430 gallon/yr of spent toluene is disposed of off-site as a hazardous waste. Annually, 240,000 lbs. of trivalent chromic sludge are disposed of in an on-site landfill. 1,800,000 lb/yr of wastewater sludge are also deposited in an on-site landfill. Approximately 47,160,000 gallons of wastewater are treated at this plant on a yearly basis.

Already, the plant is taking steps to minimize its hazardous wastes. The plant is currently transforming toxic hexavalent chrome to nontoxic chrome before off-site shipment. It is also controlling suspended and dissolved species concentrations in effluent water in an on-site wastewater treatment facility and using a water-spray fume scrubber in the anodizing area for air quality control.

Waste Minimization Opportunities

By replacing the solvent-based painting system with an electrostatic powder painting system, not only would the powder coating produce a more even coating, the spent and evaporated toluene, used air filters and plastic sheets, paint ash, and evaporated solvents would be reduced 100%. The net annual savings, which would include the lower cost of powder coatings, would be $1,084,440. The payback period for the $147,580 implementation cost would be 0.1 yr.

By installing an array of rinse spray nozzles in the anodizing line (above the detergent, etch, acid de-smut, anodizing, stannous sulfate, nickel fluoride seal, Sandoz bronze and Sandoz seal tanks) to spray water onto each parts rack as it is raised from the tank; by installing drag-out boards on all tanks in the line; and by increasing the drain time over the tanks, wastewater sludge would be reduced 0.2%, at an annual savings of $52,900. The payback period for the $28,910 implementation cost would be 0.5 yr.

By installing an automatic metering system to minimize the amount of buffing compound used, the buffing compound sludge would be reduced 20%, at an annual savings of $6,590. The payback for the $25,960 implementation cost would be 3.9 yrs.

By installing a distillation unit, 80% of the spent toluene would be recovered—toluene that could then be used to clean the paint lines. The annual savings would be $17,030. The payback period for the $37,060 implementation cost would be 2.2 yrs.

By recovering and recycling caustic and acid solutions in the rinse tanks of the anodizing process line, raw materials lost in the rinsing operations would be recovered and water purchases reduced. Wastewater sludge would be reduced 0.2% and wastewater 85%. The annual savings would be $105,480, and the payback period for the $419,160 implementation cost (for dedicated reverse osmosis solution recovery system, an electrodialysis unit, in-tank air agitation units, and a lowered flow rate of water through the rinse tanks) would be 4 yrs.

COMMERCIAL ICE MACHINE AND ICE STORAGE BIN MANUFACTURER

A plant that annually manufactures 26,000 ice machines and 12,500 ice bins was evaluated as part of EPA's University-Based Assessments Program. The raw materials used in the various processes include galvanized and rolled steel sheets; brass; copper; and metal treating and cleaning chemicals.

Waste Generation Activities

The manufacturing processes, the wastes from the processes and their methods of disposal, and the annual amount and cost to dispose of the wastes, are discussed below.

For cabinet components, galvanized and rolled steel sheets are cut, punched and drilled. These sheets are then cleaned and rinsed. The wastewater (2,569,900 gallon/yr) from this process is pre-treated and then disposed of in the sewer system at a cost of $4,320. The parts are then coated with iron phosphate/phosphoric acid spray to serve as a base for the powder coating. The wastewater (14,400 gallon/yr) from this process is pre-treated and then disposed of in the sewer system at a cost of $30. The parts are rinsed to remove excess phosphating solution. The wastewater (2,781,500 gallon/yr) from this process is pre-treated and then disposed of in the sewer system at a cost of $4,680. Before finally being coated with paint powder and baked, the parts are rinsed with chromate to provide a conversion coating. The rinse water (16,900 gallon/yr) containing hexavalent chromium is recycled and ion exchange treated. The material is then disposed of in the sewer system at a cost of $30.

For the flaker barrel, evaporator, and downchute assembly, after the brass tubes are milled, drilled, threaded, wrapped around the barrel, and then cooled, the brass shell is heated, shrink-fitted and brazed onto the assembly. The assembly is then sent through a bright dip process for corrosion protection. The spent solution (660 gallon/yr) of copper sulfate and copper nitrate crystals is sent to the supplier who reclaims the copper at a cost of $1,620. After rinsing and draining, assemblies undergo chemical sealant and rinse treatments, including baths of soda ash, rinse, and sealant with isopropyl alcohol and potassium hydride. The wastes generated from these steps (12,300 gallon/yr) are pre-treated and disposed of in the sewer system. The finished barrel/evaporator assembly and the fabricated downchutes are insulated with polyurethane foam. The spray nozzle used for the foam insulation is cleaned and the waste (170 gallon/yr) is disposed of as a hazardous waste at a cost of $1,770.

Separately, the evaporators undergo assembly. Copper parts are degreased and dipped in a 5% hydrochloric acid solution. The waste from this step (20,800 gallon/yr) is pre-treated and disposed of in the sewer system for $40. Eventually, the evaporator assemblies are cleaned in an enclosed washer. Again the wastewater from this process (3,120,000 gallon/yr) is pretreated and disposed of in the sewer system at a cost of $5,240.

General plant wastes are generated during cleaning and maintenance operations. Spent petroleum naphtha is reclaimed. Also, hydraulic and vacuum pump oil is reclaimed. Sludge (18,200 gallon/yr) is generated from on-site wastewater treatment operations and is disposed of in a sanitary landfill at a cost of $6,300.

Existing Waste Minimization Practices

Presently, the plant already has:

- Replaced solvent-based paints with powder coatings.

- Installed a high-pressure foaming system to reduce organic cleaning wastes in the bin-foaming area

- Segregated excess metal for recycling by a scrap metal dealer.

- Developed preventive maintenance for metal-forming machinery.

- Recycled waste oils and spent bright dip solutions, solvents, and ion-exchange cartridges.

- Employed a closed-loop rinse (just before powder coating) for ion-exchange treatment.

- Reduced rinse-water flow rates to the lowest possible levels.

- Employed counter flow rinsing so the initial rinse used water from the drainage collection tank.

- Employed a dead-rinse tank to collect drag-out in the bright dip process.

- Used drain boards following bright dip rinses.

- Dewatered wastewater treatment sludge.

- Replaced MEK with methylene chloride to clean foamer nozzles.

- Employed drip bars in degreasing bath and acid bath of small parts washer.

- Defined a formal waste minimization policy.

Waste Minimization Opportunities

When the metal cabinet components are cleaned and rinsed, the two rinses generate well over 5 million gallons of rinse water that are pre-treated and disposed of in the sewer system. By redirecting the rinse water overflow from the second rinse (following the iron phosphate treatment) to the first rinse, the fresh water used in the first rinse would be reduced by 2,171,520 gallon/yr. The savings would be $4,630/yr, and the payback period for the $800 implementation cost would be 0.2 yr.

The flaker barrel evaporator assemblies are cleaned in a bright dip line. By cleaning these assemblies with a plastic media blasting unit, disposal of the spent brightener and subsequent rinses could be eliminated as could the purchase cost of the brightener. (The caustic bath, the sealer, and its rinse would then follow.) The waste could be reduced by 9,960 gallon/yr; the total savings could be $3,950/yr. The payback period for the $5,000 implementation cost would be 1.3 yr.

ADDITIONAL CASE STUDIES

Four additional case studies of small and medium-sized manufacturing companies are presented here. These case studies illustrate the waste generation and management problems of smaller manufacturing companies and provide additional practical guidance on waste minimization strategies.

Printed Circuit Board Manufacturer

A manufacturing plant producing 4.3 million square feet of printed circuit boards per year was evaluated under the University-Based Assessment Program. The facility manufactures screens used for transfer ink patterns to the circuit boards, as well as the circuit boards. Presently, the plant possesses a computer-controlled regeneration system which maintains cupric chloride etch solution. Excess ink is manually scraped from the screens and returned to the reservoir, and filter presses reduce the volume of wastewater sludge.

The study identified several waste minimization opportunities for the printed circuit board manufacturer. First, WMAC staff at the University of Tennessee found that a closed-loop, chilled-water system using recirculated water to cool the UV ovens and etch tanks could reduce wastewater by 60% at an annual savings of $40,000. The payback period for the $79,640 implementation cost would be 1.9 years. Second, a steam generation system to heat spent etch solution and drive off a portion of water would reduce the volume of spent etchant by 35% at an annual savings of $33,150. The payback period for the $28,180 implementation cost would be 0.8 years. Third, WMAC staff determined that substituting reusable polymer membrane filters for the cooper- and zinc-containing paper filters in the mechanical wet scrubbing operation would reduce filter waste by 96% and save the company $3,100 annually. The payback period for the $7,700 implementation cost would be 2.5 years.

Specialty Chemical Manufacturer

WMAC staff at the University of Tennessee also evaluated the waste generation and management practices of a plant that annually produces approximately 300 million pounds of acrylic emulsions, low molecular weight resins, herbicides and other specialty chemicals. The report identifies two waste minimization opportunities for the chemical manufacturer. First, to reduce the amount of off-specification products, upgrading of redundant sensing and control devices on the reactor raw material lines would reduce burnable liquids by 75%, composited absorbed monomers by 19%, off-grade methylolacrylamide/acrylamide by 71%, and unsalable products by 15%. The annual savings would be $139,810, and the payback period for the $365,080 implementation cost would be 2.6 years. Second, installation of a gas-fired dry-off oven in the wastewater treatment system would reduce the volume of sludge hauled off-site. Annual savings of $92,730 could be realized from a $70,230 investment.

Aluminum Can Manufacturer

The waste generation activities of a plant manufacturing 1 billion 12-ounce aluminum cans per year was evaluated by WMAC staff at Colorado State University. The company already recycles its scrap aluminum, minimizes its use of water and chemicals in the washing operation, uses a filter press to reduce water content of hazardous sludge before shipment off-site, and collects waste oil from an extruder coolant system. Following study, one waste minimization opportunity was identified. WMAC staff concluded that by substituting a nonhazardous reagent for the 2-4% ammonium fluozirconate currently in use, the need to dispose of hazardous sludge at a waste disposal facility would be completely eliminated, at a cost savings of $133,060. Because there are no implementation costs, the payback period would be immediate.

Treated Wood Products Manufacturer

This case study evaluates a manufacturing plant that produces treated wood products (crossties, poles, lumber) for regional distribution. The plant annually processes approximately 1.7 million cubic feet of wood on a full-time schedule, amounting to 8,760 hours per year. The major process operations are debarking and trimming, treating lumber with a chromated copper arsenate solution in pressure cylinders, and steam-cleaning the crossties and poles in the pressure cylinders to remove excess creosote. Currently, the plant minimizes the need for periodic cleaning by fitting its large storage tanks with conical bottoms to accumulate sludge. These storage tanks are heated to maintain proper viscosity and reduce sludge formation. Also, because the plant uses Wolmanizing (a closed-loop process that does not require steam cleaning), waste containing chromated copper arsenate has been minimized. The Colorado State University WMAC recommended implementation of a waste exchange program as an additional waste minimization measure. If a waste exchange program could be arranged with a user of scrap wood, the plant's annual output of 9,750 cubic yards of bark and wood chips could be removed, and the property leased for storage purposes would no longer be necessary. This would save the company $1,200 per year and cost nothing to implement.

Chapter 8

Pollution Prevention Case Studies: Federal Facilities

INTRODUCTION

This chapter outlines various waste minimization methods that are being evaluated and implemented at federal facilities. Case studies are provided of waste reduction evaluations undertaken under the federal sites program. The Waste Reduction Evaluations at Federal Sites (WREAFS) Program consists of a series of demonstration and evaluation projects for waste reduction conducted cooperatively by the U.S. Environmental Protection Agency (EPA) and various parts of the Department of Defense, Department of Energy, and other federal agencies. The WREAFS Program focuses on waste minimization research opportunities and technical assessments at Federal sites. The objectives of the WREAFS Program include: (1) conducting waste minimization workshops; (2) performing waste minimization opportunity assessments; (3) demonstrating waste minimization techniques or technologies at Federal facilities; and (4) enhancing waste minimization benefits within the Federal community.

The WREAFS Program facilitates the adoption of pollution prevention/waste minimization practices through technology transfer. New techniques and technologies for reducing waste generation are identified through waste minimization opportunity assessments and may be further evaluated through joint research, development, and demonstration projects. The waste minimization opportunity assessments follow the procedures outlined in the EPA Waste Minimization Opportunity Assessment Manual.[145] The major phases of a WREAFS assessment are:

1. *Planning and Organization:* organization goal setting;

---

[145] EPA/625/7-88/003 (1988).

2. *Assessment*: careful review of a facility's operations and waste streams and the identification and screening of potential options to minimize waste;

3. *Feasibility Analysis:* evaluation of the technical and economic feasibility of the options selected and subsequent ranking of options; and

4. *Implementation:* procurement, installation, implementation, and evaluation (at the discretion of the facility surveyed).

SCOTT AIR FORCE BASE

As part of the WREAFS program, a waste minimization assessment of Scott Air Force Base (AFB) has been conducted. The base is part of the Military Airlift Command (MAC), and operates and maintains a fleet of C-9 medical aircraft. The assessment focuses on the non-destructive wheel inspection process. Non-destructive inspection (NDI) was of special interest to EPA because of the widespread use in the military and commercial airlines. In addition, assessments of the paint stripping/parts cleaning and printed circuit board manufacture were carried out.

Existing Waste Management Activities

As part of the preventative maintenance practices on the C-9 aircraft, landing wheels are inspected for signs of fatigue with a liquid dye penetrant method used at Scott AFB. The primary wastes produced by this method are penetrant, emulsifier, and developer. Waste penetrant is drummed and picked up by a waste handler for incineration in a cement kiln. The Defense Reutilization and Marketing Office (DRMO) currently classifies the penetrant waste as a RCRA D001 waste. The 100-gallon batch of emulsifier is changed out about every six months and sent to the sewage treatment plant. Developer batches of approximately 100 gallon are changed out on about the same frequency as the emulsifier batches. Like the emulsifier, the developer is sent to the sewage treatment plant. Due to the levels of sodium chromate present, the batches meet the criteria for a RCRA D007 waste.

The paint shop handles all aerospace ground equipment (AGE) for Scott AFB. Paint booths are normally used. Approximately 24 one-gallon kits of polyurethane paint are used each year. About 90% of the paint used at the paint shop is polyurethane. The wastes generated by painting are overspray solids, booth compound, booth wastewater, waste paint and thinner, and volatile organic compounds (VOCs). About 220 gallon of sludge and scum are placed in 55-gallon drums and hauled away each year to the appropriate facility. Booths are periodically coated with a protective film called booth compound to prevent adhesion to the metal walls. As the compound deteriorates, a new coating is then applied and the old compound is discarded in a sanitary landfill. The booth water is drained to the sewage treatment plant. Paint thinner is used to clean paint gun nozzles. The mixed thinner and paint, along with unused paint, are placed in 30-gallon drums for disposal by Safety-Kleen, Inc. VOCs are released during atomization of the paint by the spray guns.

Parts to be painted are dry-sanded or dipped into a bath containing a multi-layer stripping solvent. This solvent is used until contaminated with paint sludge; it is then drummed and hauled away as a hazardous (F002) waste. Parts requiring a clean, grease free surface for subsequent processing, such as inspection or repainting, are brought into the Cleaning Shop. The parts are wiped off and then immersed in a bath of solvent degreaser. The part is then removed, scrubbed with a brush and rinsed. The U.S. Air Force uses a contractor to recycle the contaminated solvent. The solvent is primarily mineral spirits and is classified as a RCRA D001 waste.

Waste Minimization Opportunities

The primary contaminant (the penetrant) floats on or near the surface because of its low density. This characteristic makes possible an inexpensive method of periodically skimming the top layer of fluid in these tanks. By skimming off the top layer and adding fresh make up emulsifier or developer, respectively, contaminants floating at or near the surface can be removed, and suspended contaminants can be diluted.

To eliminate the need for the wet chromate solution, new systems use a dry, nonhazardous (silica-based) developer. Changing to the silica-based developer would be technically feasible. The dry developer is technically equivalent and meets the same specifications as the current wet developer.

The plastic media blasting equipment should be used to eliminate the use of organic solvents in paint stripping. Conversion from wet operation to dry painting booth operation would result in volume reduction of wastes associated with painting. The Air Force is implementing the use of high volume, low pressure (HVLP) paint guns. The amount of overspray solids and VOCs generated can be substantially reduced.

Recommendations

The results of the study indicate that the fastest payback (0.24 year) would be from penetrant skimming option. The capital outlay needed for this option is estimated to be only $330. Switching from a wet developer to a dry developer had a payback period of 27.5 years. This option has moderate capital investment but low cost savings. If Scott AFB later determines that the wet developer should be treated as a D007 waste, the disposal costs for wet developer will increase and that option will have higher cost savings.

FITZSIMMONS ARMY MEDICAL CENTER OPTICAL FABRICATION LABORATORY

To promote waste minimization activities in accordance with the national policy objectives established under the 1984 Hazardous and Solid Waste Amendments to the Resource Conservation and Recovery Act (RCRA), the Risk Reduction Engineering Laboratory (RREL) of the USEPA Office of Research and Development is supporting WREAFS. The present project focused on a waste minimization opportunity assessment (WMOA) conducted at the Fitzsimmons Army Medical Center (FAMC) Optical Fabrication Laboratory (OFL) in Denver, Colorado.[146]

Existing Waste Management Activities

One of the sites chosen for performance of a waste minimization opportunity assessment (WMOA) under the WREAFS Program is the Fitzsimmons Army Medical Center Optical Fabrication Laboratory (FAMC/OFL) in Denver, Colorado. Glass lens fabrication operations at the OFL generate three RCRA hazardous wastes (waste lead-bearing lens blocking alloy (RCRA D008), alkaline wash water from ground and polished lens cleaning and deblocking operations (D002), and spent Stoddard solvent from the tool cleaning operations (D001) and one nonhazardous waste (ground glass fines from lens grinding and polishing operations). The waste lead-bearing blocking alloy particulates are reclaimed and recycled at the OFL (to the extent possible); the alkaline wash water is discharged to the wastewater treatment plant and ultimately used on the

---

[146] EPA/600/2-91/031 (1991).

FAMC grounds for irrigation; and spent Stoddard solvent is recycled off-site through a contractor operation. The nonhazardous ground glass fines are collected from the present on-site grinding coolant filtration operations and disposed of at a local sanitary landfill.

Results of the WMOA conducted at the OFL identified three waste minimization opportunities involving materials in use at the OFL. These options are summarized below.

Waste Minimization Opportunities

Alkaline wash water from the glass lens cleaning/deblocking operation is currently discharged from the OFL after passing through a trap to collect large particulates of the lead-bearing lens blocking alloy. This wastewater is discharged periodically from the glass lens washing machines at the rate of approximately 200 gallon/month, at a pH of about 13 to 14, and is drained to the FAMC on-site central water treatment facility. Although this waste is not discharged off-site, it is ultimately discharging lead (both as dissolved lead and submicron particulates) to the groundwater under the site. It is proposed that this possibility be avoided in one of two ways:

1. *Use of a source reduction technique*: the substitution of a non-lead-bearing blocking alloy.

2. *Use of a recycling technique*: introducing a cartridge filter in the line leaving the trap from the lens washing/deblocking operation in order to catch the submicron-size alloy particulates. This technique could recover up to 500 lb/yr of particulate material that would ultimately be recycled to the lens blocking operation.

The OFL presently generates about 37.5 ton/yr of a mixture of waste glass fines and water from the lens grinding operation. This material is not a hazardous waste under the RCRA definition. The OFL currently sends this waste to a local landfill, thereby incurring both the transportation and landfilling costs. These fines, when dry, could generate particulate emissions, thus creating possible inhalation problems, during transportation if they are transported in uncovered or improperly covered containers or at a landfill if they are improperly covered or managed. A potential use for this material is as feed stock in glass or ceramic tile production by a local facility. It is assumed that this facility would use the OFL waste material and, consequently, the land disposal cost could be eliminated.

Recommendations

Of the three waste-related opportunities developed at the OFL by the WMOA, two represent waste reduction for RCRA hazardous wastes, while the remaining option represents an opportunity to reduce or eliminate nonhazardous waste. None of these options represent substantial capital outlays or appreciable operating cost savings. In fact, one waste minimization option—substituting a nonhazardous lens blocking alloy for the present hazardous material—represents a substantial operating cost increase. The only positive value of the option is the potential elimination of an environmental pollution problem if it can be shown at FAMC that a source of lead pollution in groundwater needs to be eliminated.

FORT RILEY, KANSAS

To promote waste minimization activities in accordance with the national policy objectives established under the 1984 Hazardous and Solid Waste Amendments to the Resource Conservation and Recovery Act (RCRA), the present WREAFS project focused on a waste reduction assessment at the U.S. Army Forces Command (FORSCOM) maintenance facilities at Fort Riley, Kansas.

Results of the Fort Riley, Kansas waste minimization assessment identified two waste reduction opportunities in a multi-purpose building (Building 8100) used for automotive subassembly rebuilding, lead acid battery repair, as well as a number of other Army maintenance operations. The two waste reduction opportunities are summarized below.[147]

Existing Waste Management Activities

Battery acid (32-37% sulfuric acid) containing trace concentrations of lead and cadmium is currently drained from both dead batteries and batteries requiring repairs (e.g., replacement of battery terminals), and shipped in 15-gallon drums to the Defense Reutilization and Marketing Office (DRMO) storage facility at the installation for ultimate disposal as a hazardous waste. Instead, it is proposed that the waste acid be gathered in a holding tank, particulates removed, and the waste acid adjusted in concentration to 37% sulfuric acid (using 60 Baume commercial sulfuric acid), as needed, for

---

[147] EPA/600/S2-90/031 (1990).

reuse in reconditioned or new batteries. The buildup of dissolved metal impurities in this recycling system is prevented by purging part of the acid from the system. It is assumed in this assessment that 25% of the acid is purged and 75% is reused. The acid being purged is neutralized and treated for trace heavy metal removal to allow on-site disposal as nonhazardous waste.

Dirty aqueous alkaline detergent solution from automotive parts cleaning, which contains trace concentrations of lead, chromium and cadmium at a pH >12, as well as the oil, grease and dirt removed from the automotive parts, is currently drained to an on-site nonhazardous waste evaporation pond. This waste, heretofore regarded as nonhazardous, is currently being reclassified as a RCRA hazardous waste due to its characteristics (D007, D008) and will have to be disposed of as a hazardous waste through DRMO. The proposed waste minimization option for this waste stream would involve the use of equipment external to the automotive parts washer. The proposed process would include emulsion breaking to cause emulsified oils to float, removal of demulsified oils and other tramp oils and grease by skimming, filtration to remove particulates in an in-line cartridge filter, and addition of fresh alkaline detergent, as necessary, followed by recirculation of the cleaned wash water to the automotive parts cleaner. Buildup of impurities in the recycled wash water is prevented by purging 25% of the used alkaline detergent and recycling 75%. The material being purged is neutralized with an appropriate amount of waste battery acid and precipitated trace heavy metal impurities are removed to allow disposal of the purge stream as a nonhazardous waste.

Some in-plant experimentation will be needed to determine what type of filter elements are best suited to this operation, whether multiple cartridge filters are needed, for how many cycles the recovered wastewater is effective in cleaning automotive parts, etc. The uncertainty in the proposed procedure is reflected in a 25% contingency in the capital cost estimate.

Waste Minimization Opportunities

The battery repair shop generates 7,200 gallon/yr of RCRA hazardous waste (classifications-D002, D006, D008) at a disposal cost of $27,900/yr. Current raw material cost is $11,530. Recycling of the reformulated battery acid would require a capital investment of $15,200 but would save $36,000/yr in operating costs. This would yield a payback of 0.42 years.

Automotive parts washing generates 29,000 gallon/yr of RCRA hazardous waste (classifications-D007, D008). This waste is currently drained to an on-site evaporation pond. If it were disposed of as a RCRA hazardous waste via DRMO at the same cost per gallon as the waste battery acid, the disposal cost would be $112,000/yr. Current raw material cost is less than $100/yr. Recycling of purified alkaline detergent solution would require a capital investment of $19,800. This option would save $107,100/yr in operating costs, leading to a payback period of 0.18 years.

Recommendations

In light of the short payback periods of the two waste reduction options identified, implementation of these options should be considered. Successful application of these options at Fort Riley creates the potential for application of similar waste minimization options in at least ten other U.S. Army FORSCOM installations.

CINCINNATI DEPARTMENT OF VETERANS AFFAIRS HOSPITAL

In this study, EPA's Risk Reduction Engineering Laboratory (RREL) and the Department of Veterans Affairs (DVA-Cin) chose to look for pollution prevention alternatives for minimizing the discarded medical supply waste stream. That VA-Cin is uniquely suited to such a study is directly attributable to its cost sensitivity. The need to deliver services under a fixed budget has led DVA-Cin to both adopt environmentally clean practices on its own, and to continue clean practices that cost-reimbursement hospitals had abandoned.

According to DVA-Cin personnel, approximately 80% of the hospital's supplies are disposed of after a single use. The DVA-Cin saw an additional increase in the use of disposables in the last 2-3 years due to concern by hospitals over both patient safety and staff occupational exposure to the AIDS virus. Therefore, the increase results from greater usage of existing disposable supplies (i.e.,single-use sponges for patient surgery, and disposable gloves and masks worn to protect hospital staff) rather than from the use of newly developed disposable items.[148]

Existing Waste Management Activities

On average, hospitals generate between 0.5 and 4 pounds of infectious

---

[148] EPA/600/S2-91/024 (1991).

waste per patient each day. The DVA-Cin facility produces approximately 0.6 pounds of infectious waste per patient each day, placing it at the low end of the spectrum. However, there are inconsistencies in how hospitals from different States define what is infectious waste. For example, DVA-Cin classifies its laboratory waste as general trash after autoclaving. Inflating the DVA-Cin's quantity of infectious waste to reflect lab wastes would raise the generation rate to 0.87 pounds per patient each day - still quite low in comparison to other hospitals. DVA employs waste segregation to minimize infectious waste volume and also uses cloth gowns instead of disposable gowns.

For this study, a site assessment team was assembled with representatives from DVA-Cin, EPA, and an EPA contractor to track the flow of disposables throughout the hospital and review procedures, uses and consumption with department heads. Over a two-day period, the assessment team visited these DVA-Cin departments: Laboratory Services; Surgery; Surgical Intensive Care Unit (SICU); 5 South (a patient floor); Medical Instensive Care Unit(MICU); Hemodialysis; and Outpatient Clinic.

Research and Development Opportunities

A major concern for RREL in conducting this assessment was to look for those areas in which research and development may support advancing new alternatives. In learning of the concerns, difficulties and successes of the health care profession, RREL hopes to expand EPA's experience in the medical waste area and provide a solid basis for planning future research. Suggestions for further research in the health care industry are presented below:

- *Evaluate Reuse Potential in Single-Use Devices*: Using the rigorous investigation of Hemodialyzers as an example, a cooperative effort could be established between EPA and representatives of the health care community to undertake the research of other potential reusable single-use devices and provide substantive data to either support or reject reuse considerations for these items.

- *Quality Assurance*: Research conducted by the EPA in cooperation with health care professionals, other Federal agencies (such as the Food and Drug Administration), and trade associations can form the basis for developing a protocol for reuse, giving hospitals a standard under which to set down operating procedures and institutional policies.

- *Hidden Cost Factors*: Confusion exists in comparing the relative costs of disposables versus reusables. The EPA may wish to conduct analytical studies in conjunction with health care facilities in order to fully develop and quantify the cost of using disposable and reusable products, respectively, as an aid in decision making.

- *Development of Reprocessing Capacity*: As health care cost containment gains increasing importance, reprocessing may become cost-effective for some items. The potential for promoting some reprocessing capability should be explored, particularly in those areas exhibiting a high density of medical facilities.

- *Developing a Reusable Market*: The EPA and DVA should consider working together in developing procurement guidelines for the DVA which will stimulate the production and distribution of reusable and recyclable products.

Recommendations

DVA-Cin was very pleased with the study and considers the final report a very usable document for other VA medical facilities. For its part, the EPA hopes to learn from future cooperation with DVA, seeking the health care professional's advice and guidance in planning and implementing research programs to respond to the needs of the medical community in the areas of hazardous waste, infectious waste, and other waste streams. Opportunities to reduce these wastes do exist, and additional opportunities will be uncovered through research. Research will also provide the data on which to make operational decisions of benefit to health care facilities, while favoring environmental considerations.

AIR FORCE PLANT NO. 6

The evaluation of emulsion cleaners at Air Force Plant 6 project is part of the Waste Reduction Evaluation at Federal Sites (WREAFS) Program conducted within the Pollution Prevention Research Branch. The purpose of this project is to provide assistance to Air Force Plant 6 personnel by documenting the relevant work by other aircraft fabrication facilities to support comparison of cleaner qualification performance with trichloroethylene for the vapor degreaser operations at Air Force Plant 6.

Plant Background

Air Force Plant No. 6, located in Marietta, Georgia, is operated for the Air Force by Lockheed Aeronautical Systems Company. The facility is part of the Aeronautical Systems Division (ASD), whose headquarters is located at Wright-Patterson Air Force Base near Dayton, Ohio. There are six vapor degreaser units that utilize trichloroethylene (TCE) to prepare steel and aluminum parts for a variety of subsequent manufacturing steps in the production of C-130 aircraft. The eventual goal of the facility is to substitute water-soluble emulsion cleaners to obviate use of 650,000 pounds of TCE.

The final report has been compiled for this project. The report contains information on the evaluation of various substitute cleaners on the conformance of the emulsion cleaners to be implemented at Air Force Plant No. 6 with specific qualification test criteria. The document contains the specifications for qualification tests in 17 areas. It also contains a list of ten cleaners that were targeted for evaluation. The information for this report was developed by documenting research performed by Boeing Aircraft, Air Force Engineering Service Center (AFESC), General Dynamics, Lockheed Missile and Space Company (LMSC), Martin Marietta and Northrop. The report contains a table summarizing the status of cleaner substitute evaluations conducted by the represented companies. The document concludes with a chart that compares the performance criteria of the various companies to the criteria required by Lockheed. Also, data and information for the report was accumulated from emulsion cleaner manufacturers/suppliers and an international workshop on solvent substitution.

Cleaning Solvent Substitutions

EPA is working in cooperation with Lockheed Aeronautical Systems Company Georgia and Air Force Aeronautical Systems Division to investigate the potential for implementing emulsion cleaners as a replacement for trichloroethylene (TCE). The substitution of emulsion cleaners for TCE is currently being implemented at Air Force Plant No. 6 and EPA will be cooperating with Lockheed and Air Force personnel to document the successes, problems and costs associated with the change. The results can then be transferred to similar facilities in the Department of Defense or the Department of Energy, and can serve to expedite the use of emulsion cleaners at other facilities.

Appendix A

Pollution Prevention Act of 1990

UNITED STATES CODE
TITLE 42. THE PUBLIC HEALTH AND WELFARE
CHAPTER 133—POLLUTION PREVENTION

42 U.S.C. 13101. Findings and policy

(a) Findings
The Congress finds that:
(1) The United States of America annually produces millions of tons of pollution and spends tens of billions of dollars per year controlling this pollution.
(2) There are significant opportunities for industry to reduce or prevent pollution at the source through cost-effective changes in production, operation, and raw materials use. Such changes offer industry substantial savings in reduced raw material, pollution control, and liability costs as well as help protect the environment and reduce risks to worker health and safety.
(3) The opportunities for source reduction are often not realized because existing regulations, and the industrial resources they require for compliance, focus upon treatment and disposal, rather than source reduction; existing regulations do not emphasize multi-media management of pollution; and businesses need information and technical assistance to overcome institutional barriers to the adoption of source reduction practices.
(4) Source reduction is fundamentally different and more desirable than waste management and pollution control. The Environmental Protection Agency needs to address the historical lack of attention to source reduction.
(5) As a first step in preventing pollution through source reduction, the Environmental Protection Agency must establish a source reduction program which collects and disseminates information, provides financial assistance to States, and implements the other activities provided for in this chapter.
(b) Policy
The Congress hereby declares it to be the national policy of the United States that pollution should be prevented or reduced at the source whenever feasible; pollution that cannot be prevented should be recycled in an environmentally safe

manner, whenever feasible; pollution that cannot be prevented or recycled should be treated in an environmentally safe manner whenever feasible; and disposal or other release into the environment should be employed only as a last resort and should be conducted in an environmentally safe manner.

42 U.S.C. 13102. Definitions

For purposes of this chapter—
(1) The term "Administrator" means the Administrator of the Environmental Protection Agency.
(2) The term "Agency" means the Environmental Protection Agency.
(3) The term "toxic chemical" means any substance on the list described in section 11023(c) of this title.
(4) The term "release" has the same meaning as provided by section 11049(8) of this title.
(5)(A) The term "source reduction" means any practice which—
(i) reduces the amount of any hazardous substance, pollutant, or contaminant entering any waste stream or otherwise released into the environment (including fugitive emissions) prior to recycling, treatment, or disposal; and
(ii) reduces the hazards to public health and the environment associated with the release of such substances, pollutants, or contaminants. The term includes equipment or technology modifications, process or procedure modifications, reformulation or redesign of products, substitution of raw materials, and improvements in housekeeping, maintenance, training, or inventory control.
(B) The term "source reduction" does not include any practice which alter the physical, chemical, or biological characteristics or the volume of a hazardous substance, pollutant, or contaminant through a process or activity which itself is not integral to and necessary for the production of a product or the providing of a service.
(6) The term "multi-media" means water, air, and land.
(7) The term "SIC codes" refers to the 2-digit code numbers used for classification of economic activity in the Standard Industrial Classification Manual.

42 U.S.C. 13103. EPA Activities

(a) Authorities
The Administrator shall establish in the Agency an office to carry out the functions of the Administrator under this chapter. The office shall be independent of the Agency's single-medium program offices but shall have the authority to review and advise such offices on their activities to promote a multi-media approach to source reduction. The office shall be under the direction of such officer of the Agency as the Administrator shall designate.

(b) Functions

The Administrator shall develop and implement a strategy to promote source reduction. As part of the strategy, the Administrator shall—

(1) establish standard methods of measurement of source reduction;

(2) ensure that the Agency considers the effect of its existing and proposed programs on source reduction efforts and shall review regulations of the Agency prior and subsequent to their proposal to determine their effect on source reduction;

(3) coordinate source reduction activities in each Agency Office and coordinate with appropriate offices to promote source reduction practices in other Federal agencies, and generic research and development on techniques and processes which have broad applicability;

(4) develop improved methods of coordinating, streamlining and assuring public access to data collected under Federal environmental statutes;

(5) facilitate the adoption of source reduction techniques by businesses. This strategy shall include the use of the Source Reduction Clearinghouse and State matching grants provided in this chapter to foster the exchange of information regarding source reduction techniques, the dissemination of such information to businesses, and the provision of technical assistance to businesses. The strategy shall also consider the capabilities of various businesses to make use of source reduction techniques;

(6) identify, where appropriate, measurable goals which reflect the policy of this chapter, the tasks necessary to achieve the goals, dates at which the principal tasks are to be accomplished, required resources, organizational responsibilities, and the means by which progress in meeting the goals will be measured;

(8) establish an advisory panel of technical experts comprised of representatives from industry, the States, and public interest groups, to advise the Administrator on ways to improve collection and dissemination of data;

(9) establish a training program on source reduction opportunities, including workshops and guidance documents, for State and Federal permit issuance, enforcement, and inspection officials working within all agency program offices.

(10) identify and make recommendations to Congress to eliminate barriers to source reduction including the use of incentives and disincentives;

(11) identify opportunities to use Federal procurement to encourage source reduction;

(12) develop, test and disseminate model source reduction auditing procedures designed to highlight source reduction opportunities; and

(13) establish an annual award program to recognize a company or companies which operate outstanding or innovative source reduction programs.

42 U.S.C. 13104. Grants to States for State technical assistance programs

(a) General authority
The Administrator shall make matching grants to States for programs to promote the use of source reduction techniques by businesses.
(b) Criteria
When evaluating the requests for grants under this section, the Administrator shall consider, among other things, whether the proposed State program would accomplish the following:

(1) Make specific technical assistance available to businesses seeking information about source reduction opportunities, including funding for experts to provide on-site technical advice to business seeking assistance and to assist in the development of source reduction plans.

(2) Target assistance to businesses for whom lack of information is an impediment to source reduction.

(3) Provide training in source reduction techniques. Such training may be provided through local engineering schools or any other appropriate means.
(c) Matching funds
Federal funds used in any State program under this section shall provide no more than 50 per centum of the funds made available to a State in each year of that State's participation in the program.
(d) Effectiveness
The Administrator shall establish appropriate means for measuring the effectiveness of the State grants made under this section in promoting the use of source reduction techniques by businesses.
(e) Information
States receiving grants under this section shall make information generated under the grants available to the Administrator.

42 U.S.C. 13105. Source Reduction Clearinghouse

(a) Authority
The Administrator shall establish a Source Reduction Clearinghouse to compile information including a computer database which contains information on management, technical, and operational approaches to source reduction.
 The Administrator shall use the clearinghouse to—

(1) serve as a center for source reduction technology transfer;

(2) mount active outreach and education programs by the States to further the adoption of source reduction technologies; and

(3) collect and compile information reported by States receiving grants under section 13104 of this title on the operation and success of State source reduction programs.

(b) Public availability
The Administrator shall make available to the public such information on source reduction as is gathered pursuant to this chapter and such otherpertinent information and analysis regarding source reduction as may be available to the Administrator. The database shall permit entry and retrieval of information to any person.

42 U.S.C. 13106. Source reduction and recycling data collection

(a) Reporting requirements
Each owner or operator of a facility required to file an annual toxic chemical release form under section 11023 of this title for any toxic chemical shall include with each such annual filing a toxic chemical source reduction and recycling report for the preceding calendar year. The toxic chemical source reduction and recycling report shall cover each toxic chemical required to be reported in the annual toxic chemical release form filed by the owner or operator under section 11023(c) of this title.
This section shall take effect with the annual report filed under section 11023 of this title for the first full calendar year beginning after November 5, 1990.
(b) Items included in report
The toxic chemical source reduction and recycling report required under subsection (a) of this section shall set forth each of the following on a facility-by-facility basis for each toxic chemical:

(1) The quantity of the chemical entering any waste stream (or otherwise released into the environment) prior to recycling, treatment, or disposal during the calendar year for which the report is filed and the percentage change from the previous year. The quantity reported shall not include any amount reported under paragraph (7). When actual measurements of the quantity of a toxic chemical entering the waste streams are not readily available, reasonable estimates should be made based on best engineering judgment.

(2) The amount of the chemical from the facility which is recycled (at the facility or elsewhere) during such calendar year, the percentage change from the previous year, and the process of recycling used.

(3) The source reduction practices used with respect to that chemical during such year at the facility. Such practices shall be reported in accordance with the following categories unless the Administrator finds other categories to be more appropriate:

(A) Equipment, technology, process, or procedure modifications.
(B) Reformulation or redesign of products.
(C) Substitution of raw materials.
(D) Improvement in management, training, inventory control, materials handling, or other general operational phases of industrial facilities.

(4) The amount expected to be reported under paragraph (1) and (2) for the two calendar years immediately following the calendar year for which the report is filed.

Such amount shall be expressed as a percentage change from the amount reported in paragraphs (1) and (2).

(5) A ratio of production in the reporting year to production in the previous year. The ratio should be calculated to most closely reflect all activities involving the toxic chemical. In specific industrial classifications subject to this section, where a feedstock or some variable other than production is the primary influence on waste characteristics or volumes, the report may provide an index based on that primary variable for each toxic chemical. The Administrator is encouraged to develop production indexes to accommodate individual industries for use on a voluntary basis.

(6) The techniques which were used to identify source reduction opportunities. Techniques listed should include, but are not limited to, employee recommendations, external and internal audits, participative team management, and material balance audits. Each type of source reduction listed under paragraph (3) should be associated with the techniques or multiples of techniques used to identify the source reduction technique.

(7) The amount of any toxic chemical released into the environment which resulted from a catastrophic event, remedial action, or other one-time event, and is not associated with production processes during the reporting year.

(8) The amount of the chemical from the facility which is treated (at the facility or elsewhere) during such calendar year and the percentage change from the previous year. For the first year of reporting under this subsection, comparison with the previous year is required only to the extent such information is available.

(c) SARA Provisions

The provisions of sections 11042, 11045(c), and 11046 of this title shall apply to the reporting requirements of this section in the same manner a to the reports required under section 11023 of this title. The Administrator may modify the form required for purposes of reporting information under section 11023 of this title to the extent he deems necessary to include the additional information required under this section.

(d) Additional optional information

Any person filing a report under this section for any year may include with the report additional information regarding source reduction, recycling, and other pollution control techniques in earlier years.

(e) Availability of data Subject to section 11042 of this title, the Administrator shall make data collected under this section publicly available in the same manner as the data collected under section 11023 of this title.

42 U.S.C. 13107. EPA Report

(a) Biennial reports

The Administrator shall provide Congress with a report within eighteen months after November 5, 1990, and biennially thereafter, containing a detailed description

of the actions taken to implement the strategy to promote source reduction developed under section 13103(b) of this title and of the results of such actions. The report shall include an assessment of the effectiveness of the clearinghouse and grant program established under this chapter in promoting the goals of the strategy, and shall evaluate data gaps and data duplication with respect to data collected under Federal environmental statutes.
(b) Subsequent reports
Each biennial report submitted under subsection (a) of this title after the first report shall contain each of the following:

(1) An analysis of the data collected under section 13106 of this title on an industry-by-industry basis for not less than five SIC codes or other
categories as the Administrator deems appropriate. The analysis shall begin with those SIC codes or other categories of facilities which generate the largest quantities of toxic chemical waste. The analysis shall include an evaluation of trends in source reduction by industry, firm size, production, or other useful means. Each such subsequent report shall cover five SIC codes or other categories which were not covered in a prior report until all SIC codes or other categories have been covered.

(2) An analysis of the usefulness and validity of the data collected under section 13106 of this title for measuring trends in source reduction and the adoption of source reduction by business.

(3) Identification of regulatory and nonregulatory barriers to source reduction, and of opportunities for using existing regulatory programs, and incentives and disincentives to promote and assist source reduction.

(4) Identification of industries and pollutants that require priority assistance in multi-media source reduction

(5) Recommendations as to incentives needed to encourage investment and research and development in source reduction.

(6) Identification of opportunities and development of priorities for research and development in source reduction methods and techniques.

(7) An evaluation of the cost and technical feasibility, by industry and processes, of source reduction opportunities and current activities and an identification of any industries for which there are significant barriers to source reduction with an analysis of the basis of this identification.

(8) An evaluation of methods of coordinating, streamlining, and improving public access to data collected under Federal environmental statutes.

(9) An evaluation of data gaps and data duplication with respect to dat collected under Federal environmental statutes. In the report following the first biennial report provided for under this subsection, paragraphs (3) through (9) may be included at the discretion of the Administrator.

42 U.S.C. 13108. Savings provisions

(a) Nothing in this chapter shall be construed to modify or interfere with the implementation of title III of the Superfund Amendments and Reauthorization Act of 1986 [42 U.S.C. 11001 et seq.].

(b) Nothing contained in this chapter shall be construed, interpreted or applied to supplant, displace, preempt or otherwise diminish the responsibilities and liabilities under other State or Federal law, whether statutory or common.

42 U.S.C. 13109. Authorization of appropriations

There is authorized to be appropriated to the administrator $8,000,000 for each of the fiscal years 1991, 1992 and 1993 for functions carried out under this chapter (other than State Grants, and $8,000,000 for each of the fiscal years 1991, 1992 and 1993, for grant programs to states issued pursuant to section 13104 of this title.

Appendix B

U.S. EPA Pollution Prevention Program Contacts

EPA Regional Offices

USEPA Region 1

90 Canal Street
Boston, MA 02203
Mail Code: HER-CAN6
(617) 573-9656

USEPA Region 2

26 Federal Plaza
New York, NY 10278
Mail Code: 2AWM
(212) 264-3384

USEPA Region 3

841 Chestnut Street
Philadelphia, PA 19107
Mail Code: 3HW53
(215) 597-7938

USEPA Region 4

345 Courtland Street, N.E.
Atlanta, GA 30365
Mail Code: 4WD-RCRA
(404) 347-7603

USEPA Region 5

230 South Dearborn Street
Chicago, IL 60604
Mail Code: 5HR13
(312) 886-7598

USEPA Region 6

Interstate Bank Building
1445 Ross Avenue
Dallas, TX 75202-2733
Mail Code: 6HHS
(214) 655-6760

USEPA Region 7

726 Minnesota Avenue
Kansas City, KS 66101
Mail Code: STPG
(913) 551-7523

USEPA Region 8

999 18th Street, Suite 500
Denver, CO 80202
Mail Code: 8HWM-WM
(303) 293-1667

USEPA Region 9

75 Hawthorne Street
San Francisco, CA 94105
Mail Code: H-3-1
(415) 744-2091

USEPA Region 10

1200 6th Avenue
Seattle, WA 98101
Mail Code: HW072
(206) 553-6522

U.S. EPA POLLUTION PREVENTION PROGRAMS

Agriculture in Concert with the Environment (ACE)

The U.S. EPA and the Department of Agriculture (USDA) have joined together to undertake a grant program called Agriculture in Concert with the Environment (ACE). The primary purpose of the grant program is to promote the adoption of sustainable agriculture practices and reduce the use of highly toxic herbicides and other pesticides. Establishing a harmonious relationship between agriculture and the environment offers the opportunity for multiple gains on all sides—for farm owners, farm workers, consumers, and communities as a whole.

ACE grants are distributed from a joint pool by EPA's Office of Pollution Prevention and the USDA Cooperative State Research Service (CSRS). Host institutions in four regions of the country (northeast, south, north central, and west) manage the evaluation, project selection and distribution of funds for their regions.

ACE is jointly administered by USDA and EPA with the Sustainable Agriculture Research and Education Program (SAREP). Evaluation panels in each of the four regions include representatives from government, academic and other research institutions, the farming industry, the environmental community, and other private or public organizations.

For general information on the ACE program, contact:

Harry Wells
Office of Pollution Prevention (7409)
U.S.EPA
401 M. Street, SW
Washington, D.C. 20460
202-260-4472

G.W. Bird
Director, USDA Sustainable Agriculture
Research and Education Program
342 Aerospace Building
14th and Independence Avenues
Washington, D.C. 20250

Patrick Madden, Ph.D.
Associate Director, USDA Sustainable Agriculture
Research and Education Program
P.O. Box 10338
Glendale, CA 91209

U.S. EPA's National Industrial Competitiveness Through Efficiency Program: Energy, Environment and Economics (NICE³)

A joint project of the Department of Energy (DOE) and EPA's Office of Pollution Prevention and Toxics (OPPT), the NICE³ grant program strives to improve energy efficiency, advance industrial competitiveness, and reduce environmental emissions of industry. Large-scale research and demonstration projects are targeted at industries with the highest energy consumption and greatest levels of toxics and chemicals released.

Projects are expected to use the one-time grant funds as seed money to overcome start-up risks. It is expected that industry will finance continuation of projects past the initial grant funding period. As part of the grant-funded phase, awardees will design, test, demonstrate, and assess the feasibility of new processes and/or equipment which can significantly reduce generation of high-risk pollution.

DOE Regional Support OFFICES and EPA Regional OFFICES will work through state energy and environment OFFICES to actively seek out interested state developmental, energy, and industry organizations.

Eligible industries are in SIC codes 26 (paper), 28 (chemicals), 29 (petroleum and coal products), and 33 (primary metal industries). For more information contact:

David Bassett
Office of Pollution Prevention and Toxics
U.S. Environmental Protection Agency
401 M. Street, SW (7409)
Washington, D.C. 20460
202-260-2720

U.S. EPA's Pollution Prevention Incentives for States (PPIS) Program

The centerpiece of EPA's pollution prevention grant activities over the last several years is the ongoing Pollution Prevention Incentives for States (PPIS) program. PPIS is intended to build and support state pollution prevention capabilities and to test, at the state level, innovative pollution prevention approaches and methodologies.

Types of program activities include:

- Institutionalizing multimedia pollution prevention as an environmental management priority, establishing prevention goals, developing strategies to meet those goals, and integrating the pollution prevention ethic into state or regional institutions.

- Other multimedia pollution prevention activities, such as providing direct technical assistance to businesses; collecting and analyzing data; conducting outreach activities; developing measures to determine progress in pollution prevention; and identifying regulatory and nonregulatory barriers and incentives to pollution prevention.

- Initiating demonstration projects that test and support innovative pollution prevention approaches and methodologies.

Eligible applicants are states and federally recognized Indian tribes. Awards are made through EPA Regional OFFICES. Organizations selected for an award must support half of the total cost of the project in order to receive the 50% match required by the Pollution Prevention Act. Additional eligibility criteria include the following:

- Must be pollution prevention as defined by the Act.
- Multimedia opportunities and impacts should be identified.
- Areas for significant risk reduction should be targeted.
- Other pollution prevention efforts in the state should be leveraged and integrated into the project.
- Measures of success are identified.
- A plan for dissemination of project results should be identified.

Along with the National Eligibility Criteria, regional pollution prevention OFFICES may develop their own region-specific guidances. Interested applicants should contact their regional pollution prevention coordinator for more information.

Headquarters contact:

Lena Hann
Office of Pollution Prevention and Toxics
U.S. Environmental Protection Agency
401 M. Street, SW (TS-779)
Washington, D.C. 20460
202-260-2237

U.S. EPA's 33/50 Program

Announced early in 1991, EPA's 33/50 Program is a voluntary pollution prevention initiative seeking to achieve real reductions in pollution in a relatively short period of time.

Under this program, EPA has identified 17 high priority toxic chemicals. EPA's Administrator has set a goal of reducing the total amount of these chemicals released into the environment and transferred off-site by 33% at the end of 1992 and by 50% at the end of 1995.

See the section on EPA's Pollution Prevention Policy Statement for additional information about the 33/50 program. For copies of a brochure on the 33/50 Program or other information, fax your request to the TSCA Assistance Service at 202-554-5603. Or call the TSCA Hotline at 202-554-1404 from 8:30 a.m. to 4:00 p.m. EST.

To assist companies in participating in the 33/50 Program, EPA is developing a series of bibliographic reports for many of the industries that are major releasers of the 17 targeted chemicals. Each report will:

- Summarize the types of processes within the industrial category primarily responsible for release of the chemicals of concern.

- Describe general pollution prevention and recycling alternatives applicable to the industrial processes.

- Provide a bibliography of documents that may provide detailed technical information on pollution prevention and recycling options for the industrial processes.

The reports also will provide education to the general public, EPA staff, and State and local government employees on pollution prevention options that may be available for various industrial processes.

Design for the Environment (DfE)

Established in October 1992, EPA's Design for the Environment Program (DfE) is a voluntary cooperative program which promotes the incorporation of environmental considerations, and especially risk reduction, at the earliest stages of product design.

As part of the Office of Pollution Prevention and Toxics (OPPT), the DfE Program has initiated a number of wide-ranging projects which operate through two levels of involvement. Industry Cooperative projects work with specific industry segments to apply substitute assessment methodology, share regulatory and comparative risk information, and invoke behavioral change. Infrastructure projects are aimed at changing aspects of the general business environment which affect all industries in order to remove barriers to behavior change and provide models which encourage businesses to adopt green design strategies.

EPA's DfE Program is working closely with trade associations and individuals in three specific industry segments. These cooperative projects will develop Substitutes Assessments, which compare risk and environmental trade-offs associated with alternative chemicals, processes, and technologies, and which will provide models for other businesses to follow when including environmental objectives in their designs.

DfE Dry Cleaning Project

Two major dry cleaning associations which represent thousands of American dry cleaners have joined EPA in evaluating an alternative cleaning process and new advances in microwave dryers. This information will be part of the Substitute Assessment for Dry Cleaning.

DfE Printing Project

Over 700 printers and the major printing trade associations are assisting the development of four Substitute Assessments—one for each major printing process. Work on a draft Substitute Assessments for Screen Reclamation and for Blanket Washes in Lithography are underway.

DfE Computer Workstation Project

EPA is participating with several major computer electronics companies in an industry initiative to design a more environmentally friendly computer workstation. The project is using the Cleaner Technology Risk Scoring System and EPA information on regulatory costs to inform the design process.

Chemical Design Project

The DfE Program has awarded 6 grants to universities which fund research into alternate synthesis of important industrial chemical pathways. Results of the research will provide the chemical industry with tools for production which reduce risk and prevent pollution. The grants are providing a model for further National Science Foundation grants. For more information contact:

Pollution Prevention Information Clearinghouse
U.S. Environmental Protection Agency
401 M. Street, SW (PM-211A)
Washington, D.C. 20460
202-260-1023

U.S. EPA's Environmental Education Activities

U.S. EPA's Office of Environmental Education

The Agency has established an Office of Environmental Education as authorized by the National Environmental Education Act of 1990. The office's mandate is to foster an enhanced environmental ethic in society by improving the environmental literacy of our youth and increasing the public's awareness of environmental problems. The Office will provide national leadership in these areas, and will build upon the ongoing work of public, nonprofit, and private sector groups already pursuing these goals. Agency environmental education programs will emphasize four specific themes: wise use of natural resources, prevention of environmental problems, the importance of environmentally sensitive personal behavior, and the need for additional action at the community level to address environmental problems.

The Office's focus will be primarily on the K-12 levels, and the program will be multimedia in its approach.

Office of Environmental Education
U.S. EPA
401 M St.,SW (A-107)
Washington, D.C. 20460
202-260-4484

EPA Pollution Prevention Center for Curriculum Development and Dissemination

In 1991, the University of Michigan was awarded funding for a national pollution prevention education center. This center will develop pollution prevention curriculum modules for undergraduate and graduate engineering, business, and natural resources classes, and for broad distribution to other universities nationwide. The center conducts outreach efforts through short summer courses, offers pollution prevention internships for students at business and industrial facilities, and provides information and education for university faculty through departmental and interdepartmental seminars. The university has committed to support the center for 3 years with substantial supplemental funding . The award of this project, worth more than $300,000, grows out of an EPA "2% Set-Aside" project initiated by EPA's Office of Toxic Substances.

Dr. Gregory A. Keoleian
School of Natural Resources
University of Michigan
Dana Building
430 E. University
Ann Arbor, MI 48109-1115
313-764-1412

National Pollution Prevention Environmental Education Project

The U.S. EPA, working in partnership with State and local governments, industry, educational institutions, textbook publishers, and other entities, is embarking on a project that would ultimately produce pollution prevention education materials for students and teachers. This project will contribute to the establishment of an environmental ethic and work toward improved environmental quality. The materials to be produced will concentrate on kindergarten through grade 12, or a specific segment of this broad group, and will emphasize that preventing pollution at the source is preferable to managing pollutants after they are produced.

A short-term goal is to provide our country's youth with an appreciation and an understanding of the potential benefits of pollution prevention, including conservation and recycling. The ultimate goal of the project is to instill in our future leaders an ethic for more integrated environmental decision making, pollution prevention, and protection of human health and the environment.

This program is directed by an Agency-wide group, the National Pollution Prevention Environmental Education Task Force. This task force comprises membership from all EPA Regions, the Office of the Administrator, and the Office of Pollution Prevention.

The actual form of the materials to be produced will be determined with the assistance of an Advisory Board to the task force. The task force will also encourage the private production of complementing pollution prevention education materials, e.g., videos, films, computer software, teaching aids, textbooks, etc. Special emphasis and consideration will be given to the environmental education needs of urban and rural youths.

The National Pollution Prevention Environmental Education Task Force is co-chaired by:

Douglas Cooper
Office of Federal Activities
U.S. EPA
401 M St., SW
Washington, D.C. 20460
202-260-5052

and

Rowena Michaels
Director of Public Affairs
U.S. EPA Region VII
726 Minnesota Avenue
Kansas City, KS 66101
913-236-2803

Bibliographic Database of Educational Curriculum Material

As the first project of the National Pollution Prevention Environmental Education Project, the Office of Pollution Prevention and Toxics has developed an annotated bibliographic database of educational curriculum material. This purpose of this project is to identify the gaps in pollution prevention education materials and to avoid duplicative efforts. This bibliography contains approximately 2,500 references to pollution prevention, including recycling, resource recovery, source reduction, energy and water conservation, and composting. The database is accessible to the public on the PIES and includes, as available, the following information: date of publication, a brief abstract, cost, and information on how to obtain the publication.

U.S. EPA's Green Lights Program

The U.S. EPA's Green Lights Program was officially launched on January 16, 1991. The program's goal is to prevent pollution by encouraging major U.S. institutions - businesses, governments, and other organizations - to use energy-efficient lighting. Because lighting consumes electricity (about 25% of the national total) and more than half the electricity used for lighting is wasted, the Green Lights program offers a substantial opportunity to prevent pollution, and to do so at a profit. Lighting upgrades reduce electric bills and maintenance costs and increase lighting quality; typically, investments in energy-efficient lighting yield 20 to 30% rates of return per year.

U.S. EPA promotes energy-efficient lighting by asking major institutions to sign a Memorandum of Understanding (MOU) with the Agency; in this MOU, the signatory commits to install energy-efficient lighting in 90% of their space nationwide over a 5-year period, but only where it is profitable and where lighting quality is maintained or improved. The U.S. EPA, in sum, offers program participants a portfolio of technical support services to assist them in upgrading their buildings.

U.S. EPA commits to help Green Lights participants with technical support projects that benefit Green Lights Partners, help strengthen the infrastructure of the energy-efficient lighting industry, and lower the barriers to energy-efficient lighting.

A computerized decision supports system developed by U.S. EPA provides Green Lights corporations and governments a rapid way to survey the lighting systems in their facilities, assess their retrofit options, and select the best energy-efficient lighting upgrades. The decision support system software produces reports suitable for use by facility managers, financial staff, and senior management.

The U.S. EPA has established a national lighting product information program in conjunction with utilities and other organizations. This program provides brand name information so that purchasers will be able to choose products with confidence. In addition, it will allow innovative products to be qualified rapidly, removing a significant barrier for new technologies.

As part of the support program, U.S. EPA helps Partners identify financing resources for energy-efficient lighting. Green Lights Partners receive a computerized directory of financing and incentive programs offered by electric utilities, lighting management companies, banks, and financing companies. The database is updated and distributed on a regular basis.

The U.S. EPA has also developed Green Lights Ally programs for lighting manufacturers, service providers, and utilities to promote the environmental, economic, and quality benefits of energy-efficient lighting. Allies commit to undertake the same level of retrofits as Green Lights Partners, but also will assist in developing the technical support programs.

For more information contact:

Green Lights
U.S. Environmental Protection Agency
401 M. Street, SW (6202 J)
Washington, D.C. 20460
202-775-6650 FAX 202-775-6680

U.S. EPA Headquarters/Laboratory Pollution Prevention Contacts

The U.S. EPA OFFICES identified below are involved in various pollution prevention issues. A brief description of each office's pollution prevention program is provided. For further information regarding particular programs or functions, contact the appropriate office.

Pollution Prevention Division
Office of Pollution Prevention and Toxics
U.S. EPA
401 M Street, S.W. (7409)
Washington, D.C. 20460
202-260 3557

This office was established in 1988 to help integrate a multimedia pollution prevention ethic both inside and outside EPA. Its primary role is to support pollution prevention efforts by EPA's program OFFICES, EPA Regions, state and local governments, industry, and the public.

33/50 Program Management Staff
Office of Pollution Prevention and Toxics
U.S. EPA
401 M Street, S.W. (TS-799)
Washington, D.C. 20460
202-260 6907

This program is EPA's voluntary pollution prevention initiative to reduce national pollution releases and off-site transfers of 17 toxic chemicals by 33% by the end of 1992 and by 50% by the end of 1995. Since the program's beginning in February 1991, more than 1,000 companies have participated in the program. EPA is using the Toxic Release Inventory (TRI) data for 1988 as a baseline.

Office of Environmental Engineering and Technology Demonstration
U.S. EPA
401 M Street, S.W. (PM-681)
Washington, D.C. 20460
202-260-5747

This office has primary responsibility for EPA's pollution prevention research. In 1987, ORD initiated the Waste Minimization Research program that primarily addressed ways to minimize hazardous wastes. In 1989, ORD began to build a multi-media research program. The plan for expanding this program, described in the *Pollution Prevention Research Plan: Report to Congress*, focused on six functional areas of pollution prevention research: 1) product research; 2) process research; 3) recycling/reuse research; 4) social science research; 5) anticipatory research; and 6) technology transfer.

Over a four year period (FY 88-92), pollution prevention research grew in all six functional areas and has been expanded to air, water, pesticides, and toxic substances. Applied research to reduce hazardous/industrial wastes has continued to be a priority, and has focused primarily on "process" research related to different industry sectors.

Risk Reduction Engineering Laboratory
Office of Research and Development
U.S. EPA
26 West Martin Luther King Drive
Cincinnati, OH 45268
513-569-7418

EPA's Pollution Prevention Research Branch encourages the development and adoption of processing technologies and products in the U.S. that will reduce the generation of pollutants. The lab is involved in studies, research, and demonstration projects, including the Innovative Clean Technologies Project, the Clean Products Program, the Assessments Program, Waste Reduction Evaluation at Federal Sites (WREAFS), and support for the 33/50 Program.

Waste Minimization Branch
Office of Solid Waste
U.S. EPA 401 M Street, S.W. (OS-320W)
Washington, D.C. 20460
703-308-8402

This Branch has recently completed a multi year action plan for waste minimization; the purpose of this plan is to incorporate waste minimization throughout the RCRA program. The RCRA program will then be integrated into EPA's pollution prevention program under the Office of Toxic Substances.

Air and Energy Engineering Research Laboratory
Office of Research and Development
Research Triangle Park, NC 27711
919-541-2821

The mission of the Air and Energy Engineering Research Laboratory (AEERL) is to research, develop and demonstrate methods and technologies for controlling air pollution from stationary sources.

AEERL creates and improves air pollution control equipment, seeks means of preventing or reducing pollution through changes in industrial processes, develops predictive models and emissions estimation methodologies, identifies and assesses the importance of air pollution sources, and conducts fundamental research to define the mechanisms by which processes, equipment, and fuel combustion produce air pollution. AEERL has eight main research areas: acid rain; air toxics; hazardous waste; indoor air/radon; municipal waste combustion;

ozone non-attainment, stratospheric ozone; and global climate change. Under stratospheric ozone, AEERL is evaluating alternatives to ozone-depleting substances in the refrigeration and fire extinguishing industries.

U.S. EPA Office of Air/Small Business Assistance Program

The Clean Air Act Amendments of 1990 placed new Federal controls on small sources of air pollution that ultimately may affect hundreds of thousands of small American businesses. Section 507 of the Act is especially important to small businesses. This section requires all state governments and the Environmental Protection Agency to establish a Small Business Technical and Environmental Compliance Assistance Program to help small businesses contend with several new air pollution responsibilities.

While each state has been granted the flexibility to tailor its program to meet the needs of industries within their state, there are three mandatory components:

1. A state ombudsman;
2. Small business assistance program; and
3. A state compliance advisory panel.

As programs may vary state to state, individuals should contact their local program for specific details. (See Appendix C for a listing of state air pollution control agencies).

The EPA must establish a Federal Small Business Assistance Program which will provide state programs with several forms of guidance and assistance. EPA will also operate several technical service centers and telephone hotlines (listed below) providing support to state and local air pollution control agencies as they develop and carry out the small business assistance programs. Small businesses may also contact any of these centers for specific information and technical assistance.

EPA Technical Support Centers and Hotlines for Air/Small Business Information

The Office of Asbestos and Small Business Ombudsman Hotline
800-368-5888

-Small business pollution prevention grants
-General assistance to small business

Appendix B: U.S. EPA Pollution Prevention Program Contacts / 329

EPA Control Technology Center
919-541-0800

-General assistance and information on the Clean Air Act and its requirements
-Federal air pollution standards
-Air pollution control technologies

Emission Measurement Technical Information Center
919-541-1060

-Air emissions testing methods
-Emission monitoring guidance
-Federal testing and monitoring requirements

Emergency Planning and Community Right-to-Know Information Hotline
800-535-0202

-Accidental chemical release prevention
-Hazardous chemical emergency planning
-Toxic Release Inventory assistance

Indoor Air Quality Information Clearinghouse
800-438-4318

-Information on indoor air pollutants, sources, health affects, testing, measuring and control
-Constructing buildings to minimize indoor air pollution

Stratospheric Ozone Information Hotline
800-296-1996

-General information on stratospheric ozone depletion and its protection
-Consultation on ozone protection regulations and requirements under the 1990 Clean Air Act amendments

Technology Transfer Network (Clean Air Act Bulletin Board)
919-541-5742

-Recent EPA rules, EPA guidance documents, and updates of EPA activities

U.S. EPA Libraries

EPA libraries are excellent sources of pollution prevention information. In particular, the EPA Headquarters and Region IX Libraries have extensive collections dedicated to this topic.

EPA Headquarters Library

Lois Ramponi, Librarian
Pollution Prevention Information Clearinghouse
Craig Lelansky, Pollution Prevention Librarian
Anne Freckmann, Information Specialist
Headquarters Library
U.S. EPA 401 M Street, S.W. (PM 211A)
Washington, D.C. 20460
202-260-5921
202-260-1963 (PPIC)

EPA Laboratory Library

Stephena Harmony, Librarian
Andrew W. Briedenbach Environmental Research Center Library Risk Reduction Environmental Laboratory
U.S. EPA 26 West Martin Luther King Drive
Cincinnati, OH 45268
513-569-7707

EPA Regional Libraries

Peg Nelson, Librarian
U.S. EPA Region I - Library
John F. Kennedy Federal Building
1 Congress Street
Boston, MA 02203
617-565-3300

Eveline Goodman, Librarian
U.S. EPA Region II
26 Federal Plaza
New York, NY 10278
212-264-2881

Diane M. McCreary, Librarian
U.S. EPA Region III
841 Chestnut Building (3PM52)
Philadelphia, PA 19107
215-597-7904

Priscilla Pride, Librarian
U.S. EPA Region IV
345 Courtland Street, NE
Atlanta, GA 30365-2401
404-347-4216

Penny Boyle, Librarian
U.S. EPA Region V
12th Floor, 77 West Jackson Blvd.
Chicago, IL 60604
312-353-2022

Phyl Barris, Librarian
U.S. EPA Region VI
1445 Ross Avenue, Suite 1200
Dallas, TX 75202-2733
214-655-6444

Barbara MacKinnon, Librarian
U.S. EPA Region VII
726 Minnesota Avenue
Kansas City, KS 66101
913-551-7358

Doug Rippey, Librarian
U.S. EPA Region VIII
999 18th Street, Suite 500
Denver, CO 80202-2405
303-293-1444

Karen Sundheim, Pollution Prevention Librarian
Pollution Prevention Resource Center
U.S. EPA Region IX
75 Hawthorne Street, 13th Floor
San Francisco, CA 94105
415-744-1508

Julienne Sears, Librarian
U.S. EPA Region X
1200 Sixth Avenue
Seattle, WA 98101
206-553-1289/1259

Appendix C

State Pollution Prevention Program Contacts

Alabama

Alabama Waste Reduction and Technology Transfer (WRATT) Program
Department of Environmental Management
1715 Congressman William Dickinson Drive
Montgomery, AL 36130
(205) 260-2779

Alaska

Pollution Prevention Office
Alaska Department of Environmental Conservation
P.O. Box O
Juneau, AK 99811-1800
(907) 465-5275

Waste Reduction Assistance Program (WRAP)
Alaska Health Project
1818 West Northern Lights Boulevard—Suite 103
Anchorage, AK 99517
(907) 276-2864

American Samoa

American Samoa Environmental Protection Agency
Office of the Governor
American Samoa Government
Pago Pago, AS 96799
(684) 633-2304

Arizona

Pollution Prevention Unit
Arizona Department of Environmental Quality
3033 North Central Avenue
Phoenix, AZ 85012
(602) 207-4210

Recycling Coordinator
Conservation Programs Energy Office
1700 West Washington Street
Phoenix, AZ 85007
(502) 255-3303

Arkansas

Hazardous and Solid Waste Division
Arkansas Department of Pollution Control and Energy
P.O. Box 8913
Little Rock, AR 72219-8913
(501) 570-2861

Biomass Resource Recovery Program
Arkansas Energy Office
One State Capitol Mall
Little Rock, AR 72201
(501) 682-7322

California

California Department of Toxic Substances Control
Pollution Prevention, Public and Regulatory Assistance Division
400 P Street
P.O. Box 806
Sacramento, CA 95812-0806
(916) 322-3670

Local Government Commission
909 12th Street—Suite 205
Sacramento, CA 95814
(916) 448-1198

Integrated Waste Management Board
8800 Cal Center Drive
Sacramento, CA 95826
(916) 255-2182

Colorado

Pollution Prevention Waste Reduction Program
Colorado Department of Health
4300 Cherry Creek Drive South
Denver, CO 80202
(303) 692-3003

Connecticut

Bureau of Waste Management
Connecticut Department of Environmental Protection
165 Capitol Avenue
Hartford, CT 06106
(203) 566-5217

Connecticut Technical Assistance Program (ConnTAP)
Connecticut Hazardous Waste Management Service
900 Asylum Avenue—Suite 360
Hartford, CT 06105-1904
(203) 241-0777

Delaware

Delaware Department of Natural Resources and Environmental Control
Solid Waste Division
89 Kings Highway P.O. Box 1401
Dover, DE 19903
(302) 739-3820

Solid Waste Authority
(302) 739-5361

Pollution Prevention/Waste Minimization
(302) 739-5071

District of Columbia

Office of Recycling
D.C. Department of Public Works
65 K Street, NE
Washington, DC 20002
(202) 727-5856/5887

Department of Environmental Programs
Council of Governments
777 North Capitol St., NE
Suite 300
Washington, DC 20002-4201
(202) 962-3355

Environmental Protection Division
D.C. Department of Public Works
2000 14th Street, NW
Washington, DC 20009
(202) 939-8115

Environmental Regulation Administration
D.C. Department of Consumer and Regulatory Affairs
2100 MLK Avenue, SE
Suite 203
Washington, DC 20020
(202) 404-1136

Florida

Waste Reduction Assistance Program
Florida Department of Environmental Regulation
Twin Towers Office Building
2600 Blair Stone Road
Tallahassee, FL 32399
(904) 488-0300

Georgia

Environmental Protection Division
Georgia Department of Natural Resources
4244 International Parkway—Suite 104
Atlanta, GA 30334
(404) 362-2537

Guam

Guam Environmental Protection Agency
D-107 Harmon Plaza
130 Rojas Street
Harmon, GU 96911
(671) 646-8863

Hawaii

Waste Minimization Coordinator
Hawaii Department of Health
5 Waterfront Plaza
500 Ala Moana Blvd.
Honolulu, HI 96813
(808) 586-4226

Idaho

Division of Environmental Quality
Idaho Department of Health and Welfare
1410 North Hilton
Boise, ID 83720-9000
(208) 334-5879

Illinois

Office of Pollution Prevention
Illinois Environmental Protection Agency
2200 Churchill Road
P.O. Box 19276
Springfield, IL 62794-9276
(217) 785-0533

Solid Waste Management Section
(217) 785-8604

Indiana

Office of Pollution Prevention and Technical Assistance
Indiana Department of Environmental Management
105 South Meridian Street
P.O. Box 6015
Indianapolis, IN 46225
(317) 232-8172

Agricultural Pollution Prevention Coordinator
Environmental Management and Education Program
2129 Civil Engineering Building
Purdue University
West Lafayette, IN 47907-1284
(317) 494-5038

Iowa

Waste Management Authority Division
Iowa Department of Natural Resources
Wallace State Office Building
Des Moines, IA 50319
(515) 281-8941

Iowa Waste Reduction Center (IWRC)
University of Northern Iowa
Cedar Falls, IA 50614-0185
(319) 273-2079

Kansas

State Technical Action Plan (STAP)
Kansas Department of Health and Environment
Forbes Field, Building 740
Topeka, KS 66620
(913) 296-1630

Kentucky

Division of Waste Management
Kansas Department of Environmental Protection
Fort Boone Plaza
18 Reily Road
Frankfort, KY 40601
(502) 564-6716

Kentucky PARTNERS—State Waste Reduction Center
Ernst Hall, Room 312
University of Louisville
Louisville, KY 40292
(502) 588-7260

Louisiana

Recycling Division
Louisiana Department of Environmental Quality
P.O. Box 82178
Baton Rouge, LA 70884
(504) 765-0249

Waste Minimization Coordinator
(504) 765-0720

Maine

Bureau of Hazardous Materials and Solid Waste Control
Maine Department of Environmental Protection
State House Station #17
Augusta, ME 04333
(207) 287-2811

Maryland

Waste Management Administration
Maryland Department of the Environment
2500 Broening Highway, Building 40
Baltimore, MD 21224
(410) 631-3344

Maryland Environmental Services
2020 Industrial Drive
Annapolis, MD 21401
(301) 974-7281

Massachusetts

Division of Solid Waste
Massachusetts Department of Environmental Protection
One Winter Street, 4th Floor
Boston, MA 02108
(617) 292-5870

Office of Technical Assistance
Massachusetts Department of Environmental Protection
100 Cambridge Street
Boston, MA 02202
(617) 727-3260

Michigan

Office of Waste Reduction Services
Environmental Services Division
Michigan Department of Natural Resources
116 West Allegan Street
P.O. Box 30004
Lansing, MI 48909-7504
(517) 335-1178

Staff Specialist, Waste Reduction
(517) 335-1310

Solid Waste Coordinator, Recycling
(517) 373-4735

Minnesota

Minnesota Office of Waste Management
1350 Energy Lane—Suite 201
St. Paul, MN 55108-5272
(612) 649-5750/5744

Environmental Assessment Office
Minnesota Pollution Control Agency
520 Lafayette Road
St. Paul, MN 55155
(612) 296-8643

Minnesota Technical Assistance Program (MNTAP)
University of Minnesota
1313 5th Street, SE, Suite 207
Minneapolis, MN 55414
(612) 627-4555/4646

Mississippi

Waste Reduction/Waste Minimization Program
Mississippi Department of Environmental Quality
P.O. Box 10385
Jackson, MS 39289-0385
(601) 961-5171

Mississippi Technical Assistance Program and
Mississippi Solid Waste Reduction Assistance
P.O. Drawer CN
Mississippi State, MS 39762
(601) 325-8454

Missouri

Solid and Hazardous Waste Management Program
Missouri Department of Natural Resources
205 Jefferson Street
P.O. Box 176
Jefferson City, MO 65102
(314) 751-3176

Environmental Improvement and Energy Resources Authority (EIERA)
225 Madison Street
P.O. Box 744
Jefferson City, MO 65102
(314) 751-4919

Montana

Solid Waste Program
Montana Department of Health and Environmental Sciences
Cogswell Building
Helena, MT 59620 (406) 444-1430

Nebraska

Hazardous Waste Section
Nebraska Department of Environmental Control
301 Centennial Mall South
P.O. Box 98922
Lincoln, NE 68509
(402) 471-4217

Litter Reduction and Recycling Programs
(402) 471-2186

Nevada

Bureau of Waste Management
Nevada Division of Environmental Protection
Capitol Complex
123 West Nye Lane
Carson City, NV 89710
(702) 687-5872

New Hampshire

Waste Management Division
New Hampshire Department of Environmental Services
6 Hazen Drive
P.O. Box 95
Concord, NH 03301
(603) 271-2902

New Hampshire Waste Cap
New Hampshire Business and Industry Association
122 North Main Street
Concord, NH 03301
(603) 224-5388

New Jersey

Office of Pollution Prevention
New Jersey Department of Environmental Protection
CN-402
Trenton, NJ 08625
(609) 777-0518

New Jersey Technical Assistance Program (NJTAP)
New Jersey Institute of Technology
Hazardous Substance Research Center
323 Martin Luther King Boulevard
Newark, NJ 07102
(201) 596-5864

New Mexico

New Mexico Environment Department
Solid Waste Bureau
P.O. Box 266110
Santa Fe, NM 87502
(505) 827-2883

Municipal Water Pollution Prevention Program
(505) 827-2804

New York

Bureau of Pollution Prevention
Division of Hazardous Substance Regulation and
the Division of Solid Waste
New York Department of Environmental Conservation
50 Wolf Road
Albany, NY 12233-7253
(518) 457-7276

Erie County Office of Pollution Prevention
Erie County Office Building
95 Franklin Street
Buffalo, NY 14202
(716) 858-6231

New York State Environmental Facilities Corporation
(518) 457-4138

North Carolina

Office of Waste Reduction
North Carolina Department of Environment, Health and Natural Resources
P.O. Box 27687
Raleigh, NC 27611-7687
(919) 571-4100

North Dakota

Solid Waste Program
North Dakota Department of Health
1200 Missouri Avenue
P.O. Box 5520
Bismarck, ND 58502
(701) 221-5150

Northern Mariana Islands

Division of Environmental Quality
Dr. Torres Hospital
P.O. Box 1304
Saipan, MP 96950
(670) 322-9371

Ohio

Pollution Prevention Section
Division of Solid and Hazardous Waste Management
Ohio Environmental Protection Agency
1800 Watermark Drive
P.O. Box 1049
Columbus, OH 43266
(614) 644-3969

Division of Litter Prevention and Recycling
Fountain Square Court, Building F2
Columbus, OH 43224-1387
(614) 265-6333

Ohio Technology Transfer Organization (OTTO)
Ohio Department of Development
77 South High Street, 26th Floor
Columbus, OH 43255-0330
(614) 644-4286

Oklahoma

Environmental Health Administration—0200
Oklahoma State Department of Health
1000 N.E. 10th Street
Oklahoma City, OK 73117-1299
(405) 271-7353

Pollution Prevention Technical Assistance
Hazardous Waste Management Service, 0205
Oklahoma State Department of Health
1000 N.E. 10th Street
Oklahoma City, OK 73117-1299
(405) 271-7047

Oregon

Waste Reduction Assistance Program (WRAP)
Hazardous and Solid Waste Division
Oregon Department of Environmental Quality
Executive Building
811 S.W. Sixth Avenue
Portland, OR 97204
(503) 229-6585

Pennsylvania

Office of Air and Waste Management
Pennsylvania Department of Environmental Resources
P.O. Box 2063
Harrisburg, PA 17105-2063
(717) 772-2724

Pennsylvania Technical Assistance Program (PENNTAP)
Penn State University
110 Barbara Building II
810 North University Drive
University Park, PA 16802
(814) 865-0427

Rhode Island

Waste Reduction Section
Office of Environmental Coordination
Rhode Island Department of Environmental Management
83 Park Street
Providence, RI 02903
(401) 277-3434

South Carolina

Center for Waste Minimization
Bureau of Solid and Hazardous Waste Management
Department of Health and Environmental Control
2600 Bull Street
Columbia, SC 29201
(803) 734-4715

South Dakota

Office of Waste Management
South Dakota Department of Environment and Natural Resources
523 East Capitol Avenue
Pierre, SD 57501-3181
(605) 773-4216

Tennessee

Bureau of Environment
Tennessee Department of Health and Environment
L & C Building, 14th Floor
401 Church Street
Nashville, TN 37243-0455
(615) 741-3657

Texas

Office of Pollution Prevention and Conservation
Texas Water Commission
P.O. Box 13087, Capitol Station
Austin, TX 78711-3087
(512) 463-7869

Bureau of Solid Waste Management
Texas Department of Health
1100 W. 49th Street
Austin, TX 78756
(512) 458-7271

Utah

Division of Solid and Hazardous Waste
Utah Department of Environmental Quality
168 North 1950 West Street
Salt Lake City, UT 84114-4810
(801) 536-4480

Vermont

Pollution Prevention Division
Vermont Department of Environmental Conservation
103 South Main Street
Waterbury, VT 05676
(802) 244-8702

Virginia

Waste Minimization Program
Virginia Department of Waste Management
Monroe Building, 11th Floor
101 N. 14th Street
Richmond, VA 23219
(804) 371-8716

Division of Litter Control and Recycling
(804) 225-2667

Washington

Solid and Hazardous Waste Program
Washington Department of Ecology
P.O. Box 47600
Olympia, WA 98504
(206) 459-6316

Solid Waste Reduction and Recycling
(206) 459-6302

Litter and Recycling Information
(206) 459-6301

West Virginia

Pollution Prevention and Open Dump Program (PPOD)
West Virginia Department of Natural Resources
1356 Hansford Street
Charleston, WV 25301
(304) 558-4000

Generator Assistance Program
Waste Management Section
West Virginia Department of Natural Resources
1356 Hansford Street
Charleston, WV 25301
(304) 558-6350

Wisconsin

Bureau of Solid and Hazardous Waste Management
Wisconsin Department of Natural Resources
P.O. Box 7921 (SW/3)
Madison, WI 53707-7921
(608) 267-3763

Solid Waste Recycling/Waste Reduction Program
(608) 267-7566

Hazardous Pollution Prevention Audit Grant Program
(608) 266-3075

Wyoming

Solid Waste Management Program
Wyoming Department of Environmental Quality
Herschler Building, 4th Floor
122 W. 25th Street
Cheyenne, WY 82002
(307) 777-7752

STATE AIR POLLUTION CONTROL AGENCIES

Alabama Department of Environmental Management
Air Division
1751 Congressman Dickinson Drive
Montgomery, AL 36130
205-271-7861

Alaska Department of Environmental Conservation
Air Quality Management Section
P.O. Box O
Juneau, AK 99811-1800
907-465-5100

American Samoa Environmental Quality Commission
Governor's Office
Pago Pago, AS 96799
011-684-633-4116

Arizona Department of Environmental Quality
Office of Air Quality
P.O. Box 600
Phoenix, AZ 85001-0600
602-257-2308

Arkansas Department of Pollution Control and Ecology
Air Division
8001 National Drive, P.O. Box 9583
Little Rock, AR 72209
501-562-7444

California Secretary of Environmental Affairs
California Air Resources Board
P.O. Box 2815
Sacramento, CA 95812
916-445-4383

Colorado Department of Health
Air Pollution Control Division
4210 E. 11th Avenue
Denver, CO 80220
303-331-8500

Connecticut Department of Environmental Protection
Bureau of Air Management
165 Capitol Avenue
Hartford, CT 06106
203-566-2506

Delaware Department of Natural Resources and Environmental Control
Division of Air and Waste Management
89 Kings Highway, P.O. Box 1401
Dover, DE 19903
302-739-4791

District of Columbia Department of Consumer and Regional Affairs
Air Quality Control and Monitoring Branch
2100 Martin Luther King Avenue, SE
Washington, DC 20020
202-404-1120

Florida Department of Environmental Regulation
Air Resources Management
2600 Blair Stone Road
Tallahassee, FL 32399-2400
904-488-1344

Georgia Department of Natural Resources
Air Resources Branch
205 Butler Street, SE
Atlanta, GA 30344
404-656-6900

Guam Environmental Protection Agency
Complex Unit D-107
130 Rojas Street
Hammon, GU 96911
011-671-646-8863

Hawaii State Department of Health Laboratories
Division Air Surveillance-Analysis Branch
1270 Queen Emma Street, Suite 900
Honolulu, HI 96813
808-586-4019

Idaho Division of Environmental Quality
Air Quality Bureau
1410 North Hilton
Boise, ID 83706
208-334-5898

Illinois Environmental Protection Agency
Division of Air Pollution Control
2200 Churchill Road, P.O. Box 19276
Springfield, IL 62794-9276
217-782-7326

Indiana Department of Environmental Management
Office of Air Management
105 S. Meridan Street, P.O. Box 6015
Indianapolis, IN 46206-6015
317-232-8384

Iowa Department of Natural Resources
Air Quality Section
Henry A. Wallace Building
900 E. Grand Street
Des Moines, IA 50319
515-281-8852

Kansas Department of Health and Environment
Bureau of Air and Waste Management
Forbes Field, Building 740
Topeka, KS 66620
913-296-1593

Kentucky Department for Environmental Protection
Division for Air Quality
316 Saint Clair Mall
Frankfort, KY 40601
502-564-3382

Louisiana Department of Environmental Quality
Office of Air Quality and Radiation Protection
Air Quality Division
P.O. Box 82135
Baton Rouge, LA 70884-2135
504-765-0110

Maine Department of Environmental Protection
Bureau of Air Quality Control
State House, Station 17
Augusta, ME 04333
207-289-2437

Maryland Department of the Environment
Air Management Administration
2500 Broening Highway
Baltimore, MD 21224
301-631-3255

Massachusetts Department of Environmental Protection
Division of Air Quality Control
One Winter Street, 8th Floor
Boston, MA 02108
617-292-5593

Michigan Department of Natural Resources
Air Quality Division
P.O. Box 30028
Lansing, MI 48909
517-373-7023

Minnesota Pollution Control Agency
Air Quality Division
520 Lafayette Road
St. Paul, MN 55155
612-296-7331

Mississippi Department of Environmental Quality
Air Division, Office of Pollution Control
P.O. Box 10385
Jackson, MS 39289
601-961-5171

Missouri Department of Natural Resources
Division of Environmental Quality, Air Pollution Control
P.O. Box 176
Jefferson City, MO 65102
314-751-4817

Montana Department of Health and Environmental Science
Air Quality Bureau
Cogswell Building, Room A116
Helena, MT 59620
406-444-3454

Nebraska Department of Environmental Control
Air Quality Division
P.O. Box 98922
Lincoln, NB 68509-8922
402-471-2189

Nevada Division of Environmental Protection
Bureau of Air Quality
123 West Nye Lane
Carson City, NV 89710
702-687-5065

New Hampshire Air Resources Division
64 N. Main Street, Box 2033
Concord, NH 03301
603-271-1370

New Jersey Department of Environmental Protection
Division of Environmental Quality, Air Program
401 East State Street
Trenton, NJ 08625
609-292-6710

New Mexico Environmental Department
Air Quality Division
P.O. Box 26110
Santa Fe, NM 87502
505-827-0070

New York Department of Environmental Conservation
Division of Air Resources
50 Wolf Road
Albany, NY 12223-3250
518-457-7230

North Carolina Department of Environment, Health and Natural Resources
Air Quality Section
P.O. Box 27687
Raleigh, NC 27611-7687
919-733-3340

North Dakota State Department of Health
Division of Environmental Engineering
1200 Missouri Avenue
Bismarck, ND 58502-5520
701-221-5188

Ohio Environmental Protection Agency
Division of Air Pollution Control
1800 WaterMark Drive
Columbus, OH 43266-0149
614-644-2270

Oklahoma State Department of Health
Air Quality Service
1000 Northeast 10th Street, P.O. Box 53551
Oklahoma City, OK 73152
405-271-5220

Oregon Department of Environmental Quality
Air Quality Control Division
811 SW 6th Avenue, 11th Floor
Portland, OR 97204
503-229-5287

Pennsylvania Department of Environmental Resources
Bureau of Air Quality Control
101 South Second Street, P.O. Box 2357
Harrisburg, PA 17105-2357
717-787-9702

Puerto Rico Environmental Quality Board
Edificio Banco National Plaza
431 Avenue Ponce DeLeon
Hato Rey, PR 00917
809-767-8071

Rhode Island Department of Environmental Management
Division of Air and Hazardous Materials
291 Promenade Street
Providence, RI 02908-5767
401-277-2808

South Carolina Department of Health and Environmental Control
Bureau of Air Quality Control
2600 Bull Street
Columbia, SC 29201
803-734-4750

South Dakota Department of Environment and Natural Resources
Point Source Program
523 East Capitol Avenue
Pierre, SD 57501
605-773-3153

Tennessee Department of Environmental Conservation
Division of Air Pollution Control
701 Broadway
Nashville, TN 37243-1531
615-741-3931

State of Texas
Texas Air Control Board
12124 Park 35 Circle
Austin, TX 78753
512-908-1000

Utah Department of Environmental Quality
Division of Air Quality
1950 West North Temple
Salt Lake City, UT 84114-4820
801-536-4000

Vermont Agency of Natural Resources
Air Pollution Control Division
103 S. Main Street, Building 3
South Waterbury, VT 05676
802-244-8731

Virgin Islands Department Planning/Natural Resources
Division of Environmental Protection
Watergut Homes 1118 Christiansted
St. Croix, VI 00820-5065
809-773-0565

State of Virginia
Department of Air Pollution Control
P.O. Box 10089
Richmond, VA 23240
804-786-2378

Washington State
Department of Ecology
P.O. Box 47600
Olympia, WA 98504-7600
206-459-6632

State of West Virginia
Air Pollution Control Commission
1558 Washington Street East
Charleston, WV 25311
304-348-2275

Wisconsin Department of Natural Resources
Bureau of Air Management
Box 7921
Madison, WI 53707
608-266-7718

Wyoming Department of Environmental Quality
Air Quality Division
122 West 25th Street
Cheyenne, WY 82002
307-777-7391

Appendix D

University Pollution Prevention Research Centers

ALABAMA

University of Alabama
Environmental Institute for Waste Management Studies (EI WMS)

Activities include policy research, technology transfer, and basic research. Their Hazardous Material & Management and Resource Recovery (HAMMARR) program provides regulatory information, waste exchange and technical assistance for waste minimization, and workshops for small quantity generators and local businesses.

Dr. Robert Griffin, Director
Hazardous Materials Management and Recovery Program (HAMMMR)
University of Alabama
275 Mineral Industries Building
Box 870203
Tuscaloosa, AL 35487-0203
205-348-8403

Gulf Coast Hazardous Substance Research Center (GCHSRC)

The University of Alabama is a member of the GCHSRC, which is located at Lamar University in Beaumont, Texas (see the listing under Texas).

CALIFORNIA

University of California
Environmental Hazards Management Program

The University of California at Berkeley, Davis, Irvine, Los Angeles, Santa Cruz, Riverside, Santa Barbara, and San Diego offer post-graduate continuing education courses on toxic materials that devote some time to pollution prevention issues. Many of the courses give certificates in hazardous material management and air quality management. Some locations offer environmental auditing and other related topics.

Jon Kindschy, Statewide Coordinator
Environmental Hazards Management Program
University of California Extension
Riverside, CA 92521-0112
714-787-5804

University of California at Los Angeles
Center for Waste Reduction Technologies

The center conducts industry-supported research into waste reduction technology.

Dr. David Allen
University of California, Los Angeles
Los Angeles, CA 90024
213-206-0300

COLORADO

Colorado State University
Waste Minimization Assessment Center (WMAC)

WMAC is managed through the University City Science Center of Philadelphia. The center conducts detailed waste minimization assessments at small- to medium-sized manufacturing companies, training workshops for the Department of Health personnel, and training for EPA Region VIII RCRA inspectors. The center is also performing solvent use reduction audits at two manufacturing plants and will develop technical information on solvent use practices for small- to medium-sized manufacturing plants. In addition, the Center conducts training workshops for Department of Health personnel to develop technical expertise in pollution prevention. Contact EPA Region VIII for information on these workshops.

Dr. Harry Edwards, Director
Waste Minimization Assessment Center
Mechanical Engineering Department
Colorado State University
Fort Collins, CO 80523
303-491-5317

Marie Zanowich, Project Officer
U.S. EPA Region VIII
999 18th Street, Suite 500
Denver, CO 80202-2505
303-294-1065

CONNECTICUT

University of Connecticut
Pollution Prevention Research and Development Center (PPRDC)

An EPA-funded research center, the PPRDC will support state government and industry in reducing toxic emissions by encouraging existing and start-up companies to provide services and equipment necessary for pollution prevention technologies, and by creating new jobs to meet the demands of this industry. PPRDC's goal is to work with industry to develop pollution prevention technology and a manufacturing base in the region.

Dr. George Hoag
Director, Pollution Prevention
Research and Development Center
Environmental Research Institute
Box U-120, Route 44, Longley Building 146
University of Connecticut
Storrs, CT 06269-3210
203-486-4015 FAX 203-486-2269

Waterbury State Technical College
Industrial Environmental Management (IEM)

Waterbury State Technical College offers a waste minimization course as part of its Industrial Environmental Management certificate level and associate degree level programs. Other courses include environmental regulations, safe handling of hazardous wastes, and environmental control processes.

Cynthia Donaldson, Chairperson
Industrial Environmental Management
Waterbury State Technical College
750 Chase Parkway
Waterbury, CT 06708-3089
203-596-8703/575-8089

DISTRICT OF COLUMBIA

Howard University
The Great Lakes and Mid-Atlantic Hazardous Substance Research Center

The center is funded by EPA and focuses on the unique problems of EPA Regions III and V. Research is conducted on hazardous substances and related environmental problems. Among other projects, the center is developing materials for a hazardous waste workshop and videotapes on waste minimization information and training. The University of Michigan and Michigan State University are also members of the center.

Dr. James H. Johnson, Jr., Assistant Director
The Great Lakes and Mid-Atlantic Hazardous Substance Research Center
Department of Civil Engineering
Howard University
Washington, DC 20059
202-806-6570

FLORIDA

Florida Institute of Technology
Research Center for Waste Utilization

The center offers classroom training in waste utilization at the undergraduate and graduate levels. In addition, the center is involved in research in the areas of municipal solid waste (MSW), industrial solid waste, and pollution prevention. Specific studies include heavy metal sources in the MSW stream, uses of ash from waste-to-energy plants, biological toxicity of ash residues, and degradable plastics characteristics after disposal.

Edwin Korzun, Executive Director
Research Center for Waste Utilization
Department of Marine and Environmental Sciences
Florida Institute of Technology
150 West University Boulevard
Melbourne, FL 32901-6988
305-768-8000

University of Central Florida
Gulf Coast Hazardous Substance Research Center (GCHSRC)

The University of Central Florida is a member of the GCHSRC, which is located at Lamar University in Beaumont, Texas (see the listing under Texas).

University of Florida
Center for Training, Research, and Education for Environmental Occupations

The center's activities include developing a statewide training action plan for business, government, and the public; providing RCRA hazardous waste regulation training; developing a university-level waste reduction curriculum; sponsoring a 2-day symposium; and developing a training program for three specific industries.

Dr. James O. Bryant, Jr., Director
Center for Training, Research, and Education for
Environmental Occupations
Division of Continuing Education
University of Florida
3900 S.W. 63rd Boulevard
Gainesville, FL 32608-3848
904-392-9570

Florida Center for Solid & Hazardous Waste Management

The Center coordinates the State's solid and hazardous waste research efforts, including management practices for waste reduction, reuse, recycling, and improved conventional disposal methods.

Dr. James O. Bryant, Jr., Director
Florida Center for Solid and Hazardous Waste Management
University of Florida
3900 S.W. 63rd Boulevard
Gainesville, FL 32608-3848
904-392-9570

GEORGIA

Georgia Institute of Technology (Georgia Tech)
Environmental Science and Technology Laboratory

The institute provides continuing education workshops on a wide variety of waste reduction and pollution prevention topics, including hazardous waste reduction planning requirements. As part of a U.S. EPA grant with the Georgia Hazardous Waste Management Authority, the institute is offering workshops to help industry write proposals for grants implementing new waste minimization technologies. Within the Hazardous Materials Group of the Laboratory are the Hazardous Waste Technical Assistance Program (HWTAP) and the Pollution Prevention Program. These programs provide technical assistance to Georgia industry to encourage voluntary waste reduction and minimization, as well as compliance with hazardous waste regulations. Activities include on-site assistance, telephone consultations, information dissemination, multimedia information releases, short courses, and annual seminars. The Pollution Prevention Program is funded by EPA grants, while HWTAP is paid for through general funds.

Carol Foley
Georgia Tech Research Institute
Environmental Science and Technology Laboratory
Atlanta, GA 30332
404-894-3806

ILLINOIS

Illinois Institute of Technology
Industry Waste Elimination Research Center (IWERC)

The center's research priorities include recycling or reusing industrial by-products and developing manufacturing processes that avoid generating wastes or pollutants. In conjunction with the Department of Environmental Engineering, graduate programs are offered with an option in hazardous waste management.

Dr. Kenneth E. Noll, Director
Industrial Waste Elimination Research Center
Pritzker Department of Environmental Engineering
IIT Center
Chicago, IL 60616
312-567-3536

University of Illinois
Hazardous Waste Research & Information Center (HWRIC)

The center combines research, education, and technical assistance in a multidisciplinary approach to manage and reduce hazardous waste. HWRIC collects and shares information through its library/clearinghouse and a computerized Waste Reduction Advisory System.

Dr. David Thomas, Director
Hazardous Waste Research and Information Center
One East Hazelwood Drive
Champaign, IL 61820
217-333-8940

INDIANA

Purdue University
Pollution Prevention Program

The Pollution Prevention Program provides outreach and technical assistance efforts to industry (including on-site assessments conducted by graduate students) on pollution prevention opportunities. Purdue University and the Indiana Department of Environmental Management sponsor both general and specific workshops on pollution prevention and recycling.

Rick Bossingham, Coordinator
Pollution Prevention Program
Purdue University
2129 Civil Engineering Building
West Lafayette, IN 47907-1284
317-494-5038

IOWA

University of Northern Iowa
Iowa Waste Reduction Center

This EPA-funded center is designed to be a technology transfer center, utilizing research findings from across the globe to benefit existing and potentially new Iowa businesses and industries.

Dr. John L. Konefes
Director, Recycling and Reuse Technology
Transfer Center
Iowa Waste Reduction Center
75 BRC
University of Northern iowa
Cedar Falls, IA 50614-0185
319-273-2079 FAX 319-273-6494

KANSAS

Kansas State University
Hazardous Substance Research Center (HSRC)

This EPA-funded center provides research and technology transfer services for pollution prevention and other waste management techniques. HSRC programs include outreach to industry, assistance to government, videos, radio programs, written materials, databases, and workshops on pollution prevention and hazardous waste remediation. A pollution prevention focus of this center is on soils and mining waste.

Dr. Larry E. Erickson, Director
Hazardous Substance Research Center
Durland Hall, Room 105
Kansas State University
Manhattan, KS 66506-5102
913-532-5584

University of Kansas
Center for Environmental Education and Training

In cooperation with the Kansas Department of Health and Environment, the center offers Hazardous Waste Regulatory Training Conferences. Conference topics include waste minimization, regulatory compliance, and technology transfer components.

Lani Heimgardner
Center for Environmental Education and Training
Division of Continuing Education
University of Kansas
6330 College Boulevard
Overland Park, KS 66211
913-491-0810

KENTUCKY

University of Louisville
Kentucky PARTNERS—State Waste Reduction Center

This center conducts general and industry-specific seminars and workshops on environmental regulations and pollution prevention methods Another service is free: non-regulatory pollution prevention services for all Kentucky industries and businesses. In addition, Kentucky PARTNERS publishes a newsletter and performs on-site assessments.

Joyce St. Clair
Executive Director
Kentucky PARTNERS - State Waste Reduction Center
Ernst Hall, Room 312
University of Louisville
Louisville, KY 40292
502-588-7260

Waste Minimization Assessment Center (WMAC)

WMAC is managed through the University City Science Center in Philadelphia. The center conducts quantitative, on-site, waste minimization assessments for small- to medium-sized generators located within a 150 mile radius of Louisville. In addition, the center incorporates risk reduction and pollution prevention into the undergraduate and graduate engineering curricula. Professionals are encouraged to participate in these courses. Engineering students also conduct waste minimization projects at manufacturing plants.

Marvin Fleischman, Director
Waste Minimization Assessment Center
Department of Chemical Engineering
University of Louisville
Louisville, KY 40292
502-588-6357

LOUISIANA

Louisiana State University (Shreveport)
Hazardous Waste Research Center (HWRC)

Categories of research conducted by faculty and students include incineration and combustion, alternative methods of treatment and destruction, and transport of leachate and wastes from pits and spills.

David Constant, Director
Hazardous Waste Research Center
3418 CEBA Building
Louisiana State University
Baton Rouge, LA 70803
504-388-6770

Louisiana State University (LSU)
Gulf Coast Hazardous Substance Research Center (GCHSRC)

LSU is a member of the GCHSRC, which is located at Lamar University in Beaumont, Texas (see the listing under Texas).

Southern University at Baton Rouge
Center for Energy and Environmental Studies

The center will support individual pollution prevention, treatment technology and socio-economic policy research projects.

Dr. Robert L. Ford
Director, Center for Energy and
Environmental Studies
Southern University at Baton Rouge
Cottage #8, P.O. Box 9764
Baton Rouge, LA 70813
504-771-4723 FAX 504-771-4722

MAINE

University of Maine (UM)
Chemicals in the Environment Information Center

The Center provides courses, conferences, presentations and brochures emphasizing pollution prevention. Courses are: Issues in Environmental Pollution; Pollution Prevention-Changing Ourselves and Changing Society (Honors students); and Pollution Prevention through Understanding and Managing the Chemicals in Our Lives (teachers). Conferences are for business, e.g. Pollution Prevention in the Home, Workplace and Community. Work is carried out in cooperation with state agencies, Cooperative Extension and Maine Waste and Toxics Use Reduction Committee.

Marquita K. Hill, Ph.D., Director
University of Maine
5737 Jenness Hall
Orono, ME 04469-5737
207-581-2301

MASSACHUSETTS

Massachusetts Institute of Technology (MIT)
Center for Technology, Policy and Industrial Development

Along with the Center, the Technology, Business and the Environment Group conducts research and offers workshops in pollution prevention. Pollution prevention concepts are also included in some undergraduate and graduate courses.

John Ehrenfeld
Technology, Business and the Environment Group
Center for Technology, Policy and Industrial Development
E40-241
Massachusetts Institute of Technology
Cambridge, MA 02139
617-253-7753

Tufts University
Tufts Environmental Literacy Institute (TELI)

The Institute is conducting a demonstration project, Tufts CLEAN to analyze the energy and materials flow at the university. Funded by EPA's Office of Pollution Prevention, this project involves students in audit design, data collection and analysis, implementation, and evaluation.

Dr. Anthony Cortese
Dean of Environmental Programs
Tufts University
Office of Environmental Programs
474 Boston Avenue, Curtis Hall
Medford, MA 02155
617-627-3452

The Center for Environmental Management

The purpose of this center is to develop a multidisciplinary approach to environmental problems through health effects research, technology research, policy analysis, education and training programs, and information transfer. Pollution prevention is emphasized throughout center programs.

Dr. Kurt Fischer
Tufts University
Center for Environmental Management
474 Boston Avenue, Curtis Hall
Medford, MA 02155
617-627-3452 FAX 617-627-3084

University of Massachusetts (Lowell)
Toxics Use Reduction Institute

The Massachusetts Toxics Use Reduction Institute promotes reduction in the use of toxic chemicals or the generation of toxic by-products in Massachusetts industry. The Institute is a multi-disciplinary research, education, training and technical support center located at the University of Massachusetts (Lowell).

Dr. Jack Luskin
Associate Director for Education and Training
Toxics Use Reduction Institute
University of Massachusetts (Lowell)
Lowell, MA 01854-2881
508-934-3275 FAX 508-453-2332

MICHIGAN

Grand Valley State University
Waste Reduction and Management Program (WRMP)

The WRMP is a university-based pollution prevention program that conducts research and provides technical assistance to Michigan industry. "Design for Recycling: Solving Tomorrow's Problems Today," a 1-year waste reduction research and demonstration project, is funded by the Padnos Foundation and the Michigan Department of Natural Resources as part of the Quality of Life Bond Program. The overall objective of the project is to reduce the future generation of solid waste by infusing undergraduate engineering curricula with the concept of design for the entire product lifecycle. This project includes the following activities: identifying and prioritizing 10 products that have the greatest potential for design change to promote recycling; and developing a series of seminars to focus Michigan manufacturers, engineers, and engineering faculty on "cutting edge" design approaches; developing student awareness and skill in designing products with end-stage product management in mind.

Dr. Paul Johnson, Associate Professor
Grand Valley State University
School of Engineering
301 W. Fulton, Room 617
Grand Rapids, MI 49504
616-771-6750

Michigan Technological University
Center for Clean Industrial and Treatment Technologies (CCITT)

The emphasis of this center is pollution prevention through identification of alternatives, balanced assessment and targeted research and development. Ultimately, the goal is to develop and advocate methods to fully utilize raw materials and produce products which are highly recyclable and/or exhibit minimal lifetime environmental risk. This is to be accomplished by acting as a sort of "analytical bridge" between industry, government and academia to promote practical means of total quality management and environmental equity.

Dr. John C. Crittenden, Director
Environmental Engineering Center
Michigan Technological University
1400 Townsend Drive
Houghton, MI 49931
906-487-3143 FAX 906 487-2061

University of Detroit Mercy
Center of Excellence in Polymer Research and Environmental Study

The Center is a partnership of university, industry, and government whose purpose it is to conduct high technology research that addresses environmental problems related to polymer wastes and proposes the development of new environmentally responsible and safe polymer products. The center is also committed to the transfer of pollution prevention and waste management technologies to commercial application in products and processes through their industry partners.

Daniel Klemper
Director, Center of Excellence in Polymer
Research and Environmental Study
University of Detroit Mercy
4001 W. McNichols Road
Detroit, MI 48219-3599
313-993-1270 FAX 313-993-1409

University of Michigan
EPA Pollution Prevention Center for Curriculum Development and Dissemination

The purpose of this center is to develop pollution prevention curriculum modules for undergraduate and graduate courses in engineering business and science.

Dr. Gregory A. Keoleian, Manager
School of Natural Resources
University of Michigan
Dana Building
430 E. University
Ann Arbor, MI 48109-1115
313-764-1412

The Great Lakes and Mid-Atlantic Hazardous Substance Research Center (GLMA-HSRC)

A cooperative research consortium comprising the University of Michigan, Michigan State University, and Howard University, this center supports hazardous substance training, technology transfer, and research.

Dr. Walter Weber, Director
Hazardous Substance Research Center
University of Michigan
Suite 181 Engineering 1-A
Ann Arbor, MI 48109-2125
313-763-2274

MINNESOTA

University of Minnesota
Minnesota Technical Assistance Program

Using EPA's Toxic Release Inventory (TRI), the program provides technical transfer, workshops, and fact sheets encouraging decreased use of TRI chemicals through use of alternatives and waste minimization.

David Simmons
Public Relations Representative
Minnesota Technical Assistance Program
1315 5th St., S.E., Suite 207
University of Minnesota
Minneapolis, MN 55414-4504
612-627-4646

MISSISSIPPI

Mississippi State University
Mississippi Technical Assistance Program and Mississippi Solid Waste Reduction Assistance Program

These programs work cooperatively to provide pollution prevention research, on-site waste assessments, workshops, conferences, employee and student education materials, a waste exchange, technology databases, and a monthly newsletter.

Dr. Don Hill, Dr. Caroline Hill, or Dr. June Carpenter
Mississippi Technical Assistance Program and Mississippi Solid Waste
Reduction Assistance Program
P.O. Drawer CN
Mississippi State University
Mississippi State, MS 39762
601-325-8454

Gulf Coast Hazardous Substance Research Center (GCHSRC)

Mississippi State University is a member of the GCHSRC, which is located at Lamar University in Beaumont, Texas (see the listing under Texas).

NEVADA

University of Nevada at Reno
Nevada Small Business Development Center

The Nevada Small Business Development Center, in cooperation with the Nevada Division of Environmental Protection, offers free pollution prevention services to industry and businesses, including seminars, workshops, on-site evaluations, fact sheets, and a newsletter. The center also maintains a Hazardous Waste Information Line, assisting businesses with regulations, alternative product use, and pollution prevention.

Kevin Dick, Manager
Business Environmental Program
Nevada Small Business Development Center
University of Nevada - Reno
Reno, NV 89557-0100
702-774-1717

NEW JERSEY

New Jersey Institute of Technology
Hazardous Substance Management Research Center

Areas of research include incineration, biological/chemical treatment, physical treatment, site assessment remediation, health effects assessment, and public policy/education.

Dick Magee
Advanced Technology Center Building
323 Martin Luther King Boulevard
University Heights
Newark, NJ 07102
201-596-5864

NEW MEXICO

New Mexico State University
Waste-Management Education and Research Consortium (WERC)

WERC is a waste management education and research consortium established by New Mexico State University (NMSU) under a U.S. Department of Energy grant in 1990. Consortium members include NMSU, the University of New Mexico, the New Mexico Institute of Mining and Technology, the Navajo Community College, the Los Alamos National Laboratories, and the Sandia National Laboratories. The mission of WERC is to expand the Nation's capability to address the issues related to management of all types of waste (hazardous, solid, and radioactive). WERC activities involve all waste management options, including pollution prevention.

John S. Townsend, Assistant Director
WERC
New Mexico State University
Box 30001
Department 3805
Las Cruces, NM 88003-0001
505-646-2038

NEW YORK

Clarkson University
Hazardous Waste and Toxic Substance Research and Management Center

This center coordinates and mobilizes funding for multi-disciplinary research at Clarkson University. Projects currently being conducted include a wide range of basic research, applied engineering, and technology development topics. Many of these projects address waste minimization and pollution prevention either directly or indirectly.

Thomas L. Theis, Director
Hazardous Waste and Toxic Substance Research
and Management Center
Rowley Laboratories
Clarkson University
Potsdam, NY 13699
315-268-6542

Cornell University
Waste Management Institute

The institute coordinates interdisciplinary research on waste reduction and management options for hazardous, agricultural, solid, industrial, and sludge wastes. Numerous fact sheets and publications are made available on topics ranging from source reduction opportunities for shoppers to waste minimization opportunity assessment for communities and businesses.

Richard Schuler, Director
Waste Management Institute
313 Hollister Hall
Cornell University
Ithaca, NY 14853
607-255-8674

NORTH CAROLINA

North Carolina State University
EPA Research Center for Waste Minimization and Management

U.S. EPA is sponsoring a major university-based research center that focuses specifically on the challenge to minimize and manage hazardous substances. Located at North Carolina State University, the center involves Texas A&M University and the University of North Carolina at Chapel Hill. The mission of the center is to develop practical means for industry to eliminate the use and generation of hazardous substances, treat those wastes that cannot be eliminated, and provide secure containment for treatment residues. The major research focus at the center will be the elimination or reduction in discharge of hazardous substances to all environmental media. A strong commitment also will be made to technology transfer and training.

Dr. Michael Overcash
Dr. Cliff Kaufman
Center for Waste Minimization and Management
North Carolina State University
Box 7905
Raleigh, NC 27695-2325
919-515-2325

University of North Carolina - Chapel Hill
EPA Research Center for Waste Minimization and Management

The University of North Carolina at Chapel Hill is a member of the U.S. EPA Research Center located at North Carolina State University in Raleigh, North Carolina.

Dr. William H. Glaze
Department of Environmental Science & Engineering
University of North Carolina - Chapel Hill
Chapel Hill, NC 27514
919-966-1024

NORTH DAKOTA

University of North Dakota
Energy and Environmental Research Center (EERC)

The EERC features an integrated systems approach to energy and environmental research and technology development beginning with fundamental evaluation and characterization of earth resources, followed by research and development of innovative technologies to extract and utilize these resources in an efficient and environmentally acceptable manner, and culminating in the utilization of safe disposal of wastes generated in using natural resources.

Dr. Gerald Groenwald, Director
Energy and Environmental Research Center
Center of Excellence for Toxic Metal Emissions
University of North Dakota
15 North 23rd Street, Box 8213
University Station
Grand Forks, ND 58202-8213
701-777-5131 FAX 701-777-5181

OHIO

University of Cincinnati
American Institute for Pollution Prevention (AIPP)

Jean Boddocsi, Director
American Institute for Pollution Prevention (AIPP)
Office of the University Dean for Research
University of Cincinnati
Cincinnati, OH 45221
513-556-4532

University of Findlay
RCRA Generator Training Program

Workshops introduce U.S. EPA's Pollution Prevention Program for personnel at industries and commercial businesses that generate hazardous waste. Training courses assist generators in developing waste minimization strategies, such as source reduction, with the goal of eliminating waste generation. Regulation and compliance are also discussed. Workshops consist of 2-3 day sessions.

George Kleevic
Workshop Instructor
RCRA Generator Training Program
P.O. Box 538
St. Clairsville, OH 43950
614-695-5036

OKLAHOMA

Oklahoma State University
Center for Resource Conservation and Environmental Research (CRCER)

The goal of this Center is to establish and maintain a "center without walls" to provide Oklahoma, the Southwest region, and the nation with benefits of a coordinated, multidisciplinary, multi-institutional research, analysis, and evaluation of the technical, policy, and managerial issues related to resource conservation and reduction; reduction/disposal of municipal and industrial wastes and avoidance/correction of pollution of air, land, and water. The Center will accomplish its technical studies and policy analyses primarily through the resources of Oklahoma State University, the University of Oklahoma, and the University of Tulsa.

Mr. Robert Fulton
Vice President, Oklahoma Alliance for Public
Policy Research
2630 Northwest Expressway, Suite B
Oklahoma City, OK 73112
405-943-8989 FAX 405-840-0061

PENNSYLVANIA

University of Pittsburgh
Center for Hazardous Materials Research (CHMR)

The center conducts applied research, health and safety training, education, and international technology transfer projects involving hazardous and solid waste. It also provides technical assistance, on-site assessments, and fact sheets and manuals on pollution prevention for industries in Pennsylvania.

Dr. Edgar Berkey
Center for Hazardous Materials Research
University of Pittsburgh Trust
Applied Research Center
320 William Pitt Way
Pittsburgh, PA 15238
412-826-5320

RHODE ISLAND

University of Rhode Island
Chemical Engineering Department

Advanced students and their professors develop and evaluate pollution prevention engineering solutions for Rhode Island firms. These firms are referred by the Rhode Island Department of Environmental Management's voluntary pollution prevention technical assistance program.

Professor Stanley M. Barnett, Chairman
Chemical Engineering Department
Crawford Hall
University of Rhode Island
Kingston, RI 02881
401-792-2443

SOUTH CAROLINA

University of South Carolina
Hazardous Waste Management Research Fund

The fund sponsors research and educational programs in the area of hazardous waste reduction. Research priorities include technology transfer, assessment training, site remediation, recycling and reuse strategies, and policy issues. Topics to be covered in the educational programs include vehicle/auto service shops, textiles, metal fabrication and machine shops, painting and coating, solvent use reduction, and developing a site specific waste reduction program. The fund has also established educational programs at Clemson University in Clemson, South Carolina.

Doug Dobson, Executive Director
Institute of Public Affairs
University of South Carolina
Gambrell Hall, 4th Floor
Columbia, SC 29208
803-777-8157

TENNESSEE

University of Tennessee
Center for Industrial Services (CIS)

The center sponsors an extensive waste reduction assessment training program that includes in-depth waste reduction assessment courses. This training program was originally developed to instruct retired industrial engineers and managers who became a highly skilled waste reduction assessment team. A key program for the center has been waste reduction assessments by full-time field engineers and retired engineers.

Cam Metcalf
Center for Industrial Services
University of Tennessee
226 Capitol Boulevard Building
Suite 606
Nashville, TN 37219
615-242-2456

Waste Minimization Assessment Center (WMAC)

Managed by the University City Science Center in Philadelphia, WMAC is staffed by engineering students and faculty who have considerable expertise with process operations in manufacturing plants and who also have the skills needed to minimize waste generation. These staff members perform quantitative waste minimization assessments for small- to medium-sized generators.

Dr. Richard J. Jendrucko, Director
Department of Engineering
Science and Mechanics
University of Tennessee
310 Perkins Hall
Knoxville, TN 37996-2030
615-974-7682

TEXAS

Texas A & M University
EPA Research Center for Waste Minimization and Management

Texas A & M University is a member of the U.S. EPA Research Center located at North Carolina State University in Raleigh, North Carolina (see the listing under North Carolina).

Dr. Kirk Brown
Department of Soil and Crop Science
Texas A & M University
College Station, TX 77843
409-845-5251

Gulf Coast Hazardous Substance Research Center (GCHSRC)

Texas A & M is a member of the GCHSRC, which is located at Lamar University in Beaumont, Texas.

Texas Tech University
Center for Environmental Technologies

The center coordinates conferences, short courses, and lectures that address environmental concerns, pollution prevention, pollution controls, and Federal, State, and local regulations. Conferences and short courses are offered for State and municipal audiences, professional and civic groups, and industry. The center is also conducting at least 15 different research projects involving pollution prevention in such areas as storm water discharge, groundwater monitoring, and pesticides.

Dr. John R. Bradford
Center for Environmental Technologies
Texas Tech University
P.O. Box 43121
Lubbock, TX 79409-3121
806-742-1413

Lamar University
Gulf Coast Hazardous Substance Research Center (GCHSRC)

The GCHSRC is a research consortium of eight universities, with its center located at Lamar University. Its purpose is to conduct research to aid in more effective hazardous substance response and waste management. The center's efforts are concentrated in the areas of waste minimization and alternative technology development. The center receives funding from the U.S. EPA and the State of Texas, with a majority of those funds being pledged to pollution prevention for the petrochemical and microelectronic industries. At this time, the center has some 60 projects in progress in a joint Federal, State, and industry effort at Texas Universities, and at research centers outside the State. The other members of the consortium are Louisiana State University, Mississippi State University, University of Alabama, University of Central Florida, University of Houston, University of Texas - Austin, and Texas A & M.

Mr. Tom Pinson, Assistant Director
Gulf Coast Hazardous Substance Research Center
Lamar University
P.O. Box 10613
Beaumont, TX 77710
409-880-8707 FAX 409-880-2397

University of Houston
Gulf Coast Hazardous Substance Research Center (GCHSRC)

The University of Houston is a member of the GCHSRC, which is located at Lamar University in Beaumont, Texas.

University of Texas - Arlington
Environmental Institute for Technology Transfer (EITT)

EITT was established to facilitate research, technical assistance, and the dissemination of environmental knowledge to assist business and industry in finding cost-effective and environmentally acceptable solutions to compliance problems. In addition to offering training courses that address pollution prevention, the institute provides a forum for industry and regulators to address common concerns through workshops, seminars, and conferences.

Dr. Gerald I. Nehman, Director
Dr. Victorio Argento, Associate Director
Environmental Institute for Technology Transfer
University of Texas at Arlington

Box 19050
Arlington, TX 76019
817-273-2300

University of Texas - Austin
Gulf Coast Hazardous Substance Research Center (GCHSRC)

The University of Texas - Austin is a member of the GCHSRC, which is located at Lamar University in Beaumont, Texas.

UTAH

Weber State University
Center for Environmental Service

Environmental management training and technical assistance are available with a special emphasis on the needs of Northern Utah's small- and medium-sized businesses and manufacturers, as well as its cities and towns. Pollution prevention opportunities are among the topics covered by the Center's services.

Dianne Siegfreid, Director
Barbara A. Wachocki, Director
Center for Environmental Services
Weber State University
Ogden, UT 84408-2502
801-626-7559

WISCONSIN

University of Wisconsin - Madison
Engineering Professional Development Program

The College of Engineering offers intensive, short courses on waste minimization, environmental compliance, industrial environmental engineering, and pollution prevention from the design aspect.

Pat Eagan
Engineering Professional Development Program
College of Engineering
University of Wisconsin at Madison
432 North Lake Street
Madison, WI 53706
608-263-7429

Solid and Hazardous Waste Education Center

In cooperation with the Wisconsin Department of Natural Resources, the Extension Office offers workshops in solid waste reduction, recycling, composting, as well as general and industry-specific (electroplating and metal finishing, auto repair, local government, and schools) workshops on waste minimization and pollution prevention. The center also works directly with industry and government to provide technical assistance.

David Liebel
Wayne Pferdehirt
Solid and Hazardous Waste Education Center
University of Wisconsin - Extension
529 Lowell Hall
610 Langdon Street
Madison, WI 53703
608-265-2360

Appendix E

State Pollution Prevention Laws

MANDATORY POLLUTION PREVENTION PROGRAMS

Arizona

ARIZONA REVISED STATUTES
TITLE 49. THE ENVIRONMENT
CHAPTER 5. HAZARDOUS WASTE DISPOSAL
ARTICLE 4. POLLUTION PREVENTION

49-961. Definitions

In this article, unless the context otherwise requires:
1. "Disposal" means discharging, depositing, injecting, dumping, spilling, leaking or placing a toxic substance or hazardous waste into or on land or water so that the toxic substance or hazardous waste or any constituent of hazardous waste may enter the environment, be emitted into the air or be released into or commingled with any waters, including groundwater.
2. "Facility" means all buildings, equipment, structures and other stationary items located on a single site or on contiguous or adjacent sites and owned or operated by the same person or by any person who controls, is controlled by or is under common control with any person. Facility does not include a household hazardous waste collection facility or a facility that is primarily engaged in receiving waste from off-site and that has a permit issued or plan approved under this title for the storage, treatment or disposal of solid, special or hazardous waste.
3. "Generator" means a person who, by virtue of ownership, management or control, is responsible for causing or allowing to be caused the creation of hazardous waste.
4. "Hazardous waste" means hazardous waste as defined in 49-921.
5. "Person" means an individual, the United States, this state or a public or private corporation, local government unit, public agency, partnership, association, firm, trust or estate or any other legal entity.

Appendix E: State Pollution Prevention Laws / 385

6. "Pollution" means the disposal of a toxic substance or hazardous waste into the air, land, surface water or groundwater.

7. "Pollution prevention" means operational procedures and processes and improvements in housekeeping or management techniques that reduce potential or actual releases of pollutants to the overall environment including all air, water and land resources affected by those pollutants.

Pollution prevention includes any of the following:

(a) Toxics use reduction, source reduction, recycling of wastes or secondary materials, waste minimization, reuse, reclamation, conservation and substitution.

(b) Proportionate changes in the total volume, quantity or toxicity of a particular pollutant as prescribed in this subsection as the release of that pollutant changes as a result of production changes or other business changes.

For the purposes of this subdivision, "business changes" includes improvements in operating practices, spill and leak prevention measures, inventory control and other changes that proportionately reduce or eliminate the release of pollutants to the overall environment but does not include the transfer or relocation of an operation or process to another facility in this state with no subsequent proportionate reduction in toxics use or the release of pollutants to the overall environment.

(c) On-site or off-site treatment if that treatment can be shown to confer a higher degree of protection to the public health and safety and the environment than other technically and economically practicable waste reduction alternatives.

8. "Recycling" means a reuse, further use, reclamation or extraction through a process or activity that is separate from the process or activity that produced the waste stream but does not include combustion or incineration.

9. "Toxic substance" or "toxics" means a toxic chemical listed pursuant to the pollution prevention act of 1990 (42 U.S.C. 13102 (3)).

10. "Treatment" has the same meaning as prescribed in 40 C.F.R. 260.10 but does not include recycling.

49-962. Toxic data report; progress report; exemption

A. A person who owns or operates a facility shall file a toxic data report on July 1 for the preceding calendar year if either of the following applies:

1. During the preceding calendar year, the owner or operator was required to file an annual toxic chemical release form for the facility pursuant to Section 313 of the Superfund Amendments and Reauthorization Act of 1986 (P.L. 99-499).

2. During the preceding calendar year, the facility generated an average of one kilogram per month of acutely hazardous waste as defined in 40 C.F.R. part 261 or an average of one thousand kilograms per month of hazardous waste in a calendar year, exclusive of an episodic, accidental or remediation related release or occurrence.

B. The owner or operator of a facility shall file the report prescribed in subsection A of this section on July 1 each year for the preceding calendar year until either of the following applies:
1. The facility ceases operation.
2. The facility does not meet the requirements of both:
(a) Subsection A, paragraph 1 of this section.
(b) Subsection A, paragraph 2 of this section, for two consecutive years.

C. The toxic data report required in subsection A of this section shall include both of the following:
1. The report form required by the United States environmental protection agency pursuant to 42 U.S.C. 13106.
2. Any annual progress report required to be submitted pursuant to 49-963.

49-963. Pollution prevention plan; progress report; exemption

A. A person who owns or operates a facility that meets the reporting requirements prescribed by 49-962 shall prepare and implement a pollution prevention plan that addresses a reduction in the use of toxic substances and the generation of hazardous wastes. By January 1, 1994, the director shall establish a numeric goal for the state for waste minimization. By January 1, 1999 the director shall establish a numeric goal for the state for toxic use reduction. For purposes of this section, "toxic substance" does not include material used or produced in connection with a mining or metallurgical operation.

B. By December 31, 1992, the following shall file a plan as prescribed by this section:
1. A facility that shipped off-site for purposes other than recycling the lesser of:
(a) Twelve thousand or more kilograms cumulative total of hazardous wastes in calendar year 1991.
(b) An average of one thousand kilograms or more per month cumulative total of hazardous waste in calendar year 1991.
2. A facility that shipped off-site for purposes other than recycling an average of one kilogram or more per month cumulative total of acutely hazardous wastes, as defined in 40 C.F.R. part 261, in calendar year 1991.

C. By December 31, 1995, a facility that shipped off-site for purposes other than recycling from ten thousand kilograms to twelve thousand kilograms cumulative total of hazardous wastes in calendar year 1994 shall file a plan as prescribed by this section.

D. From and after December 31, 1994, a facility that uses in excess of ten thousand pounds in a calendar year of a toxic substance as defined in 49-961 shall file a pollution prevention plan by December 31 of the following year covering those toxic substances that exceed the threshold quantity. From and after December 31, 1998, the director may adopt by rule threshold quantities of toxic substances different from those established pursuant to this section if the director determines there is sufficient evidence to establish any one of the following:

1. The chemical is known to cause or can reasonably be anticipated to cause significant acute adverse human health effects at concentration levels that are reasonably likely to exist beyond the boundaries of a facility site as a result of continuous or frequently recurring releases, or to cause or can reasonably be anticipated to cause a significant adverse effect on the environment of sufficient seriousness to warrant inclusion in the pollution planning program because of any of the following:

(a) Its toxicity and persistence in the environment.

(b) Its toxicity and tendency to bioaccumulate in the environment.

2. The chemical is known to cause or can reasonably be anticipated to cause in humans either of the following:

(a) Cancer or teratogenic effects.

(b) Serious or irreversible reproductive dysfunctions, neurological disorders, heritable genetic mutation or other chronic health effects.

E. After December 31, 1992, a facility that meets the quantitative threshold filing requirements prescribed in subsection B of this section and that has not filed a plan with the department shall file a plan no later than December 31 of the following year.

F. Notwithstanding any other provision of law, a person required to file a toxic data report under 49-962, subsection A, paragraph 1 shall prepare, submit and begin to implement a pollution prevention plan no later than December 31, 1992.

G. A facility required to prepare a pollution prevention plan under this section shall maintain and implement that plan until the facility ceases operation or the facility no longer meets the quantitative threshold filing requirements prescribed in this section.

H. A person who is not required to prepare a pollution prevention plan may voluntarily comply with this section and shall be deemed a generator who may certify compliance as prescribed by 49-931, subsection A.

I. A person who owns or operates more than one facility that is required to prepare and implement a pollution prevention plan pursuant to subsection A of this section may prepare and implement a single pollution prevention plan that covers more than one facility.

J. The pollution prevention plan required by this section shall include all of the following:

1. The name and location of and principal business activities at the facility.

2. The name, address and telephone number of the owner or operator of the facility and of the senior official with management responsibility at the facility.

3. A certification by the senior official with management responsibility at the facility that he has read the plan and that it is to the best of his knowledge true, accurate and complete.

4. Specific performance goals for the prevention of pollution, including an explanation of the rationale for each performance goal. The plan must include a goal for the facility and may include goals for individual production processes.
5. A written policy setting forth management and corporate support for the pollution prevention plan and a commitment to implement the plan to achieve the plan goals.
6. A statement of the plan's scope and objectives.
7. An analysis identifying pollution prevention opportunities to reduce or eliminate toxic substance releases and hazardous waste generation.
8. An analysis of pollution prevention activities that are already in place and that are consistent with the requirements of this article.
9. Employee awareness and training programs to involve employees in pollution prevention planning and implementation to the maximum extent feasible.
10. Provisions to incorporate the plan into management practices and procedures in order to ensure its institutionalization.
11. A description of the options considered and an explanation of why the options considered were not implemented.

K. The pollution prevention plan, at a minimum, shall cover a two-year time period and may cover a longer time period at the discretion of the facility.

L. Each owner and operator required to prepare and maintain a pollution prevention plan under this section shall file an annual progress report. The annual progress report shall both:
1. Analyze the progress made, if any, in pollution prevention including toxics use reduction, source reduction and hazardous waste minimization relative to each performance goal established and relative to the plan contents prescribed in subsection J, paragraphs 4 through 11 of this section. Pollution prevention achieved under previously implemented activities may also be included.
2. Set forth amendments to the pollution prevention plan and explain the need for the amendments.

M. A facility that causes a one-time event that generates a hazardous waste or an acutely hazardous waste from an unused hazardous substance is exempt from filing a pollution prevention plan and an annual progress report if all of the following conditions are met:
1. The unused hazardous substance cannot lawfully be used due to changes in statute or rule.
2. A toxic data report has been filed for the event as prescribed in 49-962.
3. The toxic data report is required solely as a result of the one-time generation event.

N. This section does not apply to an episodic, accidental or remediation related release or occurrence.

O. A person who would be required to file a plan as prescribed by subsection A of this section solely due to the storage, supply, application or use of a pesticide as defined in 3-361 for agricultural application and who is subject to reporting or recordkeeping requirements pursuant to 49-305 or rules adopted pursuant to 3-363 or a person who is issued an agricultural general permit pursuant to 49-247 is exempt from the plan filing prescribed in subsection A of this section.

49-964. Review of reports and plans; enforcement; contempt

A. The department shall review the submissions required under this article, including the plan and any amendments and reports, to determine if the submission is complete and correct as prescribed in 49-962 and 49-963.

B. If a facility required to submit a plan or report under this article files an inadequate submission, the department shall notify the facility in writing of the inadequacy, identifying the specific deficiencies. In reviewing the adequacy of a plan or report, or any amendment to a plan or report, the department shall base its determination on whether the plan, report or amendment is complete and correct in accordance with the requirements of this article. If the submission is inadequate, the department shall specify a reasonable time of at least ninety days within which the facility shall file a modified submission addressing the specified deficiencies.

C. If, after the specified time, the facility has not filed a modified submission or the modified submission is otherwise inadequate, the department may enter a formal notice of inadequacy. The department shall place a copy or abstract of the notice of inadequacy in the department's annual report.

D. If a formal notice of inadequacy is entered, the department, pursuant to title 41, chapter 6, [41-1001 et seq.] may hold a public hearing at least thirty days but not more than ninety days after providing written notice to the facility.

The department may issue an administrative order requiring the facility to correct the deficiencies. If the facility fails to comply with an administrative order, the department may enforce that order in a judicial proceeding including an action for contempt.

E. In reviewing for adequacy an amendment or annual progress report, the department's review is restricted to the scope of the current submission.

Previous amendments to the plan and annual progress reports that were found to be adequate are not subject to review.

F. If a facility required under this article to submit a plan or annual progress report fails to submit the plan or report, the department shall order that facility to submit an adequate plan or report within a reasonable time period of at least ninety days. If the facility fails to develop an adequate plan or progress report in response to that order within the time period specified in that order, the department may do any of the following:

1. Under procedures established by rule, provide for inspecting the facility, gathering necessary information and preparing a plan or progress report for the facility at the facility's expense.

2. Pursuant to title 41, chapter 6, enter an administrative order for compliance that is enforceable in a judicial proceeding including an action for contempt.

G. The attorney general, at the request of the director, may bring an action in superior court to recover the department's costs incurred under subsection F of this section. The facility owner or operator may appeal the department's determination to proceed under this subsection and subsection F pursuant to title 41, chapter 6 before the department prepares the plan or progress report. Any final agency order issued pursuant to this section is subject to judicial review pursuant to title 12, chapter 7, article 6. [12-901 et seq.]

H. Failure to implement the pollution prevention plan is a violation of this article and the attorney general, at the request of the director, may bring an action in superior court to compel implementation of the provisions of an approved plan, and the director pursuant to title 41, chapter 6 may enter an administrative order for compliance that is enforceable in a judicial proceeding including an action for contempt.

I. Reports and submissions made to the department pursuant to this article shall be deemed adequate for purposes of this article unless the department notifies the facility in writing of any deficiencies within ninety days of receipt of the submission.

49-965. Pollution prevention technical assistance program

A. The department shall establish a technical assistance program designed to assist all persons in reducing to the fullest extent possible the amount and toxicity of the hazardous waste that is generated or toxic substances that are used in this state. The assistance program may include:
1. The establishment of a hazardous waste reduction clearinghouse of all available information concerning hazardous waste reduction, toxic substances minimization, recycling programs, economic and energy savings, and production and environmental improvements.
2. The production of workshops, conferences and handbooks on the topics described in paragraph 1 of this subsection.
3. Cooperation with university programs to develop hazardous waste reduction and toxic substances minimization curricula and training.
4. Presentation of on-site technical assistance for hazardous waste generators and toxic substances users.
5. Researching and recommending incentive programs for innovative hazardous waste management and toxic substances reduction.
6. Conducting a public education program to reduce emissions of ozone forming substances and accelerate pollution prevention technical assistance efforts to avoid ozone violations.

B. Presentation of on-site technical assistance by the department in accordance with subsection A of this section shall not be considered engineering practice as defined in 32-101 as long as the assistance does not involve engineering calculations or design in connection with any building, machine, equipment, process, work or project and is limited to assisting in the identification of pollution prevention opportunities and potential alternatives to existing processes or practices.

C. Persons receiving on-site technical assistance pursuant to this section are solely liable for any civil or other damages arising from on-site technical assistance. This state and the department, its officers, employees and agents and any other persons assisting in providing on-site technical assistance are not liable for any civil or other damages arising from the on-site technical assistance, including damages claimed by third parties.

D. The technical assistance program shall be funded from monies allocated from the hazardous waste management fund as prescribed in 49-927, except that the portion of the program prescribed by subsection A, paragraph 6 of this section shall be funded from monies allocated from the air quality fund as prescribed in 49-551. Notwithstanding 49-551, subsection C, paragraph 7, priority for funding shall be given to the portion of the program prescribed by subsection A, paragraph 6 of this section.

49-966. Annual report by department

On December 1 of each year, the department shall submit an annual report evaluating the pollution prevention program to the governor, the president of the senate and the speaker of the house of representatives. The department of environmental quality shall utilize information submitted to the department of agriculture pursuant to title 3, chapter 2, articles 5 and 6 [3-341 et seq. and 3-361 et seq.] and the division of emergency management pursuant to title 26, chapter 2, article 3. [26-331 et seq.] The annual report shall contain all of the following:
1. A description of the department's activities to encourage pollution prevention with particular attention to technical assistance efforts.
2. A compilation of the information received in the toxic data reports.
3. An estimate of the total number of facilities reporting and filing plans.
4. A synopsis of progress, summarized from the annual progress reports.
5. A list of facilities that have inadequate plans or reports, that have failed to make progress toward the goals established in their plans or that have made inadequate progress toward their goals.
6. A list of facilities that file an annual toxic chemical release form pursuant to 313 of the Superfund Amendments and Reauthorization Act of 1986 (P.L. 99-499) and that are not required to file a hazardous waste generator report under federal law, and recommendations regarding participation in the pollution prevention program for those facilities.

7. Specific recommendations for statutory and regulatory changes to the program to improve compliance, enforcement and progress toward the purposes of this article.

49-967. Availability of information to the public

A. Any records, reports or information obtained from any person under this article, including records, reports or information obtained or prepared by the director or a department employee, shall be available to the public, except that the information, or a particular part of the information, shall be considered confidential on either:
1. Notice from the person, accompanying the information, stating that the information, or a particular part of the information, if made public, would divulge the trade secrets of the person or other information likely to cause substantial harm to the person's competitive position.
2. A determination by the attorney general that disclosure of the information or a particular part of the information would be detrimental to an ongoing criminal investigation or to an ongoing or contemplated civil enforcement action under this chapter in superior court.

B. If the director, on his own or following a request for disclosure, disagrees with the confidentiality notice, he may request the attorney general to seek a court order authorizing disclosure. If a court order is sought, the person shall be served with a copy of the court filing and shall have twenty business days from the date of service to request a hearing on whether a court order should be issued. The hearing shall be conducted in camera, and any order resulting from the hearing is appealable as provided by law. The director may not disclose the confidential information until a court order authorizing disclosure has been obtained and becomes final. The court may award costs of litigation including reasonable attorney and expert witness fees to the prevailing party.

C. Notwithstanding subsection A, the following information obtained from any person under this article shall be available to the public:
1. The name and address of any permit applicant or permittee.
2. The types and amounts of any hazardous waste generated, stored, treated or disposed.
3. The types and amounts of any toxic substances released to the environment.

D. Notwithstanding subsection A, the director may disclose, with accompanying confidentiality notice, any records, reports or information obtained from any person under this article, including records, reports or information obtained by the director or department employees, to:
1. Other state employees concerned with administering this chapter or if the records, reports or information are relevant to any administrative or judicial proceeding under this chapter.

2. Employees of the United States environmental protection agency if such information is necessary or required to administer and implement or comply with federal statutes or regulations.

49-968. Department rules

A. The director shall adopt by rule any substance established by the administrator of the United States environmental protection agency pursuant to the pollution prevention act of 1990 (42 U.S.C. 13102(3)). Except for public notice requirements this adoption is exempt from the requirements of title 41, chapter 6. [41-1001 et seq.] The director shall give public notice pursuant to title 41, chapter 6 of the adoption of substances under this section.

B. The director may, by rule, add or delete a substance to the list of toxic substances based on the same criteria described in the Emergency Planning and Community Right to Know Act (42 U.S.C. 11023(d)).

ARIZONA REVISED STATUTES
TITLE 49. THE ENVIRONMENT
CHAPTER 5. HAZARDOUS WASTE DISPOSAL
ARTICLE 5. POLLUTION PREVENTION FOR STATE AGENCIES

49-971. Repealed by Laws 1994, Ch. 299, sec. 4

[The repealed section, added by Laws 1991, Ch. 315, sec. 32, related to pollution prevention reports.]

49-972. Pollution prevention plan for state agencies; definition

A. A state agency that produces hazardous waste or uses toxic substances in excess of the threshold quantity and time limits prescribed in 49-963 shall file a pollution prevention plan with the director. The pollution prevention plan shall have a goal of twenty percent reduction in hazardous waste within two years, fifty percent reduction in hazardous waste within five years and a seventy percent reduction in hazardous waste in ten years.

B. The pollution prevention plan shall address a reduction in the use of toxic substances and the generation of hazardous wastes. The plan shall be completed on a form published by the director and shall be filed with the director on or before January 1, 1993 and every five years thereafter.

C. A state agency required to file a pollution prevention plan may include in the report a certification that there is no reasonably available and technically feasible alternative to the current level of generation of hazardous waste at its facilities. If approved by the director, the certification shall serve as demonstration of compliance with the goals stated in subsection A of this section.

D. The state agency required to prepare a pollution prevention plan shall maintain a copy of the plan and annual summaries at the agency and at the facility, where they shall be available for inspection by the department and by the public.

E. The pollution prevention plan summary shall include a summary of all data and information in the plan, including the following:

1. A statement of the scope and objectives of the pollution prevention plan considering toxicity, volume, disposal costs and liability costs, and a numerical statement of the reductions in facility use of each hazardous waste at the facility over the next five or more years.

2. An analysis identifying pollution prevention opportunities to reduce or eliminate toxic substance releases and hazardous waste generation.

3. The name and location of all facilities associated with the state agency that are included in the plan and the name, address and telephone number of the operator and the senior official with management responsibility at the facility.

4. Identification and explanation of technology, procedures and options considered available and technically feasible for reducing the use of each hazardous waste and toxic substance at the facility, an explanation of options not implemented and a time schedule for implementing chosen options.

5. A written certification that the agency has prepared a pollution prevention plan and that a copy of the plan is available at the agency or facility for the department's inspection and for inspection by the public on request to the department.

6. Specific performance goals for the prevention of pollution, including an explanation of the rationale for each performance goal. The plan shall include a goal for the facility and may include goals for individual processes, operations, toxic substance usage and hazardous waste generation.

7. A written certification by the senior official with management responsibility that he has read the plan and that to the best of his knowledge it is true, accurate and complete.

8. A written policy setting forth management support for the pollution prevention plan and a commitment to implement the plan to achieve the plan goals.

9. An analysis of pollution prevention activities that are already in place and that are consistent with the requirements of this article.

10. Employee awareness and training programs to involve employees in pollution prevention planning and implementation to the maximum extent feasible.

11. Provisions to incorporate the plan into management practices and procedures to ensure the plan's institutionalization.

F. To the extent practicable, the information required for the preparation of a pollution prevention plan shall be based on information developed and forms completed by the state agency for the purposes of compliance with 26-347 and 26-351, the federal Pollution Prevention Act, 304(l) of the Federal Water Pollution

Control Act, pretreatment sludge permits pursuant to 40 C.F.R. part 503, or other required state and federal reports.

G. The department shall make all pollution plans and pollution prevention plan summaries available to the public.

H. If the department determines that a plan is not in compliance with the requirements of this section, the department may allow the person submitting the plan ninety days from the date of the notice of deficiency to correct the deficiency.

I. Each state agency required to prepare and maintain a pollution prevention plan shall file an annual progress report. The annual progress report shall both:

1. Analyze the progress made, if any, in pollution prevention including toxics use reduction, source reduction and hazardous waste minimization relative to each performance goal established and relative to the plan contents.

2. Set forth amendments to the pollution prevention plan and explain the need for the amendments.

J. If the threshold quantity prescribed in 49-963 is exceeded due to an accidental or remediation related release or occurrence, the requirement to file a plan pursuant to this section does not apply.

K. For purposes of this section, "state agency" includes all facilities controlled by an agency.

49-973. Toxic data report; progress report

A. A state agency shall file a toxic data report on July 1 if the agency during the preceding calendar year generated ten thousand pounds or more of hazardous waste. For purposes of this section, the hazardous waste generated shall consist of the aggregate amount of hazardous waste generated from all facilities directly controlled by the state agency.

B. The toxic data report required in subsection A of this section shall include the following:

1. An annual progress report as prescribed in 49-972, subsection H.

2. Toxic chemical information in the format of a toxic chemical release form required by 313 of the Superfund Amendments and Reauthorization Act of 1986 (P.L. 99-499) for each facility within the state agency that meets the threshold quantities as prescribed by the department and that is not subject to exemptions.

3. The address of each off-site treatment, storage or disposal facility to which each hazardous waste generated was transported and the type of treatment or disposal methods used for each hazardous waste at each off-site facility.

C. If the threshold quantity prescribed in subsection A of this section is exceeded due to an accidental or remediation related release or occurrence, the requirement to file a report pursuant to this section does not apply.

California

CALIFORNIA HEALTH AND SAFETY CODE
DIVISION 20. MISCELLANEOUS HEALTH AND SAFETY PROVISIONS
CHAPTER 6.5. HAZARDOUS WASTE CONTROL
ARTICLE 11.9. HAZARDOUS WASTE SOURCE REDUCTION AND
MANAGEMENT REVIEW ACT OF 1989

25244.12. Short title

This article shall be known and may be cited as the Hazardous Waste Source Reduction and Management Review Act of 1989.

25244.13. Legislative findings and intent

The Legislature finds and declares as follows:
(a) Existing law requires the department and the State Water Resources Control Board to promote the reduction of generated hazardous waste.
This policy, in combination with hazardous waste land disposal bans, requires the rapid development of new programs and incentives for achieving the goal of optimal minimization of the generation of hazardous wastes. Substantial improvements and additions to the state's hazardous waste reduction program are required to be made if these goals are to be achieved.
(b) It is the goal of this article to do all of the following:
(1) Reduce the generation of hazardous waste.
(2) Reduce the release into the environment of chemical contaminants which have adverse and serious health or environmental effects.
(3) Document hazardous waste management information and make that information available to state and local government.
(c) It is the intent of this article to promote the reduction of hazardous waste at its source, and wherever source reduction is not feasible or practicable, to encourage recycling. Where it is not feasible to reduce or recycle hazardous waste, the waste should be treated in an environmentally safe manner to minimize the present and future threat to health and the environment.
(d) It is the intent of the Legislature not to preclude the regulation of environmentally harmful releases to all media, including air, land, surface water, and groundwater, and to encourage and promote the reduction of these releases to air, land, surface water, and groundwater.
(e) It is the intent of the Legislature to encourage all state departments and agencies, especially the State Water Resources Control Board, the California regional water quality control boards, the State Air Resources Board, the air

pollution control districts, and the air quality management districts, to promote the reduction of environmentally harmful releases to all media.

25244.14. Definitions

For purposes of this article, the following definitions apply:

(a) "Appropriate local agency" means a county, city, or regional association which has adopted a hazardous waste management plan pursuant to Article 3.5 (commencing with Section 25135).

(b) "Hazardous waste management approaches" means approaches, methods, and techniques of managing the generation and handling of hazardous waste, including source reduction, recycling, and the treatment of hazardous waste.

(c) "Hazardous waste management performance report" or "report" means the report required by subdivision (b) of Section 25244.20 to document and evaluate the results of hazardous waste management practices.

(d) "Hazardous waste management performance report summary" or "report summary" means the summary required by subdivision (c) of Section 25244.20.

(e) (1) "Source reduction" means one of the following:

(A) Any action which causes a net reduction in the generation of hazardous waste.

(B) Any action taken before the hazardous waste is generated that results in a lessening of the properties which cause it to be classified as a hazardous waste.

(2) "Source reduction" includes, but is not limited to, all of the following:

(A) "Input change" which means a change in raw materials or feedstocks used in a production process or operation so as to reduce, avoid, or eliminate the generation of hazardous waste.

(B) "Operational improvement" which means improved site management so as to reduce, avoid, or eliminate the generation of hazardous waste.

(C) "Production process change" which means a change in a process, method, or technique which is used to produce a product or a desired result, including the return of materials or their components, for reuse within the existing processes or operations, so as to reduce, avoid, or eliminate the generation of hazardous waste.

(D) "Product reformulation" which means changes in design, composition, or specifications of end products, including product substitution, so as to reduce, avoid, or eliminate the generation of hazardous waste.

(3) "Source reduction" does not include any of the following:

(A) Actions taken after a hazardous waste is generated.

(B) Actions that merely concentrate the constituents of a hazardous waste to reduce its volume or that dilute the hazardous waste to reduce its hazardous characteristics.

(C) Actions that merely shift hazardous wastes from one environmental medium to another environmental medium.

(D) Treatment.

(f) "Source reduction evaluation review and plan" or "review and plan" means a review conducted by the generator of the processes, operations, and procedures in use at a generator's site, according to the format established by the department pursuant to subdivision (a) of Section 25244.16, and which does both of the following:

(1) Determines any alternatives to, or modifications of, the generator's processes, operations, and procedures that may be implemented to reduce the amount of hazardous waste generated.

(2) Includes a plan to document and implement source reduction measures for the hazardous wastes specified in paragraph (1) which are technically feasible and economically practicable for the generator, including a reasonable implementation schedule.

(g) "Source reduction evaluation review and plan summary" or "plan summary" means the summary required by subdivision (c) of Section 25244.19.

(h) "SIC Code" has the same meaning as defined in Section 25501.

(I) "Hazardous waste," "person," "recycle," and "treatment" have the same meaning as defined in Article 2 (commencing with Section 25110).

25244.15. Establishment of program; state agencies coordination; regulation; application of article; compliance checklist; purpose of article

(a) The department shall establish a program for hazardous waste source reduction pursuant to this article.

(b) The department shall coordinate the activities of all state agencies with responsibilities and duties relating to hazardous waste and shall promote coordinated efforts to encourage the reduction of hazardous waste. Coordination between the program and other relevant state agencies and programs shall, to the fullest extent possible, include joint planning processes and joint research and studies.

(c) The department shall adopt regulations to carry out this article.

(d) (1) Except as provided in paragraph (3), this article applies only to generators who, by site, routinely generate, through ongoing processes and operations, more than 12,000 kilograms of hazardous waste in a calendar year, or more than 12 kilograms of extremely hazardous waste in a calendar year.

(2) The department shall adopt regulations to establish procedures for exempting generators from the requirements of this article where the department determines that no source reduction opportunities exist for the generator.

(3) (A) Notwithstanding paragraph (1), all generators who, by site, routinely generate, through ongoing processes and operation, more than 5,000 kilograms per year of waste that is in any of the categories of hazardous wastes listed in paragraphs (1), (2), and (3) of subdivision (a) of Section 25179.7, shall comply

with this article for those hazardous wastes, unless they are otherwise exempted pursuant to this article. This paragraph shall become inoperative on December 31, 1997.

(B) The department shall modify the review and plan requirements of Section 25244.19 for generators who generate less than 12,000 kilograms per year by substituting a compliance checklist approach for source reduction evaluation reviews and plans. The compliance checklist shall include the specific numerical goals for reducing the generation of hazardous waste streams provided for in paragraph (9) of subdivision (b) of Section 25244.19.

The purpose of the compliance checklist is to provide a simple, understandable method for small generators to comply with the requirements of this article in an inexpensive, convenient manner. The department shall make available to all generators eligible to use the checklist a list, written in language understandable to a lay person, of the categories of hazardous wastes listed in paragraphs (1), (2), and (3) of subdivision (a) of Section 25179.7.

(e) It is the purpose of this article to reduce the generation of hazardous waste in California by 5 percent per year from the year 1993 to the year 2000. On or before January 1, 2000, the department shall recommend to the Legislature the adoption of a new annual waste reduction goal.

25244.16. Adoption of format for use by generators; data and information system for developing categories of generators

On or before January 1, 1991, the department shall do both of the following:

(a) Adopt a format to be used by generators for completing the review and plan and plan summary required by Section 25244.19, and the report and the report summary required by Section 25244.20. The format shall include at least all of the factors the generator is required to include in the review and plan, the plan summary, the report, and the report summary. The department may include any other factor determined by the department to be necessary to carry out this article. The adoption of a format pursuant to this subdivision is not subject to Chapter 3.5 (commencing with Section 11340) of Part 1 of Division 3 of Title 2 of the Government Code.

(b) Establish a data and information system to be used by the department for developing the categories of generators specified in Section 25244.18, for processing and evaluating the source reduction and other hazardous waste management information submitted by generators pursuant to Section 25244.18, and for developing the program evaluation required by Section 25244.22. In establishing the data and information system, the department shall do all of the following:

(1) Establish methods and procedures for appropriately processing or managing hazardous waste source reduction and management information.

(2) Use the data management expertise, resources, and forms of already established environmental protection programs, to the extent practicable.
(3) Establish computerized data retrieval and data processing systems, including safeguards to protect trade secrets designated pursuant to Section 25244.23.
(4) Identify additional data and information needs of the program.

25244.17. Technical and research assistance program; elements

The department shall establish a technical and research assistance program to assist generators in identifying and applying methods of source reduction and other hazardous waste management approaches. The program shall emphasize assistance to smaller businesses that have inadequate technical and financial resources for obtaining information, assessing source reduction methods, and developing and applying source reduction techniques. The program shall include at least all of the following elements, which shall be carried out by the department:

(a) The department shall encourage programs by private or public consultants, including on-site consultation at sites or locations where hazardous waste is generated, to aid those generators requiring assistance in developing and implementing the review and plan, the plan summary, the report, and the report summary required by this article.

(b) The department shall conduct review and plan assistance programs, seminars, workshops, training programs, and other similar activities to assist generators to evaluate source reduction alternatives and to identify opportunities for source reduction.

(c) The department shall establish a program to assemble, catalogue, and disseminate information about hazardous waste source reduction methods, available consultant services, and regulatory requirements.

(d) The department shall identify the range of generic and specific technical solutions that can be applied by particular types of hazardous waste generators to reduce hazardous waste generation.

25244.18. Categories of generators selected by SIC Code; generator's review and plan, plan summary, report, and report summary; enforcement requirements and actions; local agency request from generator

(a) On or before September 15, 1991, and every two years thereafter, the department shall select at least two categories of generators by SIC Code with potential for source reduction, and, for each category, shall do all of the following:
(1) Request that selected generators in the category provide the department, on a timely basis, with a copy of the generator's completed review and plan, or plan

Appendix E: State Pollution Prevention Laws / 401

summary, or both, and with a copy of the generator's completed report, or report summary, or both.

(2) Examine the review and plan or plan summary and the report or report summary of selected generators in the category.

(3) Ensure that the selected generators in that category comply with Sections 25244.19 and 25244.20.

(4) Identify successful source reduction and other hazardous waste management approaches employed by generators in the category and disseminate information concerning those approaches to generators within the category.

(b) In carrying out subdivision (a), the department shall not disseminate information determined to be a trade secret pursuant to Section 25244.23.

(c) The department may request from any generator, and the generator shall provide within 30 days of the request, a copy of the generator's review and plan, plan summary, report, or report summary. The department may evaluate any of these documents submitted to the department to determine whether it satisfies the requirements of this article.

(d) (1) If the department determines that a generator has not completed the review and plan or plan summary in the manner required by Section 25244.19, or the report or report summary in the manner required by Section 25244.20, the department shall provide the generator with a notice of noncompliance, specifying the deficiencies in the review and plan, plan summary, report, or report summary identified by the department. If the department finds that the review and plan does not comply with Section 25244.19, the department shall consider the review and plan to be incomplete. A generator shall file a revised review and plan, plan summary, report, or report summary correcting the deficiencies identified by the department within 60 days of the receipt of the notice. The department may grant, in response to a written request from the generator, an extension of the 60-day deadline, for cause, except that the department shall not grant this extension for more than an additional 60 days.

(2) If a generator fails to submit a revised review and plan, plan summary, report, or report summary complying with the requirements of this article within the required period, or if the department determines that a generator has failed to implement the measures included in the generator's review and plan or plan summary for reducing the generator's hazardous waste, in accordance with Section 25244.19, except as provided in subdivision (e), the department may impose civil penalties pursuant to Section 25187, in an amount not to exceed one thousand dollars ($1,000) for each day the violation of this article continues, notwithstanding Section 25189.2, seek an order directing compliance pursuant to Section 25181, or enter into a consent agreement or a compliance schedule with the generator.

(e) If a generator fails to implement a measure specified in the review and plan, or plan summary, pursuant to paragraph (5) of subdivision (b) of Section 25244.19, the generator shall not be deemed to be in violation of Section

25244.19 for not implementing the selected measure if the generator does both of the following:

(1) The generator finds that, upon further analysis or as a result of unexpected consequences, the selected measure is not technically feasible or economically practicable, or if the selected approach has resulted in any of the following:

(A) An increase in the generation of hazardous waste.

(B) An increase in the release of hazardous chemical contaminants to other media.

(C) Adverse impacts on product quality.

(D) A significant increase in the risk of an adverse impact to human health or the environment.

(2) The generator revises the review and plan and plan summary to comply with the requirements of Section 25244.19.

(f) When taking enforcement action pursuant to this article, the department shall not judge the appropriateness of any decisions or proposed measures contained in a review and plan, plan summary, report, or report summary, but shall only determine whether the review and plan, plan summary, report, or report summary is complete, prepared, and implemented in accordance with this article.

(g) An appropriate local agency which has jurisdiction over a generator's site may request from the generator, and the generator shall provide within 30 days, a copy of the generator's current review and plan, plan summary, report, and report summary.

25244.19. Generator source reduction evaluation review and plan; summary; multi-site review and plan and plan summary; review and certification by engineer, individual or environmental assessor; implementation or rejection of measures by generator

(a) On or before September 1, 1991, and every four years thereafter, each generator shall conduct a source reduction evaluation review and plan and a source reduction evaluation review and plan summary pursuant to subdivisions (b) and (c).

(b) Except as provided in subdivision (d), the source reduction evaluation review and plan required by subdivision (a) shall be conducted and completed for each site pursuant to the format adopted pursuant to subdivision (a) of Section 25244.16 and shall include, at a minimum, all of the following:

(1) The name and location of the site.

(2) The SIC Code of the site.

(3) Identification of all routinely generated hazardous waste streams which result from ongoing processes or operations that have a yearly volume exceeding 5 percent of the total yearly volume of hazardous waste generated at the site, or, for

extremely hazardous waste, 5 percent of the total yearly volume generated at the site. For purposes of this paragraph, a hazardous waste exceeds 5 percent of the total yearly volume, and is subject to this article, if it is routinely generated on an ongoing basis and meets either of the following criteria:

(A) It is a hazardous waste stream processed in a wastewater treatment unit which discharges to a publicly owned treatment works or under a national pollutant discharge elimination system (NPDES) permit, as specified in the Federal Water Pollution Control Act, as amended (33 U.S.C. Sec. 1251 and following), and its weight before treatment exceeds 5 percent of the weight of the total yearly volume at the site.

(B) It is a hazardous waste stream which is not processed in a wastewater treatment unit and its weight exceeds 5 percent of the weight of the total yearly volume at the site, less the weight of any hazardous waste stream identified in subparagraph (A).

(4) For each hazardous waste stream identified in paragraph (3), the review and plan shall include all of the following information:

(A) An estimate of the quantity of hazardous waste generated.

(B) An evaluation of source reduction approaches available to the generator which are potentially viable. The evaluation shall consider at least all of the following source reduction approaches:

(i) Input change.
(ii) Operational improvement.
(iii) Production process change.
(iv) Product reformulation.

(5) A specification of, and a rationale for, the technically feasible and economically practicable source reduction measures which will be taken by the generator with respect to each hazardous waste stream identified in paragraph (3). The review and plan shall fully document any statement explaining the generator's rationale for rejecting any available source reduction approach identified in paragraph (4).

(6) An evaluation, and, to the extent practicable, a quantification, of the effects of the chosen source reduction method on emissions and discharges to air, water, or land.

(7) A timetable for making reasonable and measurable progress towards implementation of the selected source reduction measures specified in paragraph (5).

(8) Certification pursuant to subdivision (e).

(9) Any generator subject to this article shall include in its source
reduction evaluation review and plan four-year numerical goals for reducing the generation of hazardous waste streams through the approaches provided for in subparagraph (B) of paragraph (4), based upon its best estimate of what is achievable in that four-year period, as follows:

(A) For those generators and waste streams subject to this program prior to January 1, 1993, the four-year numerical goals shall be included in the plan which is required to be prepared by September 1, 1995, and every four years thereafter, pursuant to subdivision (a).

(B) Any generator that is subject to this program pursuant to paragraph (3) of subdivision (d) of Section 25244.15, and was not subject to this program before January 1, 1993, shall be required to prepare its source reduction evaluation review and plan, or compliance checklist, as provided in paragraph (3) of subdivision (d) of Section 25244.15, on September 1, 1993, and every four years thereafter.

(10)(A) Any generator subject to this article shall submit to the department a progress report as part of its submittal of a biennial report on March 1 of each even-numbered year pursuant to Section 66262.41 of Title 22 of the California Code of Regulations. If the generator is not required to submit a biennial report pursuant to that regulation, it shall prepare a separate progress report on the same time schedule required for the biennial report. Generators not required to submit a biennial report shall not be required to submit their prepared progress report. However, the progress report shall be subject to the requirements applicable to reports under Section 25244.21. Progress reports shall address plan implementation activities undertaken by the generator during the two years preceding the year in which the biennial report is required to be submitted.

(B) The progress report shall briefly summarize and, to the extent practicable, quantify, in a manner which is understandable to the general public, the results of implementing the source reduction methods identified in the generator's source reduction evaluation, review, and plan for each waste stream addressed by that plan. The progress report due on March 1, 1994, and every other progress report thereafter, shall also include an estimate of the amount of reduction the generator anticipates will be achieved by the implementation of source reduction methods during the period between the preparation of the progress report and the preparation of its next progress report. This estimate shall not require any new evaluation, review, or planning, but shall be based upon the generator's existing source reduction, evaluation, review, and plan and upon the waste reduction results in the progress report. The department shall prepare a form which may be used as a convenient and expeditious method for generators to comply with the progress report requirements of this section. The department shall hold at least one public workshop concerning the development of the form.

(c) The source reduction evaluation review and plan summary required by subdivision (a) shall be completed in accordance with the format adopted pursuant to subdivision (a) of Section 25244.19 and shall include the information specified in paragraphs (1), (2), (3), and (6) of subdivision (b) and a summary of the information required pursuant to paragraphs (4) and (5) of subdivision (b).

Appendix E: State Pollution Prevention Laws / 405

(d) If a generator owns or operates multiple sites with similar processes, operations, and waste streams, the generator may prepare a single multisite review and plan and plan summary addressing all of these sites.

(e) Every review and plan and plan summary conducted pursuant to this section shall be submitted by the generator for review and certification by an engineer who is registered as a professional engineer pursuant to Section 6762 of the Business and Professions Code and who has demonstrated expertise in hazardous waste management, by an individual who is responsible for the processes and operations of the site, or by an environmental assessor who is registered pursuant to Section 25570.3 and who has demonstrated expertise in hazardous waste management. The engineer, individual, or environmental assessor shall certify the review and plan and plan summary only if the review and plan and plan summary meet all of the following requirements:

(1) The review and plan addresses each hazardous waste stream identified pursuant to paragraph (3) of subdivision (b).

(2) The review and plan addresses the source reduction approaches specified in subparagraph (B) of paragraph (4) of subdivision (b).

(3) The review and plan clearly sets forth the measures to be taken with respect to each hazardous waste stream for which source reduction has been found to be technically feasible and economically practicable, with timetables for making reasonable and measurable progress, and properly documents the rationale for rejecting available source reduction measures.

(4) The plan summary meets the requirements of subdivision (c).

(5) The review and plan and plan summary does not merely shift hazardous waste from one environmental medium to another environmental medium by increasing emissions or discharges to air, water, or land.

(f) At the time a review and plan or a plan summary is submitted to the department, the generator shall certify that the generator has implemented, is implementing, or will be implementing, the source reduction measures identified in the review and plan or the plan summary according to the implementation schedule contained in the review and plan or the plan summary. A generator may determine not to implement a measure selected in paragraph (5) of subdivision (b) only if the generator determines, upon conducting further analysis or due to unexpected circumstances, that the selected measure is not technically feasible or economically practicable, or if attempts to implement that measure reveal that the measure would result in, or has resulted in, any of the following:

(1) An increase in the generation of hazardous waste.

(2) An increase in the release of hazardous chemicals to other environmental media.

(3) Adverse impacts on product quality.

(4) A significant increase in the risk of an adverse impact to human health or the environment.

(g) If the generator elects not to implement the review and plan or plan summary, including, but not limited to, a selected measure pursuant to subdivision (f), the generator shall amend its review and plan and plan summary to reflect this rejection and include in the review and plan and plan summary proper documentation identifying the rationale for this rejection.

Minnesota

MINNESOTA STATUTES
ENVIRONMENTAL PROTECTION
CHAPTER 115D. TOXIC POLLUTION PREVENTION

115D.01. Citation

Sections 115D.01 to 115D.12 may be cited as the "Minnesota toxic pollution prevention act."

115D.02. Policy

(a) To protect the public health, welfare, and the environment, the legislature declares that it is the policy of the state to encourage toxic pollution prevention. The preferred means of preventing toxic pollution are techniques and processes that are implemented at the source and that minimize the transfer of toxic pollutants from one environmental medium to another.

(b) The legislature intends that the programs developed under sections 115D.01 to 115D.12 shall encourage and lead to a greater awareness of the need for and benefits of toxic pollution prevention, and to a greater degree of cooperation and coordination among all elements of government, industry, and the public in encouraging and carrying out pollution prevention activities.

115D.03. Definitions

Subdivision 1. Applicability. The definitions in this section apply to this chapter.

Subd. 2. Commission. "Commission" means the emergency response commission under section 299K.03.

Subd. 3. Commissioner. "Commissioner" means the commissioner of the pollution control agency.

Subd. 4. Director. "Director" means the director of the office of environmental assistance.

Subd. 5. Eligible recipients. "Eligible recipients" means persons who use, generate, or release toxic pollutants, hazardous substances, or hazardous wastes.

Subd. 6. Facility. "Facility" means all buildings, equipment, structures, and other stationary items that are located on a single site or on contiguous or adjacent sites and that are owned or operated by the same person, or by any person who controls, is controlled by, or is under common control with such person.

Subd. 7. Person. "Person" means any individual, partnership, association, public or private corporation or other entity including the United States government, any interstate body, the state and any agency, department or political subdivision of the state.

Subd. 8. Pollution prevention or prevent pollution. "Pollution prevention" or "prevent pollution" means eliminating or reducing at the source the use, generation, or release of toxic pollutants, hazardous substances, and hazardous wastes.

Subd. 9. Reduce, reducing, or reduction. "Reduce," "reducing," or "reduction" means lessening the quantity or toxicity of toxic pollutants, hazardous substances, and hazardous wastes used, generated, or released at the source. Methods of reducing pollution include, but are not limited to, process modification, inventory control measures, feedstock substitutions, various housekeeping and management practices, and improved efficiency of machinery. Decreases in quantity or toxicity are not reductions where the decrease is solely the result of a decrease in the output of the facility.

Subd. 10. Release. "Release" means any spilling, leaking, pumping, pouring, emitting, emptying, discharging, injecting, escaping, leaching, dumping, or disposing into the environment which occurred at a point in time or which continues to occur.

"Release" does not include:

(1) emissions from the engine exhaust of a motor vehicle, rolling stock, aircraft, watercraft, or pipeline pumping station engine;

(2) release of source, by-product, or special nuclear material from a nuclear incident, as those terms are defined in the Atomic Energy Act of 1954, under United States Code, title 42, section 2014, if the release is subject to requirements with respect to financial protection established by the federal Nuclear Regulatory Commission under United States Code, title 42, section 2210;

(3) release of source, by-product, or special nuclear material from any processing site designated pursuant to the Uranium Mill Tailings Radiation Control Act of 1978, under United States Code, title 42, section 7912(a)(1) or 7942(a); or

(4) any release resulting from the application of fertilizer or agricultural or silvicultural chemicals, or disposal of emptied pesticide containers or residues from a pesticide as defined in section 18B.01, subdivision 18.

Subd. 11. Toxic pollutant. "Toxic pollutant" means a chemical identified in United States Code, title 42, section 11023(c).

115D.04. Pollution prevention assistance program

Subdivision 1. Establishment. The director shall establish a pollution prevention assistance program to assist eligible recipients in preventing pollution. The program must emphasize techniques and processes that minimize the transfer of pollutants from one environmental medium to another and must focus primarily on toxic pollutants.

Subd. 2. Assistance. The pollution prevention assistance program must include at least the following:

(1) a program to assemble, catalog, and disseminate information on pollution prevention;

(2) a program to provide technical research and assistance, including on-site consultations to identify alternative methods that may be applied to prevent pollution and to provide assistance for planning under section 115D.07, excluding design engineering services; and

(3) outreach programs including seminars, workshops, training programs, and other similar activities designed to provide pollution prevention information and assistance to eligible recipients and other interested persons.

Subd. 3. Administration. (a) The pollution prevention assistance program must be coordinated with other public and private programs that provide management and technical assistance to eligible recipients.

(b) The director may make grants to public or private entities to operate elements of the program. Grantees shall provide periodic reports on their efforts to assist eligible recipients to reduce pollution.

115D.05. Pollution prevention grants

Subdivision 1. Purpose. The director may make grants to study or demonstrate the feasibility of applying specific technologies and methods to prevent pollution.

Subd. 2. Eligibility. (a) Eligible recipients may receive grants under this section.

(b) Grants may be awarded up to a maximum of two-thirds of the total cost of the project. Grant money awarded under this section may not be spent for capital improvements or equipment.

Subd. 3. Procedure for awarding grants. (a) In determining whether to award a grant, the director shall consider at least the following:

(1) the potential of the project to prevent pollution;

(2) the likelihood that the project will develop techniques or processes that will minimize the transfer of pollution from one environmental medium to another;

(3) the extent to which information to be developed through the project will be applicable to other persons in the state;

(4) the willingness of the grant applicant to implement feasible methods and technologies developed under the grant;

(5) the willingness of the grant applicant to assist the director in disseminating information about the pollution prevention methods to be developed through the project; and
(6) the extent to which the project will conform to the pollution prevention policy established in section 115D.02.
(b) The director shall adopt rules to administer the grant program. Prior to completion of any new rulemaking, the director may administer the program under the procedures established in rules promulgated under section 115A.154.

115D.06. Governor's award for excellence in pollution prevention

The governor may issue annual awards in the form of a commendation for excellence in pollution prevention. Applications for these awards shall be administered by the director.

115D.07. Toxic pollution prevention plans

Subdivision 1. Requirement to prepare and maintain a plan. (a) Persons who operate a facility required by United States Code, title 42, section 11023, or section 299K.08, subdivision 3, to submit a toxic chemical release form shall prepare a toxic pollution prevention plan for that facility. The plan must contain the information listed in subdivision 2.
(b) Except as provided in paragraphs (d) and (e), for facilities that release a total of 10,000 pounds or more of toxic pollutants annually, the plan must be completed as follows:
(1) on or before July 1, 1991, for facilities having a two-digit standard industrial classification of 35 to 39;
(2) by January 1, 1992, for facilities having a two-digit standard industrial classification of 28 to 34; and
(3) by July 1, 1992, for all other persons required to prepare a plan under this subdivision.
(c) Except as provided in paragraphs (d) and (e), facilities that release less than a total of 10,000 pounds of toxic pollutants annually must complete their plans by July 1, 1992.
(d) For the following facilities, the plan must be completed as follows:
(1) by January 1, 1995, for facilities required to report under section 299K.08, subdivision 3, that have a two-digit standard industrial classification of 01 to 50; and
(2) by July 1, 1995, for facilities required to report under section 299K.08, subdivision 3, that have a two-digit standard industrial classification of 51 to 99.
(e) For facilities that become subject to this subdivision after July 1, 1993, the plan must be completed by six months after the first submittal for the facility

under United States Code, title 42, section 11023, or section 299K.08, subdivision 3.

(f) Each plan must be updated every two years and must be maintained at the facility to which it pertains.

Subd. 2. Contents of plan. (a) Each toxic pollution prevention plan must establish a program identifying the specific technically and economically practicable steps that could be taken during at least the three years following the date the plan is due, to eliminate or reduce the generation or release of toxic pollutants reported by the facility. Toxic pollutants resulting solely from research and development activities need not be included in the plan.

(b) At a minimum, each plan must include:

(1) a policy statement articulating upper management support for eliminating or reducing the generation or release of toxic pollutants at the facility;

(2) a description of the current processes generating or releasing toxic pollutants that specifically describes the types, sources, and quantities of toxic pollutants currently being generated or released by the facility;

(3) a description of the current and past practices used to eliminate or reduce the generation or release of toxic pollutants at the facility and an evaluation of the effectiveness of these practices;

(4) an assessment of technically and economically practicable options available to eliminate or reduce the generation or release of toxic pollutants at the facility, including options such as changing the raw materials, operating techniques, equipment and technology, personnel training, and other practices used at the facility. The assessment may include a cost benefit analysis of the available options;

(5) a statement of objectives based on the assessment in clause (4) and a schedule for achieving those objectives. Wherever technically and economically practicable, the objectives for eliminating or reducing the generation or release of each toxic pollutant at the facility must be expressed in numeric terms. Otherwise, the objectives must include a clearly stated list of actions designed to lead to the establishment of numeric objectives as soon as practicable;

(6) an explanation of the rationale for each objective established for the facility;

(7) a listing of options that were considered not to be economically and technically practicable; and

(8) a certification, signed and dated by the facility manager and an officer of the company under penalty of section 609.63, attesting to the accuracy of the information in the plan.

115D.08. Progress reports

Subdivision 1. Requirement to submit progress report. (a) All persons required to prepare a toxic pollution prevention plan under section 115D.07 shall submit an annual progress report to the commissioner that may be drafted in a manner that does not disclose proprietary information. Progress reports are due on October 1 of each year. The first progress reports are due in 1992.

(b) At a minimum, each progress report must include:

(1) a summary of each objective established in the plan including the schedule for meeting the objective;

(2) a summary of progress made during the past year, if any, toward meeting each objective established in the plan including the quantity of each toxic pollutant eliminated or reduced;

(3) a statement of the methods through which elimination or reduction has been achieved;

(4) if necessary, an explanation of the reasons objectives were not achieved during the previous year, including identification of any technological, economic, or other impediments the facility faced in its efforts to achieve its objectives; and

(5) a certification, signed and dated by the facility manager and an officer of the company under penalty of section 609.63, attesting that a plan meeting the requirements of section 115D.07 has been prepared and also attesting to the accuracy of the information in the progress report.

Subd. 2. Review of progress reports. (a) The commissioner shall review all progress reports to determine if they meet the requirements of subdivision 1. If the commissioner determines that a progress report does not meet the requirements, the commissioner shall notify the facility in writing and shall identify specific deficiencies and specify a reasonable time period of not less than 90 days for the facility to modify the progress report.

(b) The commissioner shall be given access to a facility plan required under section 115D.07 if the commissioner determines that the progress report for that facility does not meet the requirements of subdivision 1. Twenty-five or more persons living within ten miles of the facility may submit a petition to the commissioner that identifies specific deficiencies in the progress report and requests the commissioner to review the facility plan. Within 30 days after receipt of the petition, the commissioner shall respond in writing. If the commissioner agrees that the progress report does not meet requirements of subdivision 1, the commissioner shall be given access to the facility plan.

(c) After reviewing the plan and the progress report with any modifications submitted, the commissioner shall state in writing whether the progress report meets the requirements of subdivision 1. If the commissioner determines that a modified progress report still does not meet the requirements of subdivision 1, the commissioner shall schedule a public meeting. The meeting shall be held in the county where the facility is located. The meeting is not subject to the requirements of chapter 14.

(d) The facility shall be given the opportunity to amend the progress report within a period of not less than 30 days after the public meeting.

(e) If the commissioner determines that a modified progress report still does not meet the requirements of subdivision 1, action may be taken under section 115.071 to obtain compliance with sections 115D.01 to 115D.12.

115D.09. Confidentiality

Information and techniques developed under section 115D.04, the reduction information and techniques under section 115D.05, and the progress reports required under section 115D.08 are public data under chapter 13. The plans required under section 115D.07 are nonpublic data under chapter 13.

Texas

TEXAS HEALTH AND SAFETY CODE
TITLE 5. SANITATION AND ENVIRONMENTAL QUALITY
SUBTITLE B. SOLID WASTE, TOXIC CHEMICALS, SEWAGE, LITTER, AND WATER
CHAPTER 361. SOLID WASTE DISPOSAL ACT
SUBCHAPTER Q. POLLUTION PREVENTION

361.501. Definitions

In this subchapter:
(1) "Acute hazardous waste" means hazardous waste listed by the administrator of the United States Environmental Protection Agency under the federal Solid Waste Disposal Act, as amended by the Resource Conservation and Recovery Act of 1976 (42 U.S.C. Section 6901 et seq.), because the waste meets the criteria for listing hazardous waste identified in 40 C.F.R. Section 261.11(a)(2).
(2) Repealed by Acts 1991, 72nd Leg., 1st C.S., ch. 3, sec. 1.029, eff. Aug. 12, 1991.
(3) "Conditionally exempt small-quantity generator" means a generator that does not accumulate more than 1,000 kilograms of hazardous waste at any one time on his facility and who generates less than 100 kilograms of hazardous waste in any given month.
(4) "Committee" means the waste reduction advisory committee established by Section 361.0215.
(5) "Environment" means water, air, and land and the interrelationship that exists among and between water, air, land, and all living things.
(6) "Facility" means all buildings, equipment, structures, and other stationary items located on a single site or on contiguous or adjacent sites that are owned or operated by a person who is subject to this subchapter or by a person who controls, is controlled by, or is under common control with a person subject to this subchapter.

(7) "Generator" and "generator of hazardous waste" have the meaning assigned by Section 361.131.

(8) "Large-quantity generator" means a generator that generates, through ongoing processes and operations at a facility:
   (A) more than 1,000 kilograms of hazardous waste in a month; or
   (B) more than one kilogram of acute hazardous waste in a month.

(9) "Media" and "medium" mean air, water, and land into which waste is emitted, released, discharged, or disposed.

(10) "Pollutant" or "contaminant" includes any element, substance, compound, disease-causing agent, or mixture that after release into the environment and on exposure, ingestion, inhalation, or assimilation into any organism, either directly from the environment or indirectly by ingestion through food chains, will or may reasonably be anticipated to cause death, disease, behavioral abnormalities, cancer, genetic mutation, physiological malfunctions, including malfunctions in reproduction, or physical deformations in the organism or its offspring. The term does not include petroleum, crude oil, or any fraction of crude oil that is not otherwise specifically listed or designated as a hazardous substance under Sections 101(14)(A) through (F) of the environmental response law, [42 U.S.C. 9601 et seq.] nor does it include natural gas, natural gas liquids, liquefied natural gas, synthetic gas of pipeline quality, or mixtures of natural gas and synthetic gas.

(11) "Source reduction" has the meaning assigned by the federal Pollution Prevention Act of 1990, Pub.L. 101-508, Section 6603, 104 Stat. 1388. [42 U.S.C. 13101 et seq.]

(12) "Waste minimization" means a practice that reduces the environmental or health hazards associated with hazardous wastes, pollutants, or contaminants. Examples may include reuse, recycling, neutralization, and detoxification.

361.502. Policy and Goals for Source Reduction and Waste Minimization

(a) It is the policy of the state to reduce pollution at its source and to minimize the impact of pollution in order to reduce risk to public health and the environment and continue to enhance the quality of air, land, and waters of the state where feasible.

(b) Source reduction is the primary goal of the state in implementing this policy because hazardous wastes, pollutants, and contaminants that are not generated or produced pose no threat to the environment and eliminate societal management and disposal costs.

(c) To further promote this policy, hazardous wastes, pollutants, and contaminants that cannot be reduced at the source should be minimized wherever possible. Waste minimization, while secondary in preference to source reduction, is an important means for achieving more effective protection of public health and the environment while moving toward source reduction.

361.503. Agency Plans

(a) Consistent with state and federal regulations, to achieve the policies stated in Section 361.502, the board by rule shall, to the maximum extent that is technologically and economically feasible, develop plans to reduce the release of pollutants or contaminants into the air.

(b) Consistent with state and federal regulations, to achieve the policies stated in Section 361.502, to the maximum extent that is technologically and economically feasible, the commission shall:

(1) develop plans to reduce the release of pollutants or contaminants into water; and

(2) establish reasonable goals for the reduction of the volume of hazardous waste generated in the state and the amount of pollutants and contaminants using source reduction and waste minimization.

(c) In order to effectively use resources and avoid duplication of effort, the commission and board by joint rule shall develop a common list of pollutants or contaminants and the level of releases of those pollutants or contaminants subject to source reduction and waste minimization planning.

361.504. Application

(a) Except as provided by Subsection (b), this subchapter applies to the following persons:

(1) all large-quantity generators of hazardous waste;

(2) all generators other than large-quantity generators and conditionally exempt small-quantity generators; and

(3) persons subject to Section 313, Title III, Superfund Amendments and Reauthorization Act of 1986 (42 U.S.C. Section 11023) whose releases exceed the levels established under Section 361.503(c).

(b) The commission and the board through joint rule making shall establish one or more schedules for the application of the requirements of this subchapter to designated classes of persons described by Subsection (a). The schedule shall provide for the inclusion of all persons described by Subsection (a) on a date to be determined by the commission and board, and until that date this subchapter applies only to those persons designated by rule of the commission or board.

(c) This subchapter does not apply to a facility regulated by the Railroad Commission of Texas under Section 91.101 or 141.012, Natural Resources Code.

361.505. Source Reduction and Waste Minimization Plans

(a) Persons identified under Section 361.504(a)(1) or (a)(3) shall prepare a source reduction and waste minimization plan. Plans developed under this section shall contain a separate component addressing source reduction activities and a separate component addressing waste minimization activities.

Appendix E: State Pollution Prevention Laws / 415

The plan shall include, at a minimum:
(1) an initial survey that identifies:
   (A) for facilities subject to Section 361.504(a)(1), activities that generate hazardous waste; and (B) for facilities subject to Section 361.504(a)(3), activities that result in the release of pollutants or contaminants designated under Section 361.503(c);
(2) based on the initial survey, a prioritized list of economically and technologically feasible source reduction and waste minimization projects;
(3) an explanation of source reduction or waste minimization projects to be undertaken, with a discussion of technical and economic considerations, and environmental and human health risks considered in selecting each project to be undertaken;
(4) an estimate of the type and amount of reduction anticipated;
(5) a schedule for the implementation of each source reduction and waste minimization project;
(6) source reduction and waste minimization goals for the entire facility, including incremental goals to aid in evaluating progress;
(7) an explanation of employee awareness and training programs to aid in accomplishing source reduction and waste minimization goals;
(8) certification by the owner of the facility, or, if the facility is owned by a corporation, by an officer of the corporation that owns the facility who has the authority to commit the corporation's resources to implement the plan, that the plan is complete and correct;
(9) an executive summary of the plan; and
(10) identification of cases in which the implementation of a source reduction or waste minimization activity designed to reduce risk to human health or the environment may result in the release of a different pollutant or contaminant or may shift the release to another medium.

(b) The source reduction and waste minimization plan may include:
(1) a discussion of the person's previous efforts at the facility to reduce risk to human health and the environment or to reduce the generation of hazardous waste or the release of pollutants or contaminants;
(2) a discussion of the effect changes in environmental regulations have had on the achievement of the source reduction and waste minimization goals;
(3) the effect that events the person could not control have had on the achievement of the source reduction and waste minimization goals;
(4) a description of projects that have reduced the generation of hazardous waste or the release of pollutants or contaminants; and
(5) a discussion of the operational decisions made at the facility that have affected the achievement of the source reduction or waste minimization goals or other risk reduction efforts.

(c) The commission shall adopt rules for the development of simplified, as appropriate, source reduction and waste minimization plans and reports for persons identified under Section 361.504(a)(2).

(d) The commission and the board shall provide information to aid in the preparation of source reduction and waste minimization plans to be prepared by a person under this section.

(e) The commission and board shall enter into memoranda of understanding as necessary to ensure that a person may meet both agencies' requirements in a single plan.

361.506. Source Reduction and Waste Minimization Annual Report

(a) A person required to develop a source reduction and waste minimization plan for a facility under this subchapter shall submit to the commission and the board an annual report and a current executive summary according to any schedule developed under Section 361.504.

(b) The annual report shall comply with rules adopted by the commission and the board through joint rule making. The report shall detail the facility's progress in implementing the source reduction and waste minimization plan and include:

(1) an assessment of the progress toward the achievement of the facility source reduction goal and the facility waste minimization goal;

(2) a statement to include, for facilities subject to Section 361.504(a)(1), the amount of hazardous waste generated and, for facilities subject to Section 361.504(a)(3), the amount of the release of pollutants or contaminants designated under Section 361.503(c) in the year preceding the report, and a comparison of those amounts with the amounts generated or released in a base year selected by agreement of the commission and the board; and

(3) any modification to the plan.

(c) The annual report may include:

(1) a discussion of the person's previous effort at the facility to reduce hazardous waste or the release of pollutants or contaminants through source reduction or waste minimization;

(2) a discussion of the effect changes in environmental regulations have had on the achievement of the source reduction and waste minimization goals;

(3) the effect that events the person could not control have had on the achievement of the source reduction and waste minimization goals; and

(4) a discussion of the operational decisions the person has made that have affected the achievement of the source reduction and waste minimization goals.

(d) The annual report shall contain a separate component addressing source reduction activities and a separate component addressing waste minimization activities.

361.507. Administrative Completeness

(a) The commission or the board may review a source reduction and waste minimization plan or annual report to determine whether the plan or report complies with this subchapter and rules adopted under Section 361.504, 361.505, or 361.506, as appropriate.

(b) Failure to have a source reduction and waste minimization plan in accordance with Sections 361.504 and 361.505 or failure to submit a source reduction and waste minimization annual report in accordance with Section 361.506 is a violation of this chapter.

361.508. Confidentiality

(a) A source reduction and waste minimization plan shall be maintained at each facility owned or operated by a person who is subject to this subchapter and shall be available to commission or board personnel for inspection. The source reduction and waste minimization plan is not a public record for the purposes of the open records law, Chapter 424, Acts of the 63rd Legislature, Regular Session, 1973 (Article 6252-17a, Vernon's Texas Civil Statutes).

(b) The executive summary and the annual report are public records. On request, the person shall make available to the public a copy of the executive summary or annual report.

(c) If an owner or operator of a facility for which a source reduction and waste minimization plan has been prepared shows to the satisfaction of the commission or board that an executive summary, annual report, or portion of a summary or report prepared under this subchapter would divulge a trade secret if made public, the commission or board shall classify as confidential the summary, report, or portion of the summary or report.

(d) To the extent that a plan, executive summary, annual report, or portion of a plan, summary, or annual report would otherwise qualify as a trade secret, an action by the commission or board or an employee of the commission or board does not affect its status as a trade secret.

(e) Information classified by the commission or board as confidential under this section is not a public record for purposes of the open records law, Chapter 424, Acts of the 63rd Legislature, Regular Session, 1973 (Article 6252-17a, Vernon's Texas Civil Statutes), and may not be used in a public hearing or disclosed to a person outside the commission or board unless a court decides that the information is necessary for the determination of an issue being decided at the public hearing.

361.509. Source Reduction and Waste Minimization Assistance

(a) The office of pollution prevention established under Section 361.0216 shall assist generators of hazardous waste and owners or operators of facilities that release pollutants or contaminants in reducing the volume, toxicity, and adverse public health and environmental effects of hazardous waste generated or pollutants or contaminants released in the state. To provide the assistance, the office may:

(1) compile, organize, and make available for distribution information on source reduction and waste minimization technologies and procedures;
(2) compile and make available for distribution to business and industry a list of consultants on source reduction and waste minimization technologies and procedures and a list of researchers at state universities who can assist in source reduction and waste minimization activities;
(3) sponsor and conduct conferences and individualized workshops on source reduction and waste minimization for specific classes of business or industry;
(4) facilitate and promote the transfer of source reduction and waste minimization technologies and procedures among businesses and industries;
(5) if appropriate, develop and distribute model source reduction and waste minimization plans for the major classes of business or industry, as identified by the office, that generate and subsequently treat, store, or dispose of hazardous waste or that release a pollutant or contaminant in the state;
(6) develop and make available for distribution recommended source reduction and waste minimization audit procedures for use by business or industry in conducting internal source reduction and waste minimization audits;
(7) provide to business and industry, as resources allow, on-site assistance in identifying potential source reduction and waste minimization techniques and practices and in conducting internal source reduction and waste minimization audits;
(8) compile and make available for distribution information on tax benefits available to business or industry for implementing source reduction and waste minimization technologies and practices;
(9) establish procedures for setting priorities among key industries and businesses for receiving assistance from the office;
(10) develop the information base and data collection programs necessary to set program priorities and to evaluate progress in source reduction and waste minimization;
(11) develop training programs and materials for state and local regulatory personnel and private business and industry regarding the nature and applicability of source reduction and waste minimization practices;
(12) produce a biennial report on source reduction and waste minimization activities, achievements, problems, and goals, including a biennial work plan;
(13) participate in existing state, federal, and industrial networks of individuals and groups involved in source reduction and waste minimization; and
(14) publicize to business and industry, and participate in and support, waste exchange programs.

(b) The commission, the department, and the board shall provide education and training to river authorities, municipalities, and public groups on source reduction and waste minimization technologies and practices.

(c) The commission and the board shall develop incentives to promote the implementation of source reduction and waste minimization, including:
(1) board and commission recommendations to the governor for awards in recognition of source reduction and waste minimization efforts;
(2) an opportunity by joint rules of the commission and the board for an owner or operator of a facility to be exempted from the requirements of this subchapter on meeting appropriate criteria for practical economic and technical completion of the source reduction and waste minimization plan for the facility; and
(3) expedited review of a permit amendment application if the amendment is necessary to implement a source reduction and waste minimization project, considering only the directly affected parts of the permit.
(d) The commission and the board shall work closely with the Gulf Coast Hazardous Substance Research Center to identify areas in which the center could perform research in the development of alternative technologies or conduct related projects to promote source reduction and waste minimization.

VOLUNTARY POLLUTION PREVENTION PROGRAMS

Colorado

COLORADO REVISED STATUTES
TITLE 25. HEALTH ENVIRONMENTAL CONTROL
ARTICLE 16.5. POLLUTION PREVENTION

25-16.5-101. Short title

This article shall be known and may be cited as the "Pollution Prevention Act of 1992".

25-16.5-102. Legislative declaration–state policy on pollution prevention

(1) The general assembly hereby finds and declares that:
(a) Colorado is blessed by natural beauty and an excellent quality of life, which should be maintained;
(b) The prevention of pollution will assist in maintaining quality of life in our state;
(c) There are resources and expertise in Colorado, including industry, government, and citizen groups, which can provide information and assistance to promote the cost-effective prevention of pollution;

(d) There are opportunities to reduce or prevent pollution through voluntary changes in procurement, production, operations, and use of raw materials throughout the state;

(e) The purpose of this article is to create a cooperative partnership among business, agriculture, the environmental community, and the department of public health and environment in which technical assistance, outreach, and education activities are coordinated and conducted to achieve pollution prevention and waste reduction and source reduction;

(f) The prevention of pollution is preferable to treatment and disposal of toxic substances and is the cornerstone of the future of environmental management.

(2) The general assembly, therefore, determines and declares that the state policy of Colorado shall be that pollution prevention is the environmental management tool of first choice. The state policy shall be that: Pollution should be prevented or reduced at the source by means including the reduction in the production or use of hazardous substances; pollution that cannot be prevented should be recycled in an environmentally safe manner; pollution that cannot be prevented or recycled should be treated in an environmentally safe manner; and disposal or other releases into the environment should be employed only as a last resort and should be conducted in an environmentally safe manner.

25-16.5-103. Definitions

As used in this article, unless the context otherwise requires:

(1) "Advisory board" means the pollution prevention advisory board created in section 25-16.5-104.

(2) "By-product" means all toxic or hazardous substances, other than a product, that are generated from production processes prior to recycling, handling, treatment, disposal, or release.

(3) "Department" means the department of public health and environment.

(4) "Federal act" means the federal "Emergency Planning and Community Right-To-Know Act of 1986", 42 U.S.C. sec. 11001 et seq., Title III of the federal "Superfund Amendments and Reauthorization Act of 1986", P.L. 99-499, as amended.

(5) "Hazardous substance" or "toxic substance" means those chemicals defined as hazardous substances under section 313 of the federal "Superfund Amendments and Reauthorization Act of 1986" (SARA Title III) [42 U.S.C. 11023.] and sections 101(14) and 102 of the federal "Comprehensive Environmental Response, Compensation, and Liability Act" (CERCLA), as amended. [42 U.S.C. 9601(14) and 9602.]

(6) "Pollution prevention" means any practice which reduces the use of any hazardous substance or amount of any pollutant or contaminant prior to recycling, treatment, or disposal, and reduces the hazards to public health and the

environment associated with the use or release or both of such substances, pollutants, or contaminants.

(7) "Production process" means a process, line, method, activity, or technique, or a series or combinations of such processes, necessary to and integral to making a product or providing a service, and does not include waste management activities.

(8) "Small and medium-sized business" means a business which has five hundred or fewer employees and which has gross annual sales of seventy-five million dollars or less.

(9) "Toxics use reduction" means changes in production processes, products, or raw materials that reduce, avoid, or eliminate the use of toxic or hazardoussubstances and the generation of hazardous by-products per unit of production,so as to reduce the overall risks to the health of workers, consumers, or theenvironment without creating new risks of concern.

(10) "Waste management" means the recycling, treatment, handling, transfer, controlled release, cleanup, and disposal of waste, and the containment of accidents and spills.

(11) "Waste reduction" and "source reduction" mean any practice which reduces the amount of any hazardous substances, pollutant, or contaminant entering any waste stream or otherwise being released into the environment (including fugitive emissions) prior to recycling, treatment, or disposal, and reduces the hazards to public health and the environment associated with the release of such substances, pollutants, or contaminants. The terms include equipment or technology modifications, process or procedure modifications, reformulation or redesign of products, substitution of raw materials, and improvements in housekeeping, maintenance, training, or inventory control. The term "source reduction" does not include any practice which alters the physical, chemical, or biological characteristics or the volume of a hazardous substance, pollutant, or contaminant through a process or activity which itself is not integral to and necessary for the production of a product or the providing of a service.

25-16.5-104. Pollution prevention advisory board—creation—repeal

(1) There is hereby created in the department of health a pollution prevention advisory board for the purposes of providing overall policy guidance, coordination, and advice to the department on pollution prevention activities and for carrying out the duties specified in section 25-16.5-105. The advisory board shall consist of fifteen members to be appointed by the governor no later than July 1, 1992. The members appointed shall include representatives of businesses, agriculture, environmental groups, academic institutions of higher education, community groups, and local governments. In addition, the governor shall appoint two representatives from state agencies to serve as ex-officio members of the

advisory board, with at least one of such appointees to be from the department of health. In making the appointments, the governor shall provide for geographic diversity. The board shall elect its own chairperson. Members of the advisory board shall serve without compensation.

(2)(a) This section is repealed, effective July 1, 1995.

(b) Prior to said repeal, the advisory board shall be reviewed as provided for in section 2-3-1203(3)(h), C.R.S.

25-16.5-105. Powers and duties of the advisory board

(1) The advisory board shall have the following powers and duties:

(a) To provide overall policy guidance, coordination, and advice in the development and implementation of the pollution prevention activities of the department;

(b) To develop pollution prevention goals and objectives;

(c) To review environmental regulatory programs, laws, and policies to identify pollution prevention opportunities and incentives;

(d) To provide direction for pollution prevention outreach, education, training, and technical assistance programs;

(e) To report to the governor and the general assembly annually on the progress of and recommendations for changes in programs on pollution prevention;

(f) To contract with a provider or providers, which may include the department, to provide pollution prevention activities as described in section 25-16.5-106.

25-16.5-106. Pollution prevention activities program

(1) The advisory board shall contract with a provider or providers, which may include the department, to develop a pollution prevention activities program. The pollution prevention activities program shall be carried out to make pollution prevention the environmental management tool of first choice, and the provider which provides the services pursuant to contract shall have the following powers and duties:

(a) To provide education and training about pollution prevention to businesses that use or produce hazardous substances and their employees, local and state governments, and the general public. Such education and training may include pollution prevention techniques, total cost analysis of toxics use and pollution prevention techniques, economic evaluation methods of such techniques, and management and employee involvement and public involvement.

(b) To expand the pollution prevention technical library and resource center providing access to information on new products, production process techniques, and raw materials for production related to pollution prevention, technical reports, fact-sheets, case studies, articles, and other reference materials;

(c) To collect and evaluate information on toxics use reduction and waste reduction and the amount of hazardous substances used in Colorado as the basis for establishing pollution prevention priorities and measuring progress in achieving pollution prevention program objectives;

(d) To conduct an evaluation of pollution prevention activities in this state analyzing existing data to determine what priority should be given to different hazardous substances and production processes;

(e) To prepare a report with data on the amount of hazardous substances, pollutants, and contaminants used in Colorado and the amount of pollution released in Colorado prior to recycling, treatment, or disposal. Such report shall be developed from existing sources and updated every two years and used as a tool to measure the success of the pollution prevention activities program and the technical assistance program in Colorado.

(f) To develop other methods to measure the success of pollution prevention projects at facilities. Methods shall be developed to measure the use of hazardous substances for production processes and the amount of waste prior to waste management practices.

(g) To cooperate with the advisory board in the performance of duties assigned to the board, including the review of environmental regulatory programs, laws, and policies for identifying pollution prevention opportunities and incentives;

(h) To coordinate with any of the academic institutions or other recipients of grants under the technical assistance program pursuant to section 25-16.5-107.

25-16.5-107. Technical assistance program

(1) The advisory board shall develop guidelines on how to allocate the portion of and shall select grant recipients for the moneys in the pollution prevention fund created in section 25-16.5-109 which is available under said section for making grants for the purpose of providing technical assistance to small and medium-sized businesses and to other generators or users of hazardous and toxic substances. Grants may be made to academic institutions, trade associations, and environmental or engineering firms with knowledge of pollution prevention techniques and processes. The advisory board shall develop guidelines for awarding grants to provide technical assistance. In developing such guidelines, the advisory board shall determine priorities for such assistance, including an emphasis on reducing the production processes that use the largest amounts of hazardous substances and an emphasis on assisting small businesses. The advisory board shall select the grant recipients and shall determine the amount of the grant awarded to each recipient. The department shall then award the grants pursuant to this section through a contract entered into between the department and the

grant recipient which details the conditions of the grant and the requirements and responsibilities of the grant recipient.

(2) For the purposes of this article, "technical assistance program" means the following types of activities:

(a) Providing technical assistance and outreach on pollution prevention to small and medium-sized businesses and to other generators or users of toxic substances;

(b) Providing on-site toxics use reduction and waste reduction assistance to small and medium-sized businesses and to other generators or users of toxic substances;

(c) Providing to businesses, trade associations, public entities, and to other generators or users of toxic substances information on the annual toxics use reduction and waste reduction that could be achieved by using pollution prevention techniques, the annual savings and implementation costs of such techniques, and the period of time necessary to recoup the money spent to implement the pollution prevention techniques;

(d) Providing on-site pollution prevention assessments of industrial plant production processes and waste generation, upon request of the affected businesses.

(3) Technical assistance programs shall be offered to small and medium sized businesses and to public generators or users of toxic substances without charge.

(4) It is the intent of the general assembly that the technical assistance program not be used to document violations of or used in the enforcement of any state law or regulation.

25-16.5-108. Pollution prevention fees—repeal

(1)(a) The department shall charge and collect pollution prevention fees from any reporting facility which is required to file a report with the department pursuant to the federal act as follows:

(I) Facilities required to report pursuant to section 11002 of the federal act shall pay a fee up to but not to exceed ten dollars per reporting facility.

(II) Each facility required to report pursuant to section 11022 of the federal act shall be required to pay a fee up to but not to exceed ten dollars for every hazardous substance located at the facility in excess of the thresholds adopted by the United States environmental protection agency.

(III) Each facility required to report pursuant to section 11023 of the federal act shall pay a fee up to but not to exceed twenty-five dollars for every extremely hazardous substance located at the facility in excess of the thresholds adopted by the United States environmental protection agency.

(b) Any retail motor fuel outlet which is required to report pursuant to the federal act shall pay one-half of the fee set forth in paragraph (a) of this subsection (1).

(c) Any single reporting organization which owns or operates multiple reporting facilities shall not be required to pay more than a total of one thousand dollars for all pollution prevention fees required by this section.

(d) Agricultural businesses which are required to report under the federal act are not required to pay the pollution prevention fees set forth in this subsection (1).

(e) It is the intent of the general assembly that the department of public health and environment collect all fees from any reporting facility required to report under the federal act, including the pollution prevention fee, in a single, centralized billing procedure.

(2) Any moneys collected pursuant to subsection (1) of this section shall be transmitted to the treasurer and credited to the pollution prevention fund created in section 25-16.5-109.

(3) This section is repealed, effective July 1, 1996.

25-16.5-109. Pollution prevention fund—created

(1) There is hereby created in the state treasury the pollution prevention fund. Any moneys collected pursuant to section 25-16.5-108 shall be credited to such fund. In accordance with section 24-36-114, C.R.S., all interest derived from the deposit and investment of moneys in the fund shall be credited to the general fund. At the end of any fiscal year, all unexpended and unencumbered moneys in the fund shall remain therein and shall not be credited or transferred to the general fund or any other fund.

(2) The moneys generated from the pollution prevention fees pursuant to section 25-16.5-108 shall be annually appropriated to the department of public health and environment for allocation to the pollution prevention advisory board created in section 25-16.5-104 for contracting for pollution prevention activities programs as set forth in section 25-16.5-106 and for the purpose of making grants under the technical assistance program as set forth in section 25-16.5-107 and as directed by the pollution prevention advisory board created in section 25-16.5-104. None of the moneys in the fund shall be used for the enforcement of any state law or regulation.

Delaware

DELAWARE CODE
TITLE 7. CONSERVATION
PART VII. NATURAL RESOURCES
CHAPTER 78. WASTE MINIMIZATION/
POLLUTION PREVENTION ACT

7801 Short title.

This chapter shall be known and may be cited as the "Waste Minimization/Pollution Prevention Act."

7802 Findings and purpose.

(a) The General Assembly finds that:
(1) Whenever possible, the generation of waste should be reduced or eliminated as expeditiously as possible, and that waste that is generated should be recovered, reused, recycled, treated or disposed of in a manner that minimizes any present or future threats to human health or the environment;
(2) There may exist many promising technologies for the reduced generation of waste, for recovery, reuse, recycling and treatment of waste; and
(3) Financial commitments by public agencies and private industry for the expeditious development and implementation of waste reduction, recovery, reuse, recycling and treatment technologies depends upon further research as well as credible and timely demonstrations of economic viability, technical feasibility, environmental acceptability and reliability of this technology.
  (b) Therefore the General Assembly declares:
(1) The purposes of this chapter are to enhance the protection of human health and the environment, and to establish a multi-media Waste Minimization/Pollution Prevention Program which will demonstrate and facilitate the potential for pollution prevention and waste minimization in Delaware through focusing on the following objectives:
  a. Targeting industries and locations for technical assistance; and
  b. Providing waste minimization/pollution prevention, education and outreach; and
  c. Developing a statewide recycling program; and
(2) That it is the policy of this State, in concurrence with the DelawareEnvironmental Legacy Report, that waste that is generated should be, in order of priority, reduced at its source, recovered, reused, recycled, treated or disposed of so as to minimize the present and future threat to human health and the environment.
  (c) It is the intent of this chapter to complement and be enforced in conjunction with other laws.

7803 Definitions.

As used in this chapter:
(1) "Department" means the Department of Natural Resources and Environmental Control.
(2) "Implementation Committee" means the Waste Minimization/ Pollution Prevention Implementation Committee.
(3) "Multi-media" means all environmental media including, but not limited to, workplaces within facilities, water, land and air.
(4) "Person" means any individual, trust, firm, joint stock company, federal agency, corporation (including a government corporation), partnership,

association, state, municipality, commission, political subdivision of a state or any interstate body.

(5) "Recycle" means to separate and to process a reusable material in order to return that material to commerce.

(6) "Secretary" means the Secretary of the referenced state department or a duly authorized designee.

(7) "Trade secret" means any information concerning production processes employed or substances manufactured, processed or otherwise used within a facility which the Secretary of the Department has determined to be propriety information.

(8) "Waste" means, but is not limited to, those substances defined as such under Delaware Codes, or the regulations promulgated thereunder: air pollution as defined in 6002 of this title; hazardous waste as defined in 6302 of this title; industrial waste as defined in 6002 of this title; liquid waste as defined in 6002 of this title; other wastes as defined in 6002 of this title; refuse as defined in 6002 of this title; rubbish as defined in 6002 of this title; sewage as defined in 6002 of this title; and solid waste as defined in 6002 of this title.

(9) "Waste minimization" means the process by which a facility conducts an analysis of a production process to determine the waste minimization techniques which could be implemented.

7804 Waste Minimization/Pollution Prevention Program.

(a) The Department shall establish the Implementation Committee and appoint the members thereof.

(1) The Implementation Committee shall be composed of, but not be limited to, the following individuals or their designee:

    a. The Secretary of the Department;
    b. The Director of the Delaware Economic Development Office;
    c. The Secretary of the Department of Agriculture;
    d. The Secretary of the Department of Administrative Services;
    e. The President of the Delaware State Chamber of Commerce;
    f. The Superintendent of the Department of Public Instruction;
    g. The General Manager of the Delaware Solid Waste Authority;
    h. A representative of the League of Local Governments;
    i. A representative of the Chemical Industry Council;
    j. A representative of Delaware State University;
    k. A representative of Delaware Technical and Community College;
    l. A representative of the University of Delaware;
    m. A representative from the private solid waste collection industry;
    n. A representative of civic organizations; and
    o. A representative of an environmental organization.

The Implementation Committee will be chaired by the Secretary of the Department and be staffed by the Department. The members of the Implementation Committee are appointed for a period of 1 ½ years and shall meet at least monthly during this period.

(2) The Implementation Committee shall select the target industry and location for the initial technical assistance program. The Committee shall evaluate the technical assistance program 1 year after the program commences or upon completion, whichever is earlier. The Committee shall then select a second target industry or location for the technical assistance program.

(b)(1) The objectives of the Waste Minimization/Pollution Prevention Program shall be:

  a. Targeting industries and locations for technical assistance;

  b. Providing waste minimization/pollution prevention education and outreach; and

  c. Developing a statewide recycling program.

(2) These objectives shall be accomplished through implementation of, but not be limited to, the programs outlined in this section.

  a. Technical assistance:

  1. The Department shall create a multi-media opportunity audit program to provide technical assistance to targeted industries, focusing on small companies within the industry because, for the most part, they do not have the economic or technical resources necessary to acquire recommendations on how to effectively minimize their wastes. The program will be made available to other industries which request assistance. The overall objective of the multi-media opportunity audit program is to reduce or recycle, to the maximum extent practicable, all waste streams within an industry.

  2. An information clearinghouse shall be established and located at the Delaware Economic Development Office. The clearinghouse will contain a database of waste minimization/pollution prevention technologies and case studies of technology applications.

  b. Waste minimization/pollution prevention education and outreach:

  1. The Department shall increase public awareness of the need for individual and community based pollution prevention and waste minimization programs by conducting educational workshops and industry seminars.

  2. The Department shall produce a pollution prevention newsletter to be distributed to industry and the public.

  c. Statewide Recycling Program:

  The Implementation Committee shall develop a Comprehensive Statewide Recycling Plan, and shall aid the Department in implementing the Plan. The Plan shall include programs which utilize appropriate collection methods in each county. These methods may include curbside collection, material recovery facilities, central drop-off collection points or other methods as deemed appropriate. The Plan shall also include programs to develop markets for recyclables and recycled products, requirements for state procurement of recycled products and incentives for processing facilities and companies which use recycled products to locate in Delaware. The Plan shall include recommendations on which programs are most feasible and beneficial, who should be responsible for implementing them, how much they will cost, and how they might be financed.

(c) An annual report on the Waste Minimization/Pollution Prevention Program shall be provided to the Governor and the General Assembly. The report shall include an evaluation of current programs and plans for implementation of any additional programs, including information on potential benefits, who should be responsible for implementing them, how much they will cost, and how they might be financed.

7805 Trade secret protection.

All trade secret information, written, verbal or observed, obtained pursuant to this chapter by the Department or any other state agency will be held as confidential unless such information is already a matter of public record or disclosure is required by law. Nothing in this section shall be construed as limiting the disclosure of information by the Department to any officer, employee or authorized representative of the state or federal government concerned with effecting this chapter. Prior to disclosure of trade secret information to an authorized representative who is not an officer or employee of the state or federal government, the person providing the trade secret information may require the representative to sign an agreement prohibiting disclosure of such information to anyone not authorized by this chapter or the terms of the agreement. Such agreement shall not preclude disclosure by the representative to any state or federal government officer or employee concerned with effecting this chapter. Any person who, by virtue of obtaining access to confidential trade secret information and knowing that disclosure is prohibited, knowingly and wilfully discloses the information to any person not entitled to receive it shall be in violation of this chapter and subject to disciplinary actions provided by the Merit Rules, including dismissal.

Florida

FLORIDA STATUTES
TITLE XXIX. PUBLIC HEALTH
CHAPTER 403. ENVIRONMENTAL CONTROL
PART I. POLLUTION CONTROL

403.072. Pollution Prevention Act

Sections 403.072-403.074 may be cited as the "Pollution Prevention Act."

403.073. Pollution prevention; state goal; agency programs; public education

(1) It is a goal of the state that all its agencies, the State University System, the State Board of Community Colleges, and all municipalities, counties, regional agencies, and special districts develop and implement strategies to prevent pollution, including public information programs and education programs.
(2) It is the policy of the state that pollution prevention is necessary for all materials and waste management activities. However, until the Pollution Prevention Council, created by s. 28, ch. 91-305, issues its report and recommendations, the department shall concentrate upon preventing pollution caused by toxic and hazardous wastes.

403.074. Technical assistance by the department

(1) To help develop effective programs to eliminate or reduce the use of materials that are toxic to humans, plants, or animals and to prevent pollution at its source, the department shall implement and administer a program of technical assistance in pollution prevention to business, industry, agriculture, and state and local government.
(2) The program shall include on-site, nonregulatory technical assistance and shall promote and sponsor conferences on pollution prevention techniques. The program may be conducted in cooperation with trade associations, trade schools, the State University System, the State Board of Community Colleges, or other appropriate entities.
(3) Proprietary information obtained by the department during a visit to provide on-site technical assistance requested under ss. 403.072-403.074 is confidential pursuant to s. 403.111, unless the confidentiality is waived by the party who requested the assistance.

Iowa

IOWA CODE
TITLE XI. NATURAL RESOURCES
SUBTITLE 1. CONTROL OF ENVIRONMENT
CHAPTER 455B. JURISDICTION OF DEPARTMENT OF NATURAL RESOURCES
DIVISION VII. TOXICS POLLUTION PREVENTION PROGRAM

455B.516. Definitions

As used in this division, unless the context otherwise requires:
1. "Commission" means the environmental protection commissionestablished pursuant to section 455A.6.
2. "Department" means the department of natural resources created pursuant to section 455A.2.
3. "Division" means the waste management assistance division created pursuant to section 455B.483.
4. "Emergency Planning and Community Right-to-know Act" or "EPCRA" means the federal Emergency Planning and Community Right-to-know Act as defined in section 30.1.
5. "Environmental waste" means a pollutant, waste, or release regardless of the type or existence of regulation and regardless of the media affected by the pollutant, waste, or release.
6. "Existing toxics user" means a toxics user installation or source constructed prior to July 1, 1991.
7. "Multimedia" means any combination of air, water, land, or workplace environments into which toxic substances or wastes are released.
8. "Release" means emission, discharge, or disposal into any environmental media including air, water, or land.
9. "Toxics pollution prevention" means employment of a practice which reduces the industrial use of toxic substances or reduces the environmental and health hazards associated with an environmental waste without diluting or concentrating the waste before the release, handling, storage, transport, treatment, or disposal of the waste. The term includes toxics pollution prevention techniques but does not include a practice which is applied to an environmental waste after the waste is generated or comes into existence on or after the waste exits a production or commercial operation.

"Toxics pollution prevention" does not include, promote, or require any of the following:
 a. Waste burning in industrial furnaces, boilers, smelters, or cement kilns for the purpose of energy recovery.

b. The transfer of an environmental waste from one environmental medium to another environmental medium, the workplace environment, or a product.

c. Off-site waste recycling.

d. Any other method of end-of-pipe management of environmental wastes including waste exchange and the incorporation or embedding of regulated environmental wastes into products or by-products.

10. "Toxics pollution prevention techniques" means any of the following practices by a toxics user:

a. Input substitution, which refers to replacing a toxic substance or raw material used in a production process with a nontoxic or less toxic substance.

b. Product reformulation, which refers to substituting for an existing end product an end product which is nontoxic or less toxic upon use or release.

c. Production process redesign or modification, which refers to developing and using production processes of a different design other than those currently in use.

d. Production process modernization, which refers to upgrading or replacing existing production process equipment or methods with other equipment or methods based on the same production process.

e. Improved operation and maintenance of existing production process equipment and methods, which refers to modifying or adding to existing equipment or methods, including but not limited to, such techniques as improved housekeeping practices, system adjustments, product and process inspections, and production process control equipment or methods.

f. Recycling, reuse, or extended use of toxic substances by using equipment or methods which become an integral part of the production process.

11. "Toxic substance" means any chemical substance in a gaseous, liquid, or solid state which is identified as a reportable substance under the federal Resource Conservation and Recovery Act, EPCRA, or defined as a hazardous air pollutant under the Clean Air Act of 1990. However, "toxic substance" does not include a chemical substance present in the article; used as a structural component of a facility; present in a product used for routine janitorial or facility grounds maintenance; present in foods, drugs, cosmetics, or other personal items used by employees or other persons at a toxics user facility; present in process water or noncontact cooling water as drawn from the environment or from municipal sources; present in air used either as compressed air or as part of combustion; present in a pesticide or herbicide when used in agricultural applications; or present in crude, fuel, or lube oils for direct wholesale or retail sale.

12. "Toxics" means toxic substances.

13. "Toxics user" means a large quantity generator as defined pursuant to the federal Resource Conservation and Recovery Act, 42 U.S.C. 6901 et seq. or a person required to report pursuant to Title III of the federal Superfund Amendments and Reauthorization Act of 1986.

14. "Waste exchange" means a method of end-of-pipe management of environmental wastes that involves the transfer of environmental wastes between

businesses or facilities owned or operated by the same business for recovery or to serve a productive purpose.

455B.517. Duties of waste management assistance division

The waste management assistance division shall do all of the following:
1. Establish the criteria for the development of the toxics pollution prevention program.
2. Develop and implement a toxics pollution prevention program.
3. Assist toxics users in the completion of toxics pollution prevention plans and inventories, and provide technical assistance as requested by the toxics user.
4. a. Seek, receive, and accept funds in the form of appropriations, grants, awards, wills, bequests, endowments, and gifts for the uses designated pursuant to section 455B.133B. The division shall also coordinate existing resources and oversee the disbursement of federal grant moneys to provide consistency in achieving the toxics pollution prevention goal of the state.
 b. Provide, through the use of moneys collected pursuant to section 455B.133A, the state matching funds for grants under the federal Pollution Prevention Act of 1990, Pub.L. No. 101-508, 6604 and 6605. 5. Develop and implement guidelines regarding assistance to toxics users to ensure that the plans are multimedia in approach and are not duplicated by the department or other agencies of the state.
6. Identify obstacles to the promotion, within the toxics user community, of toxics pollution prevention techniques and practices.
7. Compile an assessment inventory, through solicitation of recommendations of toxics users and owners and operators of air contaminant sources, of the informational and technical assistance needs of toxics users and air contaminant sources.
8. Function as a repository of research, data, and information regarding toxics pollution prevention activities throughout the state.
9. Provide a forum for public discussion and deliberation regarding toxic substances and toxics pollution prevention.
10. Promote increased coordination between the department, the Iowa waste reduction center at the university of northern Iowa, and other departments, agencies, and institutions with responsibilities relating to toxic substances.
11. Coordinate state and federal efforts of clearinghouses established to provide access to toxics reduction and management data for the use of toxics users.
12. Make recommendations to the general assembly by January 1, 1992, regarding a funding structure for the long-term implementation and continuation of a toxics pollution prevention program.
13. Work with the Iowa waste reduction center at the university of northern Iowa to assist small business toxics users with plan preparation and technical assistance.

455B.518. Toxics pollution prevention plans

1. A toxics user required to report under section 313 of EPCRA, 42 U.S.C. 11023, or a large quantity generator, as defined pursuant to the federal Resource Conservation and Recovery Act, 42 U.S.C. 6901 et seq., shall be encouraged to develop a facility-wide multimedia toxics pollution prevention plan, as described pursuant to this section.

2. The division shall adopt criteria for the information required in a multimedia toxics pollution prevention plan. To the extent possible, the plans shall coordinate reporting requirements in order to minimize unnecessary duplication. The plans shall include, but are not limited to, all of the following:

a. A policy statement which articulates upper management and corporate support for the toxics pollution prevention plan and its implementation.

b. The identification and quantities of toxic substances used and released by groups of related production processes or by processes used in producing an identifiable product.

c. An assessment of the applicability of the approaches designated as toxics pollution prevention techniques including the following: input substitution; production reformulation; production process redesign or modification; production process modernization; improved operation and maintenance of existing production process equipment and methods; and recycling, reuse, or extended use of toxic substances, to the toxics users' production processes as identified in paragraph "b".

d. A description of current and previous techniques used to reduce or eliminate toxics used or released.

e. An economic analysis of the proposed toxics pollution prevention plan. The economic analysis shall also include an evaluation of the impact upon the toxics user's existing labor force by division or department, and the projected impact upon future expansion of the toxics user's labor force.

f. A clear statement listing specific reduction objectives.

g. A method for employees of a toxics user to provide input and to be involved in the development of the plans. If the employees are represented by a labor union, organization, or association, a representative of the union, organization, or association shall be included in the development of the plans.

3. The plans developed under this section shall not promote the use of pollution control or waste management approaches that address waste or pollution after the creation of the waste or pollution.

4. A toxics pollution prevention plan developed under this section shall be reviewed by the authority for completeness, adequacy, and accuracy.

5. A toxics user shall maintain a copy of the plan on the premises, and shall submit a summary of the plan to the department.

Glossary of Terms

Aboveground Storage Tank: A device situated so that the entire surface area of the tank is completely above the plane of the adjacent surrounding surface and the entire surface area of the tank (including the tank bottom) is able to be visually inspected.

Acutely Hazardous Waste: Commercial chemical products and manufacturing intermediates having the generic names listed in 40 CFR 261.33; off-specification commercial chemical products and manufacturing chemical intermediates which, if they met specifications, would have the generic names listed; and any residue or contaminated soil, water, or other debris resulting from the cleanup of a spill of any of these substances.

Administrative Order: A legal document signed by EPA directing an individual, business, or other entity to take corrective action or refrain from an activity. It describes the violations and actions to be taken, and can be enforced in court. Such orders may be issued, for example, as a result of an administrative complaint whereby the respondent is ordered to pay a penalty for violating a statute.

Aquifer: A geologic formation, group of formations, or part of a formation capable of yielding a significant amount of groundwater to wells or springs.

Applicable or Relevant and Appropriate Requirements (ARARs): ARARs include the federal standards and more stringent state standards that are legally applicable or relevant and appropriate under the circumstances. ARARs include cleanup standards, standards of control, and other environmental protection requirements, criteria, or limitations.

RCRA has frequently been used as an ARAR for cleanup of Superfund sites.

Baler: A machine used to compress recyclables into bundles to reduce volume. Balers are used often on newspapers, plastic, corrugated cardboard, and other sorted paper products.

Best Demonstrated Available Technology (BDAT): The technology EPA establishes for a land-banned hazardous waste to reduce overall toxicity or mobility of toxic constituents in the waste. BDAT must be applied to such a waste prior to land disposal unless one can successfully demonstrate the validity of an equivalent treatment method.

Best Management Practice (BMP): Schedules of activities, prohibitions of practices, maintenance procedures, and other management practices to prevent or reduce the pollution of waters of the United States. BMPs also include treatment requirements, operating procedures, and practices to control facility site run off, spillage or leaks, sludge or waste disposal, or drainage from raw material storage.

Bioremediation: A hazardous waste site remediation technique that utilizes microorganisms to metabolize hazardous organic constituents in waste to nonhazardous compounds. Environmental conditions are carefully controlled in an attempt to create optimum growth conditions for the organisms.

CERCLIS or CERCLA Information System: EPA database which identifies hazardous waste sites that require investigation and possible remedial action to mitigate potential negative impacts on human health or the environment.

Characteristic Hazardous Waste: A solid waste that is a hazardous waste because it exhibits one or more of the following hazardous characteristics: ignitability, corrosivity, reactivity, or toxicity.

Clean Air Act (CAA): The law that authorizes regulations governing releases of airborne contaminants from stationary and non-stationary sources. The regulations include National Ambient Air Quality Standards for specific pollutants.

Clean Water Act (CWA): The law that authorizes establishment of the regulatory program to restore and maintain the physical and biological integrity of the nation's waters. The CWA established, among other things, the National Pollutant Discharge Elimination System (NPDES) to regulate industrial and municipal point-source discharges.

Closure (per RCRA): The act of securing a solid waste landfill or hazardous waste management facility according to specific technical and administrative EPA requirements.

Compactor: Power-driven device used to compress materials to a smaller volume.

Compost: The stable, decomposed organic material resulting from the composting process. Also referred to as humus.

Composting: The controlled biological decomposition of organic materials in the presence of oxygen into a stable product that may be used as a soil amendment or mulch.

Comprehensive Environmental Response, Compensation, and Liability Act (CERCLA): Also known as "Superfund," it is a program to identify sites where hazardous substances have been or might have been released into the environment and to ensure that they are cleaned up. CERCLA is primarily concerned with abandoned sites.

Conditionally Exempt Small Quantity Generator (CESQG): Those who generate no more than 100 kilograms of hazardous waste per month. Other than the hazardous waste determination requirement in 40 CFR 262.11, CESQGs are exempt from RCRA provided they do not exceed certain quantity limits for hazardous waste storage or generation.

Consent Decree: A legal document that specifies your obligations when you enter into a settlement with the government. The decree is entered as all order by a federal district court or a state court. A consent decree is mandatory in settlements about remedial action. The decree must be filed with and approved by a court. Either a consent decree or on administrative order may be used in settlements involving RI/FSs. The consent decree may be subject to a public comment period.

Container: Any portable device in which a material is stored, transported, treated, disposed of, or otherwise handled.

Contaminant: Foreign material that makes a recyclable or compostable material impure; for instance, food scraps or paper products.

Contingency Plan: A document setting out an organized, planned, and coordinated course of action to be followed in case of a fire, explosion, or release of hazardous waste constituents which could threaten human health or the environment.

Critical or Unique Habitat: The essential segment(s) of habitat that contain the unique combination of conditions (e.g., vegetation, topography, soil, species niches, etc.) necessary for the continued survival of any endangered species as listed in 50 CFR 17 or 226.

Designated Facility: An authorized hazardous waste treatment, storage, or disposal facility that has been designated on the manifest by the generator to receive a shipment of waste.

Disposal: The discharge, deposit, injection, dumping, spilling, leaking, or placing of any hazardous waste or hazardous substance into or on any land or water so that such waste or substance may enter the environment or be emitted into the air or discharged into any waters, including groundwater.

Effluent: Any discharge flowing from a conveyance.

Emergency Planning and Community Right-to-Know Act (EPCRA): The primary goal of EPCRA is to facilitate public awareness and emergency response planning for chemical hazards. To fulfill this purpose, EPCRA requires that certain companies that manufacture, process, and use chemicals in specified quantities must file written reports, provide notification of spills/releases, and maintain toxic chemical inventories. EPCRA is also referred to as SARA Title III, 42 U.S.C. 11001-11050.

Emission: Pollution discharged into the atmosphere from smokestacks, other vents, and surface areas of commercial or industrial facilities and from motor vehicle, locomotive, or aircraft exhausts.

EPA Hazardous Waste Number: A number assigned by EPA to waste that is hazardous by definition; to each hazardous waste listed in 40 CFR 261 Subpart D from specific and nonspecific sources identified by EPA (F, K, P, U); and to each characteristic waste identified in 40 CFR 261 Subpart C, including wastes with ignitable (D001), reactive (D002), corrosive (D003), and EP toxic (D004D017) characteristics.

EPA Identification Number: A number assigned by EPA to each generator; transporter; and treatment, storage, or disposal facility. Identification numbers are facility-specific, except for the transporter who has one number for all his operations.

Extraction Procedure (EP) Toxicity: One of the characteristics, along with ignitability, reactivity, and corrosivity, to make a waste a characteristic hazardous waste. The EP toxic list includes maximum concentrations for 14 constituents which, if exceeded, would make a waste hazardous. Effective September 1990, EP Toxicity was replaced by the Toxicity Characteristic.

Facility: All contiguous land, and structures, other appurtenances, and improvements on the land used for treating, storing, or disposing of hazardous waste. A facility may consist of several treatment, storage, or disposal operational units (e.g., one or more landfills, surface impoundments, or combinations of them). Under CERCLA 101(9), (1) any building, structure, installation, equipment, pipe or pipeline (including any pipe into a sewer or publicly owned treatment works), well, pit, pond, lagoon, impoundment, ditch, landfill, storage container, motor vehicle, rolling stock, or aircraft; or (2) any site or area where a hazardous substance has been deposited, stored, disposed of or placed, or has otherwise come to be located. Does not include any consumer product in consumer use or any vessel.

Fate and Transport Modeling: A mathematical process for simulating the behavior of contaminants in various environments to predict contaminant concentration and mobility. Models range from relatively simple analytical solutions to complex numerical models.

Floodplain: Lowland and relatively flat areas adjoining inland and coastal waters and other flood prone areas such as offshore islands, including at a minimum that area subject to a one percent or greater chance of flooding in any given year. The base floodplain shall be used to designate the 100-year floodplain (one percent chance floodplain).

General Permit: A permit issued under the NPDES program to cover a certain class or category of storm water discharges. These permits allow for a reduction in the administrative burden associated with permitting storm water discharges associated with industrial activities.

Generator: Any person whose process produces a hazardous waste in excess of 100 kg/month or acutely hazardous waste in excess of 1 kg/month, or whose actions first cause a hazardous waste to become subject to regulation.

Groundwater: Water in a saturated zone or stratum beneath the surface of land or water.

Hauler: A garbage collection company that offers a complete refuse removal service. Many haulers now offer to serve as collectors of recyclables as well.

Hazard Ranking System (HRS): The method EPA uses to determine which sites should be listed on the National Priorities List (NPL) under CERCLA. The HRS ranks sites by means of a mathematical rating scheme that combines the potential of a release to cause hazardous situations with the severity/magnitude of these potential impacts and the number of people who may be affected. Using the numerical scores from this scheme, EPA and the states list sites by priority and allocate resources for site investigation, enforcement, and cleanup. Sites receiving high HRS scores appear on the National Priorities List. Under SARA, the HRS must be revised by EPA to determine whether it is adequately identifying sites for the NPL (CERCLA 105(c)). Citizens may now petition EPA to conduct a preliminary assessment of a site near them. If the assessment indicates that a release may pose a threat to health or the environment, EPA will do an HRS scoring.

Hazardous and Solid Waste Amendments of 1984 (HSWA): The HSWA greatly increased the complexity of the RCRA regulatory program by imposing restrictions on land disposal of hazardous wastes, authorizing EPA to require corrective action for releases from hazardous waste management facilities, and instituting requirements for underground storage tanks containing petroleum and hazardous substances

Hazardous Substance Superfund: The fund, largely financed by taxes on petroleum and chemicals, and by an "environmental tax" on corporations, which provides operating money for government-financed actions under CERCLA. The fund is a revolving fund in the sense that it enables the government to take action and then seek reimbursement later, or to clean up sites when responsible parties with sufficient cleanup funds cannot be found. Money recovered from PRPs is returned to the fund rather than to the U.S. Treasury.

Hazardous Substance: Under CERCLA 101(14), any element, compound, mixture, solution, or substance which, when released into the environment, may present substantial danger to public health/welfare or the environment. Also includes (1) any substance designated pursuant to section 311(b)(2)(A) of the Federal Water Pollution Control Act; (2) any element, compound, mixture, solution, or substance designated pursuant to section 102 of CERCLA; (3) any hazardous waste having the characteristics identified under or listed pursuant to section 3001 of the Solid Waste Disposal Act; (4) any toxic pollutant listed under section 307(a) of the Federal Water Pollution Control Act; (5) any hazardous air pollutant listed under section 112 of the Clean Air Act; and (6) any imminently hazardous chemical substance or mixture so identified pursuant to the Toxic Substances Control Act. Excludes petroleum (including crude oil not otherwise specifically listed or designated as a hazardous substance under any of the above laws), natural gas, natural gas liquids, liquefied natural gas, or synthetic gas usable for fuel (or mixtures of natural gas and such synthetic gas). The definition of hazardous substances in CERCLA is broader than the definition of hazardous wastes under RCRA.

Hazardous Waste (RCRA): A solid waste which because of its quantity, concentration or physical, chemical, or infectious characteristics may (1) cause or contribute to an increase in mortality or an increase in serious irreversible or incapacitating reversible illness; or (2) pose a substantial present or potential hazard to human health or the environment when improperly treated, stored, transported or disposed of, or otherwise managed. EPA hazardous waste regulations (40 CFR 261) specify that if a material qualifies as a solid waste, and does not qualify for an exemption, it is a hazardous waste if it is listed by 40 CFR Part 261, Subpart D, or if it exhibits any of the four hazardous waste characteristics (ignitability, reactivity, corrosivity and toxicity).

Hazardous Waste Constituent: A constituent that caused the waste to be listed as a hazardous waste under 40 CFR Part 261 Subpart D.

Hazardous Waste Facility Permit: Companies that treat, store or dispose (TSD) of hazardous waste are considered hazardous waste facilities and must abide by extensive EPA regulations, which govern all aspects of the facility including location, design, operation and closure. TSD facilities must apply to the EPA for a hazardous waste facility permit. The application places the facility on Interim Status during EPA review, and identifies in detail how the facility will comply with EPA facility standards. If the application is approved, the company receives a Part B permit. The Part B is a final operating document, based on the application as revised by EPA, which authorizes the company to operate the facility subject to the permit provisions.

Hazardous Waste Generator: Any person whose act or process produces hazardous waste identified or listed in 40 CFR 261 or whose act first causes hazardous waste to become subject to regulation. Generators are subject to specific hazardous waste management regulations, which apply only to the particular site of generation. The regulations vary by the volume of waste annually generated but include reporting, testing, recordkeeping, storage, and shipping and disposal requirements.

Hazardous Waste Management Unit: A contiguous area of land on or in which hazardous waste is placed, or the largest area in which there is significant likelihood of mixing hazardous waste constituents in the same area. A unit may be a surface impoundment, waste pile, land treatment area, landfill cell, incinerator, tank and its associated piping and underlying containment system, or container storage area. A container alone does not constitute a unit.

Incinerator: Any enclosed device using controlled flame combustion that neither meets the criteria for classification as a boiler nor is listed as an industrial furnace.

Incompatible Waste: A hazardous waste which is unsuitable for (1) placement in a particular device or facility because it may cause corrosion or decay of containment materials; or (2) commingling with another waste or material under uncontrolled conditions because commingling might produce heat or pressure; fire or explosion; violent reaction; toxic dusts, mists, fumes or gases; or flammable fumes or gases.

Industrial Toxics Project: The Pollution Prevention Act of 1990 required that EPA develop a pollution prevention strategy that reduces pollution at the source. In February 1991, EPA published its pollution prevention strategy, commonly referred to as the Industrial Toxics Project or the "33/50" Initiative. Under the plan, EPA is seeking to reduce releases and off-site transfers of 17 high-volume EPCRA Section 313 toxic chemicals. These industrial chemicals include known and potential carcinogens, developmental toxins, chemicals that bioaccumulate, ozone-depleting chemicals, and chemicals contributing to ozone at ground level. EPA set a goal to reduce the releases of these chemicals by 33% by the end of 1992, and 50% by the end of 1995.

Industrial Waste: Solid waste generated from an industrial process or from a commercial/manufacturing facility and may or may not be hazardous. Hazardous industrial waste is defined in RCRA and includes wastes that are listed as hazardous or defined by characteristic.

Injection Well: A well into which fluids are injected.

Land Disposal: Includes, but is not limited to, placement in a landfill, surface impoundment, waste pile, injection well, land treatment facility, salt dome formation, salt bed formation, underground mine or cave, or concrete vault or bunker intended for disposal purposes. Land disposal facilities are a subset of TSD facilities. Groundwater monitoring is required at all land disposal facilities.

Leachate: Liquid that has percolated through solid and/or hazardous waste and has extracted dissolved or suspended materials from the waste.

Leak-Detection System: A system capable of detecting the failure of a primary or secondary containment structure or the presence of a release of hazardous waste or accumulated liquid in the secondary containment structure. Detection is based on systemic operational controls (such as visual monitoring) or continuous automatic monitoring of any releases from the containment areas.

Markets: Generally, a recycling business (i.e., a buyer) or municipal recycling facility that accepts recyclable materials for processing and final sale to an end user, either for their own use or for resale.

Material Safety Data Sheet (MSDS): Fact sheets required by OSHA on every commercial chemical that must be prepared by the chemical's manufacturer or importer and must include specific information, including its ingredients (one percent or more), known or suspected health risks associated with its use or exposure, proper safety precautions and waste disposal, and other information necessary to prevent or minimize a health and safety risk to employees or consumers.

Materials Exchange: A mutually beneficial relationship whereby two or more organizations exchange materials that otherwise would be thrown away. In some areas, computer and catalog networks are available to match up companies who wish to participate in exchanging their materials.

Maximum Contaminant Level (MCL): The maximum permissible level of a contaminant in water delivered to any user of a public water system. MCLs are enforceable standards.

Maximum Contaminant Level Goal (MCLG): The maximum level of a contaminant in drinking water at which no known or anticipated adverse effect on human health would occur, and which includes an adequate margin of safety. MCLGs are nonenforceable health goals.

Municipal Solid Waste: Municipal solid waste includes all materials typically disposed of in dumpsters by businesses and removed for off-site disposal by private or municipal haulers. Includes wastes such as durable and nondurable goods, containers and packaging, food scraps, yard trimmings, and miscellaneous inorganic wastes from residential, commercial, institutional, and industrial sources. Municipal solid waste does not include wastes from other sources, such as municipal sludges, combustion ash, and industrial nonhazardous process wastes that might also be disposed of in municipal waste landfills or incinerators.

National Contingency Plan (NCP): The basic policy directive for federal response actions under CERCLA Section 105. It sets forth the Hazard Ranking System and procedures and standards for responding to releases of hazardous substances, pollutants, and contaminants. The plan is a regulation (40 CFR Part 300) subject to regular revision.

National Priorities List (NPL): A list of sites designated as needing long-term remedial cleanup. The purpose of the list is to inform the public of the most serious hazardous waste sites in the nation. EPA revises the list periodically to add new sites or delete sites following cleanup. Sites on the list are generally slated for EPA enforcement or cleanup. Note that many elements of the CERCLA/SARA program apply to sites regardless of whether they are on the NPL.

National Response Center (NRC): The federal operations center that receives notifications of all releases of oil and hazardous substances into the environment. The Center, open 24 hours a day, is operated by the U.S. Coast Guard, which evaluates all reports and notifies the appropriate agency.

Navigable Water: Water which by itself, or by uniting with other waters navigable, forms a continuous highway over which interstate or international commerce may be conducted in the customary mode of trade and travel on water.

Notice Letter: EPA's formal notice to PRPs that CERCLA-related action is to be undertaken at a site for which those PRPs are considered responsible. Notice letters arc generally sent at least 60 days prior to scheduled obligation of funds for an RI/FS at a designated site. The intent is to give PRPs sufficient time to organize and to contact the government. A notice letter is sent again prior to implementing the remedy.

Notice of Intent (NOI): An application to notify the permitting authority of a facility's intention to be covered by a general permit; exempts a facility from having to submit an individual or group application.

National Pollutant Discharge Elimination System (NPDES): The program used by EPA to control the discharge of pollutants to waters of the United States. See 40 CFR 122.2 for further guidance.

NPDES Permit: A permit for the discharge of pollutants into navigable waters under the National Pollutant Discharge Elimination System pursuant to the federal Clean Water Act.

Operator: The person responsible for the overall operation of a facility.

Owner: The person who owns a facility or part of a facility.

Pallet: A wooden platform used with a forklift for moving bales or other large items. Also called a "skid."

Person: An individual, trust, firm, joint stock company, federal agency, corporation (including a government corporation), partnership, association, state, municipality, commission, political subdivision of a state, or any interstate body.

Pesticides: Any substance or mixture of substances intended for preventing, destroying, repelling, or mitigating any pest, including insecticides, fungicides, rodenticides, plant regulators, defoliants, and desiccants.

Point Source: Any discernible, confined, and discrete conveyance, including but not limited to any pipe, ditch, channel, tunnel, conduit, well, discrete fissure, container, rolling stock, concentrated animal feeding operation, or vessel or other floating craft, from which pollutants are or may be discharged. This term does not include return flows from irrigated agriculture or agricultural storm water run off.

Pollutant (Clean Water Act): Dredged spoil, solid waste, incinerator residue, filter backwash, sewage, garbage, sewage sludge, munitions, chemical wastes, biological materials, some radioactive materials, heat, wrecked or discarded equipment, rock, sand, cellar dirt, and industrial, municipal, and agricultural waste discharged into water.

Pollutant or Contaminant (CERCLA): Any element, substance, compound, or mixture, including disease-causing agents that, after release into the environment and upon exposure, ingestion, inhalation, or assimilation into any organism, either directly from the environment or indirectly by ingestion through food chains, or may reasonably be anticipated to cause death, disease, behavioral abnormalities, cancer, genetic mutation, physiological malfunctions or physical deformations.

Pollution Prevention: Any source reduction or recycling activity that results in reduction of total volume of hazardous waste, reduction of toxicity of hazardous waste, or both, as long as that reduction is consistent with the goal of minimizing present and future risks to public health and the environment. Transfer of hazardous constituents from one environment medium to another does not constitute waste minimization.

Pollution Prevention Act of 1990 (PPA): The Pollution Prevention Act, 42 U.S.C. 13101 et seq., was passed in 1990 to focus national attention on reduction of volume and toxicity of wastes at the source. Congress declared it a national policy to prevent or reduce pollution at the source whenever feasible. The PPA sets forth a hierarchy of waste management options in descending order of preference: prevention/source reduction, recycling, treatment, and disposal. Pollution should be prevented or reduced at the source wherever feasible, while pollution that cannot be prevented should be recycled in an environmentally safe manner. In the absence of feasible prevention or recycling opportunities, pollution should be treated. Disposal or other release into the environment should be used as a last resort.

Potentially Responsible Parties (PRPs): Those identified by EPA as potentially liable under CERCLA for cleanup costs. PRPs may include generators and present or former owners/operators of certain facilities or real property where hazardous wastes have been stored, treated, or disposed of, as well as those who accepted hazardous waste for transport and selected the facility.

Processing: The operations performed on recycled materials to render them reusable or marketable. Processing can include grinding glass, crushing cans, or baling newspaper. Processing has two distinct functions: a separation function and a processing or beneficiation function. Processing generally results in adding value to a particular material.

Publicly Owned Treatment Works (POTW): Any device or system used in the treatment (including recycling and reclamation) of municipal sewage or industrial wastes of a liquid nature which is owned by a state or municipality. This includes sewers, pipes or other conveyances if they convey wastewater to a POTW providing treatment.

Quality Assurance/Quality Control: A system of procedures, checks, audits, and corrective actions to ensure that all EPA research design and performance, environmental monitoring and sampling, and other technical and reporting activities are of the highest achievable quality.

Rebuilding: Modifying a component of municipal solid waste by repairing or replacing certain parts and reusing it again for its original purpose (e.g., refillable or rebuild able toner cartridges, wooden cable reels, or plastic wire reels). Rebuilding of solid waste components is most often done by a middleman.

Recyclables: Materials that still have useful physical or chemical properties after serving their original purpose. Such materials can be remade into new products.

Recycling: The process by which materials are collected and used as raw materials for new products.

Release: Any spilling, leaking, pumping, pouring, emitting, emptying, discharging, injecting, escaping, leaching, dumping, or disposing into the environment (See CERCLA 101(22)). Includes the abandonment or discarding of barrels, containers, and other closed receptacles containing any hazardous substance, pollutant, or contaminant.

Remedial Action: Actual construction and implementation of a Superfund remedial design that results in long-term site cleanup.

Removal, Remove, or Removal Action: Under CERCLA 101(23), generally short-term actions taken to respond promptly to an urgent need. With regard to hazardous substances, the cleanup or removal of released substances from the environment; actions in response to the threat of release; actions that may be necessary to monitor, assess, and evaluate the release or threat; disposal of removed material; or other actions needed to prevent, minimize, or mitigate damage to public health or welfare or to the environment. Removal also includes, without being limited to, security fencing or other measures to limit access; provision of alternative water supplies; temporary evacuation and housing of threatened individuals not otherwise provided for; and any emergency assistance provided under the Disaster Relief Act.

Reportable Quantity (RQ): The quantity of a hazardous substance or oil that triggers reporting requirements under CERCLA, the Clean Water Act, or other environmental laws. If a substance is released in amounts exceeding its RQ, the release must be reported to the National Response Center, the State Emergency Response Commission, and/or other federal, state, and local authorities.

Response Action: Any remedial action, removal action, or cleanup at a site under CERCLA 101(25). Includes enforcement-related activities.

Reuse: Taking a component of municipal solid waste (possibly with slight modification) and using it again for its original purpose (e.g., refillable beverage bottles, foam peanuts, or pallets).

Risk Assessment: A qualitative and quantitative evaluation performed to define the risk posed to human health and/or the environment by the presence or potential presence and/or use of specific pollutants.

Septic Tank: A watertight, covered receptacle designed to receive or process, through liquid separation or biological digestion, the sewage discharged from a building sewer. The effluent from such a receptacle is distributed for disposal through the soil and settled solids and scum from the tank are pumped out periodically and hauled to a treatment facility.

Shelf Life: The time for which chemicals and other materials can be stored before becoming unusable due to age or deterioration.

Small Quantity Generator (SQG): A regulated facility that generates more than 100 kilograms and less than 1,000 kilograms (about 1 ton) of hazardous waste in a calendar month. However, even if a small quantity generator avoids the requirements of full generator status, the facility may still be subject to certain RCRA conditions (e.g., if the quantity of acutely hazardous wastes generated in a calendar month exceeds quantities specified under RCRA).

Sole Source Aquifer: An aquifer which is the sole or principal drinking water source for an area and which if contaminated would create a significant hazard to public health. Sole source aquifers may receive special protective status.

Solid Waste (per RCRA): Garbage, refuse, sludge from a waste treatment plant, water supply treatment plant or air pollution control facility and other discarded material including solid, liquid, semi-solid, or contained gaseous materials resulting from industrial, commercial, mining and agriculture activities and from community activities, but does not include solids or dissolved materials in domestic sewage or solid or dissolved materials in irrigation return flows or industrial discharges that are point sources subject to permits under the federal Clean Water Act or source, special nuclear or byproduct material as defined by the Atomic Energy Act. EPA defines hazardous waste as a subset of solid waste.

Solid Waste Disposal Act (SWDA): The SWDA was amended in 1976 by The Resource Conservation and Recovery Act (RCRA). SWDA has since been amended by several public laws, including the Used Oil Recycling Act of 1980 (UORA). the Hazardous and Solid Waste Amendments of 1984 (HSWA), and the Medical Waste Tracking Act of 1988 (MWTA).

Solid Waste Management Unit (SWMU): Any unit in which wastes have been placed at any time, regardless of whether the unit was designed to accept solid or hazardous waste. Units include areas from which solid wastes have been routinely released.

Source Reduction: Source reduction means the reduction or elimination of waste at its source.

Source Separation: Separating waste materials such as paper, metal, and glass by type at the point of discard so that they can be recycled.

Spill Prevention Control and Countermeasures Plan (SPCC): Plan consisting of structures, such as curbing, and action plans to prevent and respond to spills of oil and hazardous substances as defined in the Clean Water Act. See 40 CFR 112.3.

Storage (per RCRA): The holding of hazardous waste for a temporary period, at the end of which the hazardous waste is treated, disposed of, or stored elsewhere. Facilities are required to have a RCRA permit for storage of hazardous waste for more than 90 days; storage for less than 90 days does not require a RCRA permit.

Storm Water: Run off from a storm event, snow melt run off, and surface run off and drainage.

Superfund Amendments and Reauthorization Act (SARA): Signed by the President on October 17, 1986, expanded the scope of CERCLA. SARA is a 5-year extension of the program to clean up toxic releases at uncontrolled or abandoned hazardous waste sites. CERCLA is due for reauthorization again in 1995.

Surface Impoundment: A natural topographic depression, manmade excavation, or diked area formed primarily of earthen materials (may be lined with manmade materials) but is not an injection well. Examples of surface impoundments are holding, storage, settling, and aeration pits, ponds, and lagoons.

Surface Water: All water which is open to the atmosphere and subject to surface run off.

Tank: A stationary device, designed to contain an accumulation of hazardous waste which is constructed primarily of non-earthen materials (e.g., wood, concrete, steel, plastic) which provide structural support.

Threatened/Endangered Species: Any plant or animal species so listed in section 4 of the Endangered Species Act.

Toxic Release Inventory (TRI): Under EPCRA Section 313, facilities that handle certain types of chemicals must report to EPA annually the quantities of these chemicals released to the environment during the year. These releases may be allowed under air or water permits, may be spills of waste or product materials, or may be fugitive smoke stack emissions or product/process tank/oil line losses. EPA is beginning to evaluate TRI data to determine whether facilities have environmentally significant releases that must be addressed under RCRA corrective action or Superfund authorities.

Toxicity Characteristic (TC) Rule: This rule replaced the Extraction Procedure (EP) toxicity test with the TC test to determine whether or not a waste is a characteristic waste based on toxicity. The TC test requires analysis of 25 organic compounds in addition to the eight metals and six pesticides that were subject to the EP test.

Toxicity Characteristic Leaching Procedure (TCLP): The analytical method one must use to determine whether or not a waste is a characteristic hazardous waste based on toxicity. The TCLP is also necessary to comply with provisions of land disposal restrictions as well.

Transporter: A person transporting hazardous waste within the United States, which requires a manifest. On-site movement of hazardous waste does not apply. Transporters must comply with 40 CFR Part 263.

Trash: Material often considered worthless, unnecessary, or offensive that is usually thrown away. Generally defined as dry waste material; but in common usage, it is a synonym for garbage, rubbish. or refuse.

Treatment: Any method, technique, or process, including neutralization, designed to change the physical, chemical, or biological character or composition of any hazardous waste so as to neutralize such waste, or so as to recover energy or material resources from the waste, or so as to render such waste nonhazardous, or less hazardous; safer to transport, store, or dispose of; or amenable for recovery, amenable for storage, or reduced in volume.

Treatment Standards: Standards that hazardous wastes must meet prior to land disposal. A treatment standard generally expresses a treatment technology as concentration limits to give generators flexibility in choosing treatment options. Note that concentration limits are based upon the use of best demonstrated available technology (BDAT) for a particular waste or a similar waste.

Treatment, Storage and Disposal (TSD) Facility: A facility that treats, stores, and/or disposes of hazardous wastes .

Underground Storage Tank (UST): *RCRA definition*—Any one or combination of tanks (including its connecting underground pipes) used to contain an accumulation of regulated substances, and the volume of which (including the volume of the underground pipes) is 10% or more beneath the surface of the ground. Regulated substances include hazardous chemical products regulated under CERCLA and petroleum products. Some tank uses are exempt from regulation, including septic tanks, residential/agricultural fuel or heating oil tanks, and wastewater collection systems. (See 40 CFR 280.12 for the complete RCRA definition).

Waste Minimization: Generators and TSD facilities operating under RCRA permits are required to certify annually that they have waste minimization plans in place and that the plans are being implemented at their facilities.

Waste Prevention: The design, manufacture, purchase or use of materials or products to reduce their amount or toxicity before they enter the municipal solid waste stream. Because it is intended to reduce pollution and conserve resources, waste prevention should not increase the net amount or toxicity of wastes generated throughout the life of a product.

Waste Reduction: Preventing and/or decreasing the amount of waste being generated either through waste prevention, recycling, composting, or buying recycled and reduced-waste products.

Water Quality Standards: State-adopted and EPA-approved ambient standards for water bodies. The standards cover the use Or the water body and the water quality criteria that must be met to protect the designated use or uses.

Index

ARIZONA
    Pollution Prevention Law, 42, 384

AUTOMOTIVE REFINISHING INDUSTRY
    Body Repair, 253
    Paint Application, 253
    Shop Cleanup Wastes, 259
    Waste Minimization Options, *Table 6.7,* 252

AUTOMOTIVE REPAIR INDUSTRY
    Automotive Maintenance, 183
    Parts Cleaning, 177
    Shop Cleanup, 175
    Waste Minimization Options, *Table 6.2,* 173

BEST MANAGEMENT PRACTICES, *See* CLEAN WATER ACT; WATER POLLUTION PREVENTION PROGRAMS

CALIFORNIA
    Hazardous Waste Source Reduction and Management Review Act, 45, 396

CASE STUDIES, *See* POLLUTION PREVENTION CASE STUDIES

CLEAN WATER ACT, *See also* WATER POLLUTION PREVENTION PROGRAMS
    Best Management Practices, 33
    NPDES Permits, 33
    Stormwater Pollution Prevention Plans, 37
        Permit Requirements for Industrial Facilities, 38

COMMERCIAL PRINTING INDUSTRY
    Image Processing, 217

Makeready, 221
Material Handling and Storage, 216
Plate Processing, 220
Printing and Finishing, 223
Waste Minimization Options, *Table 6.5,* 215

DESIGN FOR THE ENVIRONMENT (DfE) PROGRAM
Chemical Design Project, 321
DfE Computer Workstation Project, 321
DfE Dry Cleaning Project, 321
DfE Printing Project, 321
Generally, 320

EMERGENCY PLANNING AND COMMUNITY RIGHT-TO-KNOW ACT (EPCRA)
EPCRA Toxic Chemical Release Inventory, 32
Industrial Toxics Project, 10, 320
Target Chemicals, 11

ENVIRONMENTAL PROTECTION AGENCY (U.S. EPA)
EPA Libraries, 330
EPA Pollution Prevention Program Contacts, 317
EPA Regional Offices, 315
Office of Environmental Education, 322
Office of Prevention, Pesticides and Toxic Substances, 14
Pollution Prevention Policy, 6
Five-Year Strategic Plan, 14
Pollution Prevention Policy Statement, 12
Pollution Prevention Programs, 317
Agriculture in Concert with the Environment (ACE), 317
Design for the Environment (DfE), 320
Chemical Design Project, 321
DfE Computer Workstation Project, 321
DfE Dry Cleaning Project, 321
DfE Printing Project, 321
Green Lights Program, 324
Industrial Toxics Project (33/50 Program), 10, 320
National Industrial Competitiveness Through Efficiency Program (NICE), 318
Pollution Prevention Incentives for States (PPIS) Program, 318
Waste Minimization Assessment Centers (WMACs), 269

FIBERGLASS AND PLASTICS INDUSTRY
　　Air Emissions, 198
　　Empty Bags and Drums, 197
　　Equipment Cleaning Wastes, 187
　　Gelcoat Resin and Solvent Overspray, 193
　　Miscellaneous Waste Streams, 199
　　Rejected and/or Excess Raw Materials, 196
　　Scrap Solvated and Partially Cured Resins, 192
　　Waste Minimization Options, *Table 6.3*, 186

GLOSSARY OF POLLUTION PREVENTION TERMS, 435

HAZARDOUS WASTE MINIMIZATION PROGRAMS
　　Agent Regeneration Case Study, 101
　　　　Air Oxidation Option, 108
　　　　Electrochemical Option, 106
　　　　Option Selection; Results, 109
　　　　Ozone Gas Option, 107
　　　　Problem Identification, 102
　　　　Strategic Alternatives, 105
　　Benefits of, 95
　　Program Elements, 96
　　　　Cost Allocation, 100
　　　　Management Support, 97
　　　　Program Implementation and Evaluation, 101
　　　　Technology Transfer, 100
　　　　Waste Generation and Management Costs, 99
　　　　Waste Minimization Assessments, 99
　　RCRA Waste Minimization Regulatory Guidance, 24
　　Waste Minimization Assessment Centers (WMACs), 269

INDUSTRIAL TOXICS PROJECT (33/50 PROGRAM), 10, 320

MANDATORY STATE PROGRAMS
　　Arizona, 42, 384
　　California, 45, 396
　　Generally, *Table 2.1*, 42
　　Minnesota, 49, 406
　　Texas, 51, 412

MARINE MAINTENANCE AND REPAIR YARDS
　　Abrasive Blast Wastes, 204
　　Chemical Stripping Wastes, 202

Engine Repair Shop Wastes, 211
Equipment Cleaning Wastes, 210
Machine Shop Wastes, 210
Paint and Solvent Wastes, 206
Spill Control, 212
Vessel Cleaning Wastes, 212
Waste Minimization Options, *Table 6.4,* 201

MINNESOTA
Minnesota Toxic Pollution Prevention Act, 49, 406

PAINT MANUFACTURING INDUSTRY
Air Emissions, 170
Bags and Packages, 169
Equipment Cleaning Wastes, 164
Filter Cartridges, 171
Obsolete Products/Customer Returns, 172
Off-Site Specification Paint, 169
Spills, 171
Waste Minimization Options, *Table 6.1,* 163

PESTICIDE FORMULATION INDUSTRY
Air Emissions, 266
Containers, 264
Equipment Cleaning Wastes, 262
Miscellaneous Waste Streams, 268
Off-Specification Products, 264
Spills and Area Washdowns, 263
Waste Minimization Options, *Table 6.8,* 261

POLLUTION PREVENTION ACT OF 1990
Generally, 8
Pollution Prevention Act of 1990, *Appendix A,* 307

POLLUTION PREVENTION CASE STUDIES - FEDERAL FACILITIES
Air Force Plant No. 6, 305
Cleaning Solvent Substitutions, 306
Plant Background, 306
Cincinnati Department of Veterans Affairs Hospital, 303
Existing Waste Management Activities, 303
Recommendations, 305
Research and Development Opportunities, 304

Fitzsimmons Army Medical Center Optical Fabrication
  Laboratory, 299
    Existing Waste Management Activities, 299
    Recommendations, 301
    Waste Minimization Opportunities, 300
  Fort Riley, Kansas, 301
    Existing Waste Management Activities, 301
    Recommendations, 303
    Waste Minimization Opportunities, 302
  Scott Air Force Base, 297
    Existing Waste Management Activities, 297
    Recommendations, 299
    Waste Minimization Opportunities, 298

POLLUTION PREVENTION CASE STUDIES - INDUSTRIAL FACILITIES
  Aluminum Can Manufacturer, 294
  Aluminum Extrusions Manufacturer, 288
    Existing Waste Minimization Practices, 289
    Waste Generation Activities, 289
    Waste Minimization Opportunities, 290
  Automotive Air Conditioning Condenser Manufacturer, 280
    Waste Generation and Management Activities, 280
    Waste Minimization Opportunities, 281
  Finished Leather Manufacturer, 272
    Existing Waste Management Activities, 272
    Waste Minimization Opportunities, 273
  Ice Machine and Ice Storage Bin Manufacturer, 290
    Existing Waste Minimization Practices, 292
    Waste Generation Activities, 291
    Waste Minimization Opportunities, 293
  Local School District, 274
    Additional Options Identified, 275
    Existing Waste Management Activities, 274
    Waste Minimization Opportunities, 275
  Metal Bands, Clamps, and Tools Manufacturer, 286
    Existing Waste Minimization Practices, 288
    Waste Generation Activities, 287
    Waste Minimization Opportunities, 288
  Permanent-Magnet DC Electric Motor Manufacturer, 284
    Waste Generation and Management Activities, 284
    Waste Minimization Opportunities, 285
  Printed Circuit Board Manufacturer, 293
  Printer of Legal Supplies, 278

> Engraving Process, 279
> Printing Process, 280
> Waste Reduction Opportunities, 278
Railcar Refurbisher, 282
> Waste Generation and Management Activities, 282
> Waste Minimization Opportunities, 283
Specialty Chemical Manufacturer, 294
State DOT Maintenance Facility, 276
> Antifreeze Waste Reduction Opportunities, 277
> Freon/CFC Waste Reduction Opportunities, 277
> Oil Waste Reduction Opportunities, 276
> Paint Waste Reduction Opportunities, 277
> Recycling of Used Tires, 278
Treated Wood Products Manufacturer, 295

POLLUTION PREVENTION, GENERALLY
> Benefits of Pollution Prevention, 4
> EPA's Five-Year Strategic Plan, 14
> EPA's Pollution Prevention Policy Statement, 12
> EPA's Pollution Prevention Strategy, 10
> Importance of Pollution Prevention, 1
> Pollution Prevention Regulation, Federal, 23
>> CWA Best Management Practices, 33
>> EPCRA Toxic Chemical Release Inventory, 32
>> Pollution Prevention Act of 1990, 8
>> RCRA Regulation of Recycled Materials, 25
>> RCRA Waste Minimization Regulatory Guidance, 24
>> Stormwater Pollution Prevention Plans, 37
> Pollution Prevention Regulation, State, 41
>> Mandatory Pollution Prevention Programs, *Table 2.1,* 42, 384
>> Voluntary Pollution Prevention Programs, *Table 2.2,* 54, 419
> Regulatory Policy Shift to Pollution Prevention, 6
> Types of Pollution Prevention, 1
>> Recycling, 3
>> Source Reduction, 1
>> Treatment, 4

POLLUTION PREVENTION REGULATION - FEDERAL
> CWA Best Management Practices, 33
> EPCRA Toxic Chemical Release Inventory, 32
> Pollution Prevention Act of 1990, 8, 307

RCRA Regulation of Recycled Materials, 25
RCRA Waste Minimization Regulatory Guidance, 24
Stormwater Pollution Prevention Plans, 37

POLLUTION PREVENTION REGULATION -STATE
Mandatory State Requirements, 41, 384
Arizona, 42, 384
California, 45, 396
Minnesota, 49, 406
Texas, 51, 412
State Pollution Prevention Laws, *Appendix E,* 384
State Pollution Prevention Program Contacts, *Appendix C,* 333
Voluntary State Requirements, 53, 419
Colorado, 419
Delaware, 426
Florida, 430
Iowa, 431

PRINTED CIRCUIT BOARD INDUSTRY
Cleaning and Surface Preparation, 232
Electroplating and Electroless Plating, 236
Etching, 247
Pattern Printing and Masking, 235
Product Substitution, 231
Wastewater Treatment, 248
Waste Minimization Options, *Table 6.6,* 229

RECYCLABLE MATERIAL, 25, 28

RECYCLING ACTIVITIES
Generally, 3
Common Recyclable Materials, 88
RCRA Regulation of Recycled Materials, 25
Proof Against Claims of "Sham" Recycling, 27
"Inherently Waste-Like" Materials, 28
"Recyclable Material", 28
Recycling of Used Tires, 278
Recycling Options, Solid Waste, 59, 76

RESOURCE CONSERVATION AND RECOVERY ACT (RCRA)
Regulation of Recycled Materials, 25
Proof Against Claims of "Sham" Recycling, 27
"Inherently Waste-Like" Materials, 28

"Recyclable Material", 28
Waste Minimization Regulatory Guidance, 24

SOLID WASTE REDUCTION PROGRAMS
    Benefits of Waste Reduction, 55
    Common Solid Waste Reduction Practices, 84
    Common Recyclable Materials, 88
    Conducting Waste Assessments, 64
        Documenting the Waste Assessment, 71
        Facility Walk-Through, 68
        Purpose, 64
        Records Examination, 68
        Sample Waste Assessment, 72
        Waste Assessment Approaches, 65
        Waste Sort, 70
    Evaluating Waste Reduction Options, 73
        Analyzing and Selecting Options, 75
        Compiling and Screening Options, 74
        Composting Options, 79
        Purchasing Options, 80
        Recycling Options, 76
        Waste Prevention Options, 76
    Implementing the Waste Reduction Program, 81
        Employee Participation, 82
        Employee Training and Education, 81
        Program Evaluation, 83
    Program Planning and Organization, 60
        Employee Involvement, 63
        Management Support, 61
        Program Objectives, 62
        Waste Reduction Team, 61
    Waste Reduction Approaches, 56
        Composting, 59
        Purchasing, 60
        Recycling, 59
        Waste Prevention, 56

STORMWATER POLLUTION PREVENTION PLANS
    General Requirements, 37
    Permit Requirements for Industrial Facilities, 38

TEXAS
    Pollution Prevention Law, 51, 412

U.S. EPA POLLUTION PREVENTION PROGRAMS, *See* ENVIRONMENTAL PROTECTION AGENCY (U.S. EPA)

UNIVERSITY POLLUTION PREVENTION RESEARCH CENTERS,
  Appendix D, 358

VOLUNTARY STATE PROGRAMS
      Colorado, 419
      Delaware, 426
      Florida, 430
      Generally, *Table 2.2,* 54
      Iowa, 431

WASTE MINIMIZATION, *See* HAZARDOUS WASTE MINIMIZATION PROGRAMS; SOLID WASTE REDUCTION PROGRAMS

WASTE MINIMIZATION ASSESSMENT CENTERS (WMACs), 269

WATER POLLUTION PREVENTION PROGRAMS
      Best Management Practices, 111
      Factors Affecting BMP Selection, 114
      Types of BMPs, 112
      Components of BMP Plans, 115
      Development Phase, 132
            Employee Training, 149
            Good Housekeeping, 133
            Inspections, 142
            Preventive Maintenance, 137
            Recordkeeping and Reporting, 154
            Security, 146
      Plan Evaluation and Reevaluation, 159
      Planning Phase, 116
            BMP Committee, 116
            BMP Policy Statement, 121
            Release Identification and Assessment, 124

About Government Institutes

Government Institutes, Inc. was founded in 1973 to provide continuing education and practical information for your professional development. Specializing in environmental, health and safety concerns, we recognize that you face unique challenges presented by the ever-increasing number of new laws and regulations and the rapid evolution of new technologies, methods and markets.

Our information and continuing education efforts include a Videotape Distribution Service, over 200 courses held nation-wide throughout the year, and over 250 publications, making us the world's largest publisher in these areas.

Other related books published by Government Institutes:

Environmental Law Handbook, 13th Edition — The recognized authority in the field, this invaluable text, written by nationally-recognized legal experts, provides practical and current information on all major environmental areas. *Hardcover/550 pages/Apr '95/$79 ISBN: 0-86587-450-6*

Environmental Statutes, 1995 Edition — All the major environmental laws incorporated into one convenient source.
Hardcover/1,200 pages/Feb '95/$67 ISBN: 0-86587-451-4
Softcover/1,200 pages/Feb '95/$57 ISBN: 0-86587-452-2
3.5" Floppy Disks for DOS (#4054) or Windows (#4055) $125
Statutes Package, hardcover w/DOS disk, (#4056)
or w/Windows disk (#4057) $192

Environmental Regulatory Glossary, 6th Edition — This glossary records and standardizes more than 4,000 terms, abbreviations and acronyms, all compiled directly from the environmental statutes or the U.S. Code of Federal Regulations. *Hardcover/544 pages/June '93/$68 ISBN: 0-86587-353-4*

Environmental Guide to the Internet — Find the Environmental Information You Need Quickly and Easily on the Internet! The new Environmental Guide to the Internet is your key to a wealth of electronic environmental information. From environmental engineering to hazardous waste compliance issues, you'll have no problem finding it with this easy-to-use guide. Includes: 94 Internet Mailing Lists, 11 Usenet News Groups, 45 Electronic Journals and Newsletters, and 29 Bulletin Board Systems. *Softcover/250 pages/Feb '95/$49 ISBN: 0-86587-449-2*

Publications (cont.)

The Product Side of Pollution Prevention:
Evaluating the Potential for Safe Substitutes
By Gary A. Davis and Lori Kincaid, et. al.

This book is an in-depth study of the potential for substituting safer or less toxic chemicals for seventeen of the priority 33/50 chemicals from the Toxic Release Inventory. Seven priority product categories have been developed--each of these categories provides timely information on current releases, status of the manufacturing and projections of possibilities for safer chemical substitutions. This report focuses on safe substitutes for products that contain or use toxic chemicals in their manufacturing process. Identifying products for substitution and evaluating the feasibility of safe substitutes for those products will help prevent toxic chemical pollution at the source. Softcover, 240 pages, Sept '95, 0-86587-479-4 $69

Federal Facility Pollution Prevention:
Planning Guide and Tools for Compliance
By Environmental Protection Agency

This EPA guide presents pollution prevention tools and provides a step-by-step approach to develop a pollution prevention program plan for Federal Facilities. However, any facility will find this quick reference guide valuable for developing plans to reduce waste generation and emission or release rates in order to meet their compliance goals. It includes important information on pollution prevention opportunity assessments; training and outreach; energy conservation and efficiency; cost/benefit analysis; life-cycle costing; total cost assessment; life cycle analysis; and pollution prevention planning. Key contacts, technical resources, and pollution prevention publication lists are some of the subjects covered in the appendices. Softcover, 240 pages, July '95, ISBN: 0-86587-476-X **$75**

Facility Pollution Prevention Guide
This comprehensive manual takes you from the initial assessment phase (including collecting information, and selecting assessment targets and teams) through technical and economic feasibility analysis to implementation of the options best suited to your operation. Softcover/143 pages/Oct '92/$36 ISBN: 0-86587-314-3

Case Studies in Waste Minimization
This book contains 68 actual success stories in pollution prevention from the experts in the field! With Case Studies in Waste Minimization, you'll learn first hand from the experience of other companies which have made the most of their pollution prevention programs. These case studies provide proven waste reduction strategies that you can use for planning your own successful program from several successful organizations. Softcover/290 pages/Oct '91/$62 ISBN: 0-86587-267-8

Revised RCRA Inspection Manual
By Environmental Protection Agency

You'll see EPA's step-by-step instructions to its inspectors on how to actually conduct the inspection, beginning to end! In addition, you can tailor your own pre-inspection audit with the 144 pages of Inspection Checklists that are provided! And finally, you'll have a ready resource of over 1,000 checklist items—specific yes/no questions on compliance with regulatory requirements, waste code information, effective dates, length of storage, occurrences (e.g., "number of unplanned incineration stack emissions"), detailed record review questions (e.g., updated contingency plan information), treatment standards, and more. *Softcover/689 pages/Mar '94/$125 ISBN: 0-86587-395-X*

RCRA Hazardous Wastes Handbook, 11th Edition

Clear, Factual explanation of the law from the law firm of Crowell & Moring-who bring you up-to-date on important developments in the Resource Conservation and Recovery Act(RCRA). This completely revised new 11th Edition includes new technical amendments, major regulatory changes, and recent case law developments that have taken place since the publication of the last edition. The manual provides you with careful analysis of the impact of RCRA on your business and suggestions for how you can cost-effectively and efficiently comply with the law. Chapters include: Overview of Subtitle C of RCRA: The Hazardous Waste Program; the Identification of Hazardous Waste; Generators of Hazardous Waste; Transporters of Hazardous Waste; The Used Oil Recycling Act of 1980; and Underground Storage Tanks. *Softcover/466 pages/Oct '95/$115   ISBN: 0-86587-503-0*

Waste Analysis: *EPA Guidance Manual for Facilities that Generate, Treat, Store and Dispose of Hazardous Wastes*
By Environmental Protection Agency

With the EPA guidance provided in this manual, you will be able to • determine whether your waste is defined as a hazardous waste under RCRA • identify and classify the waste according to RCRA • and ensure that your waste is managed properly. The contents are divided into four parts: Identifying Waste Analysis Responsibilities; Documenting and Conducting Waste Analysis; Detailed Checklist; and Sample Waste Analysis Plans (WAPs). *Softcover/198 pages/Aug'94/$69   ISBN:0-86587-414-X*

Educational Programs

■ Our **COURSES** combine the legal, regulatory, technical, and management aspects of today's key environmental, safety and health issues — such as environmental laws and regulations, environmental management, pollution prevention, OSHA and many other topics. We bring together the leading authorities from industry, business and government to shed light on the problems and challenges you face each day. Please call our Education Department at (301) 921-2345 for more information!

■ Our **TRAINING CONSULTING GROUP** can help audit your ES&H training, develop an ES&H training plan, and customize on-site training courses. Our proven and successful ES&H training courses are customized to fit your organizational and industry needs. Your employees learn key environmental concepts and strategies at a convenient location for 30% of the cost to send them to non-customized, off-site courses. Please call our Training Consulting Group at (301) 921-2366 for more information!